AF796661

BEHAVIORAL FOUNDATIONS OF SYSTEM DEVELOPMENT

WILEY SERIES IN HUMAN FACTORS

Edited by David Meister

Human Factors in Quality Assurance

Douglas H. Harris and Frederick B. Chaney

Man-Machine Simulation Models

Arthur I. Siegel and J. Jay Wolf

Human Factors Applications in Teleoperator Design and Operation

Edwin G. Johnsen and William R. Corliss

Human Factors: Theory and Practice

David Meister

Human Factors in Highway Traffic Safety Research

T. W. Forbes, Editor

Applied Maintainability Engineering

Clair E. Cunningham and W. Cox

Prediction and Development of Industrial Work Performance

Gavriel Salvendy and W. Douglas Seymour

Perception and Performance Under Water

John Adolfson and Thomas Berghage

Engineering Anthropometry Methods

J. A. Roebuck, Jr., K. H. E. Kroemer, and W. G. Thomson

Behavioral Foundations of System Development

David Meister

TA 167
.M44

BEHAVIORAL FOUNDATIONS OF SYSTEM DEVELOPMENT

DAVID MEISTER

U.S. Navy Personnel Research and Development Center

A WILEY-INTERSCIENCE PUBLICATION

JOHN WILEY & SONS New York • London • Sydney • Toronto

SOLOMON R. BAKER LIBRARY
BENTLEY COLLEGE

Copyright © 1976 by John Wiley & Sons, Inc.

All rights reserved. Published simultaneously in Canada.

No part of this book may be reproduced by any means, nor transmitted, nor translated into a machine language without the written permission of the publisher.

Library of Congress Cataloging in Publication Data

Meister, David.
Behavioral foundations of system development.

(Wiley series in human factors)
"A Wiley-Interscience publication."
Includes bibliographies and index.
1. Man-machine systems. 2. Human engineering.

I. Title.
TA167.M44 620.8′2 76–1834
ISBN 0–471–59195–5

Printed in the United States of America

10 9 8 7 6 5 4 3 2 1

FOR SHIRLEY

SERIES PREFACE

Technology is effective to the extent that men can operate and maintain the machines they design. Equipment design which consciously takes advantage of human capabilities and constrains itself within human limitations amplifies and increases system output. If it does not, system performance is reduced and the purpose for which the equipment was designed is endangered. This consideration is even more significant today than in the past because the highly complex systems that we develop are pushing human functions more and more to their limits of efficient performance.

How can one ensure that machine and machine operations are actually designed for human use? Behavioral data, principles, and recommendations—in short, the Human Factors discipline—must be translated into meaningful design practices. Concepts like ease of operation or error-free performance must be interpretable in hardware and system terms.

Human Factors is one of the newer engineering disciplines. Perhaps because of this, engineering and human-factors specialists lack a common orientation with which their respective disciplines can communicate. The goal of the Wiley Human Factors Series is to help in the communication process by describing what behavioral principles mean for system design and by suggesting the behavioral research that must be performed to solve design problems. The premise on which the series is based and on which each book is written is that Human Factors has utility only to the degree that it supports engineering development; hence the Series emphasizes the *practical application* to design of human-factors concepts.

Because of the many talents on which Human Factors depends for its implementation (design and systems engineering, industrial and experimental psychology, anthropology, physiology, and operations research, to name only a few), the Series is directed to as wide an audience as possible. Each book is intended to illustrate the usefulness of Human Factors to some significant aspect of system development, such as human factors in design or testing or simulation. Although cookbook answers are not provided, it is hoped that this pragmatic approach will enable the many specialists concerned with problems of equipment design to solve these problems more efficiently.

DAVID MEISTER
Series Editor

PREFACE

The reader of a book that includes a review and evaluation of the behavioral literature is entitled to know the author's biases, particularly those that underly his selection of studies and the limitations of the conclusions he derives.

This book attempts to describe what the available behavioral literature—very broadly defined as including anything that describes personnel functioning—has to say in guiding the development of large man-machine systems (which I refer to as MMS, or simply as "systems"). The systems to which the data are to be applied are relatively *complex* (they perform multiple functions, often simultaneously; they often require crew interactions), and they emphasize *information processing* (primary stress being placed on the operator's perceptual and cognitive, rather than psychomotor, functions).

Books dealing with human factors are frequently criticized for unduly emphasizing the military side of the house. Most of the system examples I have used are military: air defense, air surveillance, command and control, sonar, radar, communications and weapons delivery; but there are also civilian parallels (e.g., air traffic control, taxi dispatching). Despite the desirability of including many more examples from the civilian sector, I have found that the available literature provides few of them.

A further qualification about the coverage in this book is that more attention might have been paid to vehicular and aircraft systems (with particular emphasis on their manual control functions). This reflects my own relative lack of interest in these functions. An even more important qualification is that I have made no attempt to summarize all or even the greatest part of the vast number of behavioral studies in the literature. Since the orientation of this book is to system development (analysis, design/development, test and evaluation, and system operation), only the studies that appear to me to be most directly relevant to these functions are described.

The studies reviewed are organized around what I have termed "cases." These are studies I consider to be especially relevant. The concept of the case arose from the hypothesis that certain studies illustrate trends in the way in which the discipline has approached certain problems, and by examining the cases it might be possible to clarify the approaches. Studies not denoted as cases are described in somewhat less detail as supporting evidence. I cannot, however, lay claim to complete consistency in this organization by cases.

Moreover, in describing individual cases it has been impossible to provide all the detail the reader might wish. Only enough material has been supplied to enable the reader to understand the major points in the study. To have described each study in the detail it deserves would have at least doubled the size of the present volume.

Because system behavior is largely interrelated, the reader will find that certain topics are discussed under several headings. For example, material relating to inputs may be found in both Chapter 2 and Chapter 4; team effects in Chapters 4 and 5, crew organization in Chapters 3 and 5, and so on.

The conclusions I have derived are in some cases based on only a few studies, because these are all the literature provides, and the reader must be more cautious in accepting such conclusions than those that are supported by a mass of evidence. I have attempted to find all the evidence available on specific questions, but unfortunately when one breaks the literature down in terms of specifics, too often the data are sparse. In generalizing the available data, I have sometimes combined data secured from different functions (e.g., vigilance, tracking), although functional differences may make this a dubious proposition.

A basic assumption I have made is that since the operator is a subsystem of and is controlled by the MMS, one cannot understand his behavior outside the system context. Explanation of system and subsystem behavior must be appropriate to the level of that behavior. It is therefore inappropriate to explain operator and MMS behavior in terms of molecular variables valid only for the individual. Principles derived from studying the individual can be applied only with great caution to the operator and the system.

Studies that purport to describe the MMS should ideally possess the following characteristics:

1. The tasks performed by subjects must involve the transformation of individual responses to a system output. The transformation should be accomplished by means of an actual or simulated environmental interface between the individual subsystems and between the subsystems and the total system.

2. The tasks performed by subjects should involve a complete function, not merely part of one. An example of the kind of task that would *not* satisfy this criterion is reaction time to the onset of a single indicator. Such a task is meaningful in individual but not system terms, because almost no system involves a single indicator alone. Because of this criterion I have

omitted many studies that deal with human engineering or operation of individual controls and displays.

3. As the previous item suggests, the task performed should be one found in the real world environment or should be a recognizable analogue of such a task. Work load and subject characteristics should also approximate those found in the real world.

4. Along with the individual response, the system output should be measured, and the two should be related.

5. The significance of the variables being tested should be determined by their effect on system outputs.

6. To study complex systems, the transformation referred to in item 1 should occur over at least two subsystem levels, as, for example, between two operators, each of whom performs distinctly different system functions.

The studies included in this book were selected to correspond as much as possible with the foregoing criteria. Unfortunately, comparatively few of the experiments reported in the behavioral literature satisfy these criteria completely. To piece out the narrative, therefore, it was necessary to include studies that only partially met the criteria. Since the use of the criteria drastically reduces the number of relevant studies, in some cases it was necessary to say that evidence for or against a given hypothesis is lacking.

In outline form, the arrangement of this book (except for Chapters 1 and 7) is as follows:

1. Introduction (definitions of terms, variables, and research questions)
2. Findings
   A. Conclusions
   B. Illustrative studies (one or more experiments illustrating the preceding conclusions)
   C. A and B repeated as necessary
   D. Summary of conclusions

3. Developmental implications (applications of study findings). The purpose of describing these implications is to tie as best one can the information from the studies described in each chapter to the *uses* that can be made of this information.

The book was written in the hope that it would prove useful to a varied audience: (1) the investigator, as a review of what has been learned from previous research and as a possible source of new hypotheses, (2) the student and teacher, as supplementary background readings for the student and as a guide to direct his attention to further reading, (3) the psychologist, who can see here where his research applies to a critical area of real world activity, and (4) the system developer, for the design guidance behavioral research supplies.

It is necessary to dampen any unreasonable expectations on the part of the reader, since the book does not pretend to be a comprehensive review of every topic discussed. Because of the commonality between individual and system variables, the following pages deal with many variables found in individual psychology, but *only from a system standpoint.* This means that most of the individual (psychological) literature is not discussed in detail.

Originally I had intended to integrate the relevant data into summary tables and graphs of "standard" values, such as one might find in an engineering reference book. This would make the book much more useful to the engineer, and the approach works with variables describing individual capabilities such as visual acuity. For example, the graph relating visual angle to probability of target detection summarizes many studies.

This form of organization has not been possible with the variables described in this book, primarily because of their macro- nature. These variables interact with so many factors (particularly task characteristics) that combining data (as distinct from generalizing conclusions) from studies with different tasks or measures makes little sense. It has therefore been necessary to describe principles and data in terms of the studies from which they were derived because the data can be understood only in terms of the original experimental conditions.

In concluding, I must acknowledge my indebtedness to the American Psychological Association and the publishers of the following journals, for permission to reproduce graphic material: *American Psychologist, Behavioral Science, Human Factors, Journal of Abnormal and Social Psychology, Journal of Applied Psychology, Journal of Engineering Psychology, Journal of Experimental Psychology, Organizational Behavior and Human Performance, Perceptual and Motor Skills,* and *Psychonomic Science.*

Finally, much of this volume was written while I was a staff member at the Army Research Institute for the Behavioral and Social Sciences (ARI), Arlington, Virginia. I must therefore express my appreciation for the use of the ARI library and the helpfulness of the library staff. Miss Marie McDowell of the Navy Personnel R&D Center library was most helpful in securing "AD" numbers. Needless to say, the opinions expressed here are those of the author alone and have not been endorsed by either the U.S. Army or Navy.

DAVID MEISTER

San Diego, California
January 1976

CONTENTS

BEHAVIORAL FOUNDATIONS OF SYSTEM DEVELOPMENT

I
INTRODUCTION TO THE MAN-MACHINE CONCEPT

SUMMARY. This chapter provides the theoretical structure needed to understand subsequent chapters. It describes the man-machine system (MMS) concept and the assumptions on which it is based, defines the variables affecting system performance, and introduces the concept of indeterminacy as a dimension running across all systems.

THE MAN-MACHINE SYSTEM CONCEPT

The man-machine system is essentially a concept based on certain assumptions (to be described later); it is an abstraction, not a physical configuration or a type of organization. It is primarily a framework for analyzing systems.

The system is a concept because it is organized around transformations (from inputs to outputs, stimuli to responses) that are invisible; all one sees are the products of these transformations. What goes on within the system, like human behavior in general, can be inferred only from antecedent inputs and consequent outputs.

The construct nature of the system is suggested by the various ways in which it has been defined. Not surprisingly, these are quite general. For example, Miller (1953) defines the MMS as "machines and men plus the processes by which they interact within an environment"; Morgan et al. (1963) prefer to say that it is "any group of men and machines that operates as a unit to carry out an assigned task or tasks." In its most general sense, the system is defined by Hall (1962) as a set of objects with relationships between the objects and their attributes.

Any definition comprehensive enough to encompass the spectrum of system characteristics becomes impractically abstruse. Jordan (1968) has suggested that an attempt to define the MMS rigorously is doomed to failure. The term is meaningful only if one knows the particular system to which reference is made. Nevertheless, despite our inability to define the MMS precisely, an examination of its characteristics is useful. (For other definitions and an examination of systems in general terms, see Van Gigch, 1974.)

The common element in all definitions of the MMS is the concept of *purposiveness.* Since the MMS is an artificial creation, its characteristics depend on the purpose of its creator. (This is a welcome change from the ambiguity that characterizes the notion of purpose in traditional psychological thinking.) The analogue of human purpose is the system requirement from which the system configuration, its functions, and its operations can be logically deduced.

Certain common characteristics are possessed by all systems, but by biological systems in particular (and the MMS is a biological system, since it is controlled by the operator). The most important of these are as follows: (1) all the elements of a system interact, (2) each element has an effect, however slight, on other elements and on the system as a whole, and (3) system functioning involves the transformation of energy and matter from one form to another (Miller, 1965). What most distinguishes the MMS from other systems is that this transformation is accomplished by means of a mechanical/electronic interface with the system environment. This interface not only permits the system to perform more extensive, far-reaching transformations, it also expands the individual's environment tremendously. As an extreme example, men have landed on the moon. Taking these characteristics as a starting point, it is possible to distinguish between systems and individuals in the following way.

The *individual's* elemental interactions and energy transformations occur within himself and between him and his immediate environment, without an intermediary. The individual has no tools (other than his hands) to interface with his environment and therefore cannot easily modify that environment.

On a somewhat more complex level, the individual can extend himself by interacting with other individuals. This interaction is direct and almost always verbal, occasionally with mechanical/electronic interfaces (e.g., telephones). This is the *group* or man-man system.

The *operator (sub)system* is qualitatively more complex than the individual or the group because the operator interacts with his environment through an interface, which means that there is a transformation of energy or matter across that interface.

Once the individual is given a tool to manipulate his environment, he becomes the operator. The tool can be as simple as a stick with which he can draw in the sand, or as complex as a spaceship he can pilot to the moon. Simple instruments like the stick sometimes make it difficult to tell whether one is dealing with an operator or an individual. The basic criterion, however, is clear; it is whether the individual can exercise an effect on his environment through some mechanism other than his body.

At a still more complex level there is the *multiman-machine system* (MMMS), more simply termed the *multiman system,* which consists of two or more operators interacting with each other and with their environment through a mechanical/electronic interface. Again, the multiman system varies in complexity from the minimal two-man operation to something as extensive as the NASA space launch complex in Houston.

It is with the operator and the MMMS in particular that this book is concerned.

There is a very natural temptation to equate the individual with the operator, and the group with the multiman system; but the comparison is inaccurate. The operator with his equipment is much more than an individual. The purposiveness of the MMS, which is derived from a goal specified by the system developer, immediately forces the operator into a controlling mold that cannot be found in the individual. Because of this system purpose can be specified more precisely and can be more deliberately measured and modified than can the individual's purpose.

The behavior of the individual and that of the system operator may therefore be discontinuous. When the individual is embedded in the system, many factors affect his behavior that cannot be studied in the context of individual psychology. To that extent, at least, human factors (HF) is a distinctive discipline. It has its own rationale and makes use of psychological principles, but in its application of these principles it goes well beyond their source.

MAN-MACHINE SYSTEM ASSUMPTIONS

Table 1-1 lists the assumptions underlying the MMS concept. The following paragraphs expand on those assumptions.

Since the operator is a subsystem of the MMS, the overall goals of the MMS must control his behavior. He is effective when he implements these goals, ineffective when he does not.* The significance of his behavior in terms of an effect on the overall system can be determined only in relation to these goals. It follows therefore that only studies that relate operator behavior to a system requirement (expressed as system output) are meaningful for system description.

*Needless to say, this assumption has profound ethical and humanistic implications. It might suggest, for example, that individual goals and desires need not be considered in developing and operating the system. The author rejects this point of view; system requirements in conflict with personnel needs are almost certain to produce an inefficient system.

Table 1-1 Fundamental MMS Assumptions

1. The man-machine relationship forms a system (MMS) whose elements—the man, the machine, and the environment—are subsystems organized in distinctive ways and controlled by overall system requirements.
2. MMS elements interact with and influence each other and the total MMS.
3. As an artificial entity, the MMS is purposefully directed (by means of requirements imposed on it) to produce specified outputs (on the basis of anticipated inputs).
   (a) Overall system requirements define subsystem, input, and output characteristics.
   (b) System activity is energized and channeled by the need to accomplish these requirements.
   (c) The system functions adequately only when these requirements are accomplished.
   (d) Failure to accomplish system requirements produces changes in system functioning.
4. Like other biological systems, the MMS and its subsystems function in both space and time and are affected by changes occurring in these dimensions.
5. All subsystem outputs must contribute to the total system output; those that do not are inefficient.
6. To the extent permitted by its structure, the system acts to adjust its functioning to optimize input/output relationships in accordance with overall system requirements.

Development of an efficient MMS necessitates the construction of the operator subsystem in accordance with the requirements of the overall system. This is done in various ways: by designing hardware and procedures that call for specific operator behaviors, by selecting operators who have appropriate capabilities, and by training the chosen personnel to perform these behaviors. In any event, the MMS is constructed by forcing individuals to behave in ways that implement system purposes.

The system is an entity developed by humans to perform in accordance with specified purposes; thus the processes by which it is developed become not only a legitimate but actually the most critical area for investigation by HF researchers. Behavioral data and principles have limited value unless they answer questions addressed to system development needs.

The purposes for which the system is developed are specified as *requirements* that the system, once developed, must satisfy (e.g., to fly at *X* mph or to process *N* items of information per hour). The overall system requirement levies requirements on lower-order subsystems and elements and also in part defines their characteristics.* This stems from the logical nature of system development. To develop X, we need $X_1, X_2, X_3, \ldots, X_n$ elements, attributes, or characteristics.

Since the system requirement is by definition phrased in system output (i.e., nonbehavioral) terms, the system developer's primary problem is that of translating the requirement into behavioral (i.e., operator) equivalents and correspondingly inferring physical design principles from behavioral data.

The MMS functions in both space and time. We are familiar with the movement of vehicular systems over space and the input changes that this movement provides. Of potentially greater significance, however, is the movement of the system in time. Both men and machines change their characteristics over time. Two factors are involved in time changes:

1. The performance of a task or mission *in a given time* may be a system requirement (e.g., "fire in 3 minutes"). Goals and subgoals may be significant therefore only in relation to time. Consequently, the operator's anticipation of failure to accomplish the mission *in time* may severely stress him. The task situation may also change, because inputs often change their characteristics over time. It has even been suggested (Alvares and Hulin, 1972) that operator capabilities change over time spent performing a task.
2. The operator's responses also change over time as a consequence of input changes, practice, fatigue, and motivational variability. Whereas the changes that occur in equipment characteristics and outputs are very slow (requiring sometimes thousands of operational cycles to manifest themselves), changes in operator behavior occur quite rapidly, often within a single mission or operational cycle.

Operator/system history is therefore a critical element in understanding system behavior. Unfortunately, little attention has been given to this aspect except in a training context.

The need to accomplish system requirements ordinarily energizes and directs the system. The existence of a mission requirement that has not been accomplished motivates the operator (assuming he is willing to function within the system) to exercise the system in specified ways to accomplish that requirement. Consequently inability to satisfy system

*We say "in part defines" because there are usually alternative ways of satisfying a higher-order requirement. For example, one can perform a function automatically, semiautomatically, or manually. The desired mechanism is selected by trading off costs against utility.

performance criteria adequately represents partial or complete system failure.

As a purposeful entity, the MMS attempts to regulate its own performance in an effort to achieve its goal. Feedback in the form of equipment displays and information are critical elements in that regulation. Attempts at self-regulation also produce changes in system performance. For example, an operator's anticipation of mission failure and his attempts to avoid that failure may result in his working harder, changing system criteria, or reinterpreting inputs by filtering out information for transmission.

MMS CHARACTERISTICS

The structure of a multiman system contains a hierarchy of subsystems, each higher level being composed of systems at lower levels. The concept of *system level* is important because higher-order subsystems may have properties (e.g., complexity, dependency) not found in lower-level subsystems. The interaction between different system levels is obviously of considerable consequence to overall system functioning.

Since the functioning of the subsystem is determined by the next higher subsystem of which it is an element, it is necessary during system development to translate higher-order system requirements into lower-order subsystem terms. However, the translation process admits of various alternative configurations, and one of these is selected on the basis of tradeoffs among competing considerations. Since the tradefoff process is often judgmental, a less than optimal subsystem configuration may be designed. Consequently, if the translation process is inadequate, the resultant system may display inefficiencies.

The transformation of energy/matter, which is the outstanding characteristic of the MMS, occurs across an environmental *boundary* that differentiates one subsystem or system from another. Each individual subsystem and system has its own environment. Thus, there are multiple boundaries and environments between one's own system and any other, between one's system and the physical environment, and between the system and its component subsystems and their elements.

Since all transmissions of energy and matter pass through a boundary that defines a particular environment, and since the environment is the source of inputs to the system, it is of some practical (i.e., design) importance to determine precisely where that boundary should be or is and what the environment consists of. The design of the equipment interface depends on the characteristics of the environmental boundary conceptualized by the designer. The boundary for an operator may be a display surface (e.g., a CRT), for a maintenance man, a circuit.

In the same way, if one is to measure operator/system performance, it is necessary to determine the interface and the boundary across which transformations take place. Often this determination is obvious; in other cases, such as that of the maintenance function, it is extremely difficult. Where multiple functions are performed, there may be multiple interfaces.

The effect of the environment must also be determined. Environments of molecular components (e.g., piece-parts) have little direct influence on the operator except when the component fails and the system malfunctions. Other environments (e.g., terrain) have a much greater direct influence on operators of systems such as tanks, ships, and aircraft.

Transformations of energy or matter occur by means of *inputs* and *outputs*. An input or output is the result or product of the *transmission* of energy or matter across an environmental boundary. These occur at three levels: (1) within a subsystem, (2) between subsystems at the same or different levels, and (3) between subsystems and the total system. The last transformation produces what has been referred to as the system output. Because transmissions occur at various levels, all must be measured to describe system performance correctly.

Inputs and outputs *do not* correspond to stimuli and responses. The former describe subsystem and system processes; the latter, internal human processes. The two sets of concepts are related, however, because the mechanical/electronic interface is designed to transform an input into an operator stimulus and an operator response into an output.

Inputs and outputs refer to events occurring within or outside the system. For example, on the input side, the appearance of an enemy tank on a distant ridge is an event; the image of that tank in the operator's binoculars (the interface) is the stimulus. On the output side, the report of the detection of the tank is the operator's response, and one's own tank opening fire is the output.

It may appear that these distinctions are arbitrary and theoretical, since inputs and stimuli, and outputs and responses, often coincide; but the differences are important. In HF research we study the process by which inputs are transformed to stimuli, and responses to outputs. (The processes that interrelate stimuli and responses are the province of the psychologist, since they occur solely within the human.) Failure of the designer to examine closely the characteristics of the input event will lead to inefficiencies in his design of the most effective transformation mechanism.

From the standpoint of the HF researcher, the transformation process has significant implications for MMS measurement methods. To describe the MMS correctly and completely, it is necessary to measure transformation effects that are reflected in the output of the subsystem and system as a whole, as well as in operator responses. This means that measures of two types (individual and system) are required to describe the system meaningfully; neither measure alone can supply adequate data.

The implication of the requirement to measure both the individual and the system is that data describing only individual responses cannot be utilized efficiently as a behavioral foundation for system design.

All system processes incur *costs* in terms of the expenditure of resources, including energy, time, and power, to achieve their end product. The costs of these processes differ from one another and from time to time. They may be immediate or delayed, short term or long term. Certain types of systems have the capability of selecting (deciding) among response mechanisms, thus improving performance by decreasing costs. The efficiency with which such systems function is determined by the strategies the operator employs in selecting its outputs. System performance is evaluated by the accomplishment of the mission as related to the costs incurred in achieving goals.

SYSTEM VARIABLES

Because of the complexity of system processes, the number of variables that may affect system output is very great. This section presents a taxonomy of these variables. It is interesting that although there have been a number of attempts to develop a generally acceptable task taxonomy (see Chapter 3), there has been no extensive effort to arrive at a system taxonomy (although see Finley et al., 1975).

The system variable taxonomy developed for this book is a mixed bag. Some variables, like system organization, communication structure, and performance criteria, are more or less fixed by the manner in which the system is developed. Others, like inputs and environmental characteristics, are more dynamic; they may be partly under the control of the system developer and user, or not at all regulated by him. Personnel variables are in between, partially modifiable for system purposes in terms of selection criteria and training, partially fixed by biological constraints.

System Organization

Subsystem arrangement: hierarchical, series, parallel, independent/dependent, redundant

Subsystems may be arranged in terms of a hierarchy of levels. A subsystem may be wholly or partly dependent for its operation on inputs from another subsystem (in which case the subsystems are in series); or it may be independent of other subsystems, in which case the subsystems operate in parallel. When a subsystem performs the same functions as others, the subsystems are redundant. Most complex systems combine all these characteristics.

A series organization has a greater potential for failure, since one subsystem is uniquely dependent on the effectiveness of another. Highly interdependent subsystems provide more opportunity for contingent responses. This increases system flexibility but also the amount of uncertainty about how the system will respond.

Number of subsystems

System size or complexity can be defined in terms of the number of subsystems it contains and the number of states each can assume. It is possible that the larger or more complex the system, the greater its inertia in adapting to changed input conditions. The point at which the size of the system becomes maladaptive may depend in part on the stringency of the requirements laid on the system.

Subsystem function: type, number

A subsystem may perform only a single function or it may perform several simultaneously or sequentially. Presumably the larger the number of different functions a subsystem performs, the more complex it must be; and greater complexity tends to increase the probability of function inefficiency or breakdown.

Procedural Variability

Procedural variability is the extent to which operating procedures may be varied by personnel in nonprogrammed ways during system operation. Procedural flexibility implies contingent responses and a higher degree of uncertainty about whether the correct response will be performed. As system personnel gain operational experience, they tend to reduce procedural variability.

System Requirements: Qualitative/Quantitative, Specific/General

The nature of system requirements and particularly their precision are tremendously important to the design of the system. Qualitative, general requirements make it difficult to develop the system efficiently. Meister (1971) gives a more extended discussion of this point.

System Type: Information-Sensing Display, Information Display and Processing, Communications, Maintenance, Production, Weapons Selection and Delivery, Command-Control, Transportation, Construction, Legal/Penal, Health, Administrative, Entertainment, Social

System types are defined by the purpose they serve. However, systems of different types often employ the same behavioral mechanisms. Since the goals of military and civilian systems differ, it is often assumed implicitly and explicitly that their structure, organization, and method of functioning differ also. This is incorrect: for *every* civilian system there is a military counterpart. For example, production systems exist in the military as well as in civilian life, and both have legal and health systems. If one looks at systems in terms of the function they perform (e.g., information processing, monitoring) rather than their context, differences between systems become minimal; different types of complex systems often perform identical functions. Military and civilian systems are influenced by almost identical variables.

Man-machine systems do differ in terms of the extent to which

1. They seek to control other systems (as in the military).
2. There are societal benefits received from system operations.
3. They investigate (information sensing) and/or seek to control/modify (construction) their environment.
4. They control their own or other personnel (e.g., legal/penal systems).
5. They produce/maintain their own or other systems.

Systems therefore differ with respect to their interaction with the environment, other systems, and their own personnel.

Task Characteristics: Difficulty, Patterning, Complexity, Organization

Task characteristics (defined in Chapter 3) are assumed to be one of the major factors influencing system performance. The task is isomorphic with the system; that is, it contains all system elements for a given function at any moment in system operation. The system, however, is more than any single task because the system represents the transformations of individual task responses to a higher-order output. Nevertheless, it is possible that the only practical way to understand system processes is by examination of task characteristics as these impact system outputs.

Personnel

Number: Required, Available

If an insufficient number of personnel are provided, it is obvious that system efficiency will suffer greatly, and the system developer attempts to avoid this mistake. In small, relatively simple systems there is little difficulty in determining the appropriate number of personnel because functional requirements are easily associated with individual operators. For large systems requiring many people to perform many different functions (e.g., a ship's complement), the problem of determining the required number of personnel becomes more difficult, since personnel interactions (which are not easy to determine from system/task requirements) play a greater role. System developers are more likely to commit too many than too few personnel to system operations.

Background: Intrinsic (Age, Sex, Race, Aptitude, Personality, Motivation); Extrinsic (Experience, Training); Crew Characteristics (Duties, Organization, Distribution of Responsibility)

With the exception of aptitude, which can seriously constrain operator/system functioning, the variables inherent in the individual —that is, the intrinsic variables—probably play a minimal role in system functioning except as they impact on aptitude and when mission requirements are highly demanding. Within limits, all ages and both sexes can perform most jobs. However, we know very little about the effects of these intrinsic variables on system performance. Some research has been performed on age and aptitude, very little on sex, race, and personality. Almost no studies tie these variables directly to system functioning. Personality and motivational factors presumably influence system performance most where system operations are relatively unprogrammed and heavily dependent on the operator's self-regulation for their output (e.g., a research laboratory).

With regard to extrinsic personnel variables, it is obvious that training is essential to effective operator/system functioning. A great deal of research has been done on the effects of training on the individual; comparatively little has been done in relation to system variables and the system as a whole. Very little is known about the effect of various types of experience on operator and team performance.

With regard to crew characteristics, the organization of crew responsibilities may be important as it relates to overall system organization (particularly in systems where the crew must adapt to changing inputs). Subsystem arrangement may determine crew responsibilities.

Communications Structure: Number and Arrangement of Channels

Communications can be viewed in a number of ways: *(a)* as a first-order system output (in a communications system), *(b)* as an implementing mechanism in a noncommunications system (i.e., a means of accomplishing first-order outputs such as carrier landings, in which case it can be considered as a second-order output), *(c)* as a correlate of system operations, in which case it reflects those operations. In all these cases communications can serve as a measure of that aspect of system performance. In systems with much procedural variability, the arrangement of communication channels is likely to be flexible.

Inputs: Number, Frequency, Duration, Characteristics, Mode of Presentation

The input attributes of number, frequency, duration, and so on, should be familiar to anyone who has studied individual performance. All affect operator and perhaps system performance. Special attention, however, should be given to input characteristics.

Structured inputs (e.g., alphanumerics) have a high degree of regularity and redundancy; their meaning is apprehended with relative ease. *Unstructured* inputs (e.g., raw video) are irregular and require analysis and interpretation. Input *variability* refers to the frequency with which inputs change over time. *Patterning* refers to the repetition over time of identical inputs. *Relevant* inputs can be related directly to the mission goal, *irrelevant* ones cannot. Input *distribution* refers to the extent to which operators in a team have access—direct or indirect, immediate or delayed—to all the information needed to perform their tasks (e.g., all radar operators view the same display versus one operator viewing azimuth information, another elevation data). Operators receiving unstructured, unpatterned and irrelevant inputs have a higher probability of error in interpreting and responding to these inputs. The impact of such inputs on system outputs is therefore greater.

Outputs: Number, Frequency, Duration, Characteristics, Mode of Implementation (equipment/operator)

A distinction must be drawn between operator behaviors and operator outputs. Many operator behaviors are only tenuously related to task accomplishment; most outputs are relevant to mission accomplishment but a few, such as social communications among team members, may not be. Certain outputs are directed solely to accomplishment of the mission requirement, whereas others serve an internal regulatory function. Many outputs have been programmed in advance of system operation, but others represent responses to purely contingency events (e.g., emergencies).

Performance Criteria: System-Descriptive, Mission-Descriptive, Personnel Performance

System-descriptive criteria include reliability (equipment/human), maintainability, survivability/vulnerability, cost, acceptability (to operators), effectiveness (output), and efficiency (output divided by cost). Mission-descriptive criteria include output quantity and accuracy, reaction time, and queues and delays. Personnel performance criteria are associated with individual operator and crew responses: reaction time, accuracy, response number, speed, variability, and so on.

Performance criteria may act as independent or dependent variables. As independent variables (e.g., the requirement to produce *N* units) they impose on the operator a requirement that serves as a forcing function for operator/system performance. As dependent variables, they describe operator/system performance and can be used to measure that performance. In the same way that system requirements determine subsystem configuration, system criteria are translated into personnel criteria and regulate the operator's performance.

Because these criteria define three different system aspects, the investigator may secure different measures of performance, depending on which criterion he uses to describe performance. In fact, much of the difficulty in interpreting behavioral effects results from the investigator's use of multiple criteria. However, use of a single criterion may produce an erroneous picture of system performance.

Environmental Factors: Locus, Characteristics

The environment as a source of inputs is of great interest to the HF specialist, but aside from its structural characteristics (e.g., temperature) and its impact on the physical well-being of the operator, little thinking and research have gone into this aspect of the system.

Each element of the system (the system as a whole, the subsystem, equipment, component) has its own environment, the medium within which transformations occur (across the environmental boundary). The locus of the environment is the level at which one designs or measures. The environment conceptualized here is a physical one; yet it is clear that as a source of inputs that are transformed into operator stimuli, one can also think about a psychological (operator) environment correlated with the physical one. Since the psychological environment is often highly individual, it is difficult to understand and control. For this reason very little is said in this book about the psychological environment.

The physical environment as a source of inputs has many of the same characteristics as do the inputs themselves; some of these are (1) variability (the rate at which the environment changes its state), (2) the extent of environmental change, and (3) passivity, the extent to which the environment is determined *only* by the action of one's own system (Alexander and Cooperband, 1965) as opposed to an environment that produces its own inputs. One dimension that perhaps underlies differences among systems is the extent to which and how systems seek to control their environment. The environment as the source of inputs is the primary element determining the nature of the system.

THE INDETERMINACY DIMENSION

Since so many variables affect it, the system can be characterized in different ways. The most common characterization is probably in terms of the functions performed by the system and/or the operator. However, systems that differ widely in certain respects often perform the same functions, as do their operators.

The system also can be characterized in terms of *indeterminacy,* or the extent to which input, procedural, and response characteristics are permitted by system structure to *vary.* Another way of saying the same thing is that systems possesss greater or lesser amounts of uncertainty (roughly equivalent to the predictability of any individual response).

Inputs may vary as a function of the probabilistic nature of the system environment producing them. That environment may inherently have stochastic properties, or adversary systems within that environment may intentionally vary inputs to disguise them. When input variability is high, the system possesses a large amount of uncertainty. Under these circumstances, the responses permitted to the operator may also have to be highly contingent (variable) to accommodate that variability. The procedures available to the system for making this accommodation also must be flexible.

The indeterminacy factor is described by three dimensions:

1. *Input characteristics.* Inputs may vary in terms of their structure and variability over time. For example, unprocessed plan position indicator (PPI) imagery is highly unstructured (i.e., without definite patterning) and may therefore be interpreted in different ways. The same is true of inputs with many interrelated dimensions. Inputs that change frequently over time are highly variable. Both aspects contribute to the overall uncertainty of the system.

2. *Procedural variability.* This dimension is defined by the extent to which system operations can be modified during system functioning. If operating procedures permit little or no deviation or require little or no operator selection among alternatives, the procedure is relatively inflexible.

3. *Response preprogramming.* This dimension involves the degree to which the response required of the operator can be specified *in advance* of the input that calls the response out. As an example of a response-programmed situation, consider the sonar operator who must classify his inputs (however they may vary) in terms of two categories (i.e., submarine or nonsubmarine). If, on the other hand, the particular response to be made is dependent on individual input characteristics, the response is less programmed.

Any system that has highly structured inputs, relatively invariable procedures, and highly preprogrammed responses can be said to be "determined" before system operations begin. On the other hand, the functioning of any system in which inputs are unstructured and variable, and whose responses are highly dependent on the immediate input situation, cannot be predicted in advance, hence is "indeterminate."

The theme that runs through the various systems is one of probability; in determinate systems the probability of occurrence of certain events (inputs, procedures, outputs) is high; in indeterminate systems the probability is lower. Where probability is low, the operator must make choices among responses; hence indeterminate systems require the use of complex decision-making processes, and determinate ones do not.

Katz (1974) has pointed out that indeterminacy is a necessary and useful characteristic of many systems. It is hypothesized that indeterminacy is not only a major dimension differentiating among systems but is also responsible for much of the variability in operator behavior. Indeterminate systems and tasks are obviously of special interest to the HF specialist because processes and tasks in these systems are more highly dependent on the operator and can be affected by self-regulating mechanisms.

Indeterminacy in MMS has both positive and negative aspects. On the one hand, a greater number of options produces higher probability of operator error in selection of the correct option. On the other hand, the wider range of potential responses may permit the indeterminate system to respond more effectively to a wider range of inputs.

By their nature, indeterminate systems permit (and require if they are to be maximally effective) *adaptation* to changing inputs and development of operating procedures through learning and experience. Where the system is highly indeterminate, operators may be required to develop procedures to organize system inputs and outputs. This may be necessary because often there is no unique way of handling inputs and outputs.

Nevertheless, the goal of the system developer and the task of the HF specialist is to reduce indeterminacy to the minimum that permits the system to function most effectively. As the preceding paragraphs suggest, a degree of indeterminacy is frequently necessary and acceptable. Katz (1974) refers to "bounded determinacy." Given, for example, that the operator has the capability of recognizing N dimensions in his input, it may be desirable to allow him N contingent responses, provided also that the system can use these responses.

SYSTEM RESEARCH QUESTIONS

There are three such questions:

1. What is the relationship between the operator and the system? This

breaks down into, first, what is the effect of the operator on the system? In answering this question the operator and his characteristics are the independent variables, and the dependent (measurement) variable is the system output (e.g., fuel consumed, units produced). Second, what is the effect of the system on the operator? In answering this question the type of system and its characteristics are the independent variables, and the dependent (measurement) variable is the operator's performance (e.g., accuracy, response time).

Because of the primary role of the operator in the system, the relationship between individual variables and system outputs may appear to be obvious. Given the existence of an individual effect on the system, is that effect significant? The effect may be and often is minimal. It cannot simply be assumed that every variable influencing individual performance will have a critical effect on the system.

The importance of a factor in system operations can only be evaluated in terms of its effect on the system output. This does not mean that if, for example, excessive noise affects the operator's comfort, but not his production, we should ignore noise as a variable. It does mean, however, that on a priority scale of factors for system improvement or modification, noise is less important than some other variable that affects both the operator and his production.

2. To what extent do data, conclusions, and principles derived from studies of the individual and the group (i.e., when they function outside the system context) apply to the operator and the system? How are principles changed because of the system context and what mechanisms are responsible for the changes?

The individual in the system has become an operator subsystem and must perform in accordance with system goals and requirements; yet he functions also simultaneously as an individual, although his individual functioning must be subordinated to the system and may sometimes be at variance with system goals. The same is true of groups in such multiman systems as production lines and ships; groups are formed that involve direct verbal contact and these, although perhaps irrelevant to the multiman system mission, probably have some effect on that system's performance.

Despite the common elements among the individual, the group, the operator, and the multiman system, it cannot be assumed that principles applicable to the individual or the group apply also *without change* to the MMS.

Since one of the system assumptions referred to earlier is that all elements affect one another, hence the total system, it follows that principles of individual functioning should change under operator and multiman system conditions. How much of the data and principles available from the large behavioral literature can then be accepted as valid in a system orientation?

3. How can behavioral research results be transformed into system design principles? Since a primary goal of HF research is to assist in system design and development, behavioral research outputs must be capable of application. If developmental applications cannot be drawn from that research, the value of the latter is reduced (this is not to say that its value is nil).

Traditional scientific wisdom suggests that if one can explain phenomena, one can manipulate or modify these phenomena. However, hard experience teaches us that explanation does not necessarily lead to engineering design; knowledge of how a function is accomplished may not permit one to create that function.

GENERAL REFERENCES

Alexander, L. T. and Cooperband, A. S. System training and research in team behavior. Report TM-2581, System Development Corp., Santa Monica, Calif., August 16, 1965. (AD 620 606.)*

Alvares, K.M. and Hulin, C.L. Two explanations of temporal changes in ability-skill relationships: a literature review and theoretical analysis. *Hum. Factors,* 1972, **14,** 295–308.

Finley, D. L. et al. *An Analysis and evaluation methodology for command and control: Final technical report.* Contract N00014-74-C-0324, Naval Analysis Program, Office of Naval Research, Arlington, Va., 1975 (in press).

Hall, A. D. *A Methodology for Systems Engineering.* Princeton, N.J.: Van Nostrand, 1962.

Jordan, N. *Theses in Speculative Psychology.* London: Tavistock, 1968.

Katz, F. E. Indeterminacy in the structure of systems. *Behav. Sci.,* 1974, **19,** 394–403.

Meister, D. *Human Factors: Theory and Practice.* New York: Wiley, 1971.

Miller, J. G. Living systems: Basic concepts. *Behav. Sci.,* 1965, **10**(3), 193–237.

Miller, R. B. Some working concepts of systems analysis. American Institute for Research, Pittsburgh, February 1953.

Morgan. C. T. et al. *Human Engineering Guide to Equipment Design.* New York: McGraw-Hill, 1963.

Van Gigch, J. P. *Applied General Systems Theory.* New York: Harper and Row, 1974.

*AD-numbered documents can be secured from the National Technical Information Service, Springfield, VA 22151.

II
INPUT LOAD EFFECTS ON SYSTEM PERFORMANCE

SUMMARY. Inputs are examined in terms of how they produce a load on the operator and the system. Structural variables reviewed include number of stimuli, presentation rate, attention sharing in multitask situations, number of channels, and amount of information. Operator strategy, stimulus value and payoff, training and experience, and origanization (e.g., crew size, flexibility, division of responsibility) are also shown to affect operator performance. Conclusions from the 59 cases investigated indicate that although excessive load degrades operator performance, an effect that can be observed in all functions and tasks, these effects can be mitigated by the operator's strategies. Load effects are highly variable, often disproportionate to inputs, and sometimes dependent on the particular measure employed. Load exhibits the characteristics of additivity and modifiability; in multiman systems its effects on the operator and on the system, respectively, may differ. Organizational variables appear to be less important than structural ones in producing and controlling load effects. Display format can affect the way in which the operator handles his information load.

INTRODUCTION

The author has chosen to discuss input characteristics in relation to the concept of *work load* or, more simply, *load.*

Inputs—primarily, and to a lesser extent response requirements, which also serve as inputs to the operator—impose a demand (a superordinate input) on the MMS to which the system responds by modifying its performance. This demand, which the concept of load in its most general sense assumes, differs from the system requirement. Although load ultimately derives from system requirements (since the more demanding these are, the more likely it is that load will build up), the requirement imposed by load stems from the immediate task situation.

Thus inputs can be examined in terms of the load-inducing effects they produce, and this is the tack taken in this chapter. Despite its many disadvantages, the load concept organizes a large number of input variables in terms of a common framework. For example, load has been defined as " . . . *any* aspect of the task situation which requires that the subjects attend to and/or process increased amounts of input information" (Moore, 1961, p. 24; italics not in original). As we shall see in the following section, load is an intervening variable that can be used for both the operator and the system to relate inputs to outputs in a stimulus-response framework.

Definitions

Load is multidimensional and may be conceptualized in various ways.

1. Load may be caused by a discrepancy between an input attribute (often amount, frequency, or rate) and the system's capability to handle the input.
2. It may be expressed as the amount of physical effort required by a task or performed by the operator independent of task performance itself (Senders, 1970).
3. It may be experienced and expressed both as a subjective operator state (e.g., the task is difficult) and as an objective consequence (e.g., reduced performance).
4. It may be defined in terms of how much attention is demanded by a task and the consequent reduction in attention available for other tasks (Knowles, 1963).
5. It may be related to a concept of stress—the time required to execute a series of tasks divided by the time available to perform those tasks (Siegel and Wolf, 1969).
6. It may be viewed as both an independent and a dependent variable, as the input cause of certain system effects, and as the effects themselves.
7. It may be simply a special case of the classic problem of attention or an aspect of a general drive state.
8. It may be considered as an intervening variable between inputs and stimuli and outputs and responses, therefore inferrable only by the relationship between them.

If one accepts the validity of item 8, load in any individual case can be defined only in terms of specific antecedent conditions that lead to specified operator/system responses. In accordance with this concept, the individual sections in this chapter are phrased in terms of input variables, although for convenience we continue to use the term "load."

The work load concept presents special problems. For example, load is correlated with a number of physiological, psychological, and performance phenomena. In addition to objective indices of performance (e.g., a change in number of errors, an increase in response time, an increase in blood pressure), load is often experienced in terms of the operator's feeling of task "difficulty." Any task that is "harder" to perform or requires more physical effort to accomplish may be thought of as imposing a load. What happens then if two load measures describe the same task in different ways?

Load is difficult in define in terms of independent stimulus variables; one must discern a change in performance (response) before it can be said that the load (stimulus) condition exists or has existed. The stimulus must therefore be defined in terms of the response to that stimulus, which is inherently tautological and undesirable.

What this means is that if an experimenter imposes a set of conditions that he defines as load inducing, and the operator/system fails to react to these in the sense that no performance change results, either the system has withstood the load and his hypothesis (that load affects performance) was incorrect or the inputs were not load inducing at all and the hypothesis is still tenable. Neither alternative is very satisfactory.

The existence of load can be determined only by *comparison* with some baseline performance. Thus load can be inferred only when performance *changes* significantly because a changed input has been introduced. In the real world, as differentiated from a formal experiment, it may be difficult to distinguish a performance change resulting from a load input from "normal" operator variability, particularly when changes in input conditions are not highly visible.

The concept of load in terms of inability to keep up with inputs requires moreover the conception of a *capacity* of the operator or system to respond and a *demand* imposed by inputs exceeding that capability (almost like a jug filled to overflowing). Capacity represents a potential, a hypothetical maximum that is never observed unless the operator/system is exercised to the point of failure, which rarely occurs. Demand is similarly difficult to objectify and measure.

Despite these difficulties, the advantages of the concept (in tying together disparate input variables) more than outweigh its disadvantages.

Research Questions

It hardly seems necessary to prove that abnormal input conditions will reduce system efficiency and will lead, unless reversed, to system breakdown. Many of the studies described in subsequent pages demonstrate this. More specifically, however, one wishes to know the following.

1. What types of antecedent conditions and input characteristics are most likely to induce load effects?

2. How much load input produces how much load effect and how is that effect manifested?

3. Are there ways of ameliorating the load effect? What mechanisms can the operator/system use to adjust to and reduce load effects?

4. Do load effects vary as a function of certain system characteristics?

5. Are load effects for the system different from those for the individual operator?

Independent Variables

It has already been stated that any factor may impose a load on the operator/system. In this section we describe certain input conditions (e.g., number, frequency, and intensity) to which the load-inducing property has commonly been attributed. Note that nothing about the nature of the input itself (whether it be verbal report, Morse code, or radar blips) defines load, which is always relative to some set of operator, task, or system conditions.

Number of Inputs

A load can be said to exist when the total number of inputs or stimuli at any one time is greater than the operator can efficiently process. (For simplicity, "inputs" and "stimuli" are used here interchangeably, but the distinctions made in Chapter 1 are still valid.) A load may arise at the operator level because of increased demands on the operator for distributing his attention among these inputs. At the system level, for the same reason, an input queue may develop.

Inputs can be defined either as *event occurrences* or as *input sources.* For example, Conrad (1951) defines load as the number of separate streams of signals that occur independently of each other but require simultaneous consideration. In another context (Conrad, 1955), the same author uses the term "load" to mean the number of stimulus sources (channels or instruments) that can present critical signals for response. The "number of inputs" category can be broken down further.

Number of Relevant Critical Stimuli. Those to which the operator is cued to respond because of his instructions or other system criteria.

Number of Irrelevant Stimuli. Those which the operator must discriminate from relevant stimuli (e.g., "noise" or "clutter").

Number of Stimulus Dimensions. The more variations in stimulus dimensions, the greater the discrimination task required of the operator and consequently (all other things being equal), the greater the load. For example, stimuli differing simultaneously in size, shape, color, and brightness would pose greater difficulty for the operator than would stimuli differing only in size.

Stimulus Similarity

The more each stimulus resembles another, the greater the task load.

Stimulus Intensity

The weaker the stimulus, the more difficult it is to respond to it effectively (e.g., as in signal/noise relationships).

Stimulus Presentation Rate

Inputs may be presented (occur) so rapidly that the operator has difficulty keeping up with them. Tied in with this is exposure time; when presentation rate is increased, the exposure duration of any single input is likely to be decreased, which means that the operator's opportunity to scan the input is reduced.

Frequency of Stimulus Change

When the operator is exposed to a large number of stimuli and the frequency with which they change from one to another is high, the operator may have difficulty keeping up with the changes. On the other hand, when stimulus change is very slow (as in the traditional vigilance task), the operator's performance tends to degrade.

Number of Sequential Functions/Tasks To Be Performed

Quite apart from (although possibly related to) number of inputs and stimuli, the number of individual functions or tasks required of the operator, relative to the time available to perform them all, may impose a load on the operator. This is the measure utilized by Siegel and Wolf (1969).

Number of Concurrent Functions or Tasks

This situation is similar to the one just described, except that the operator's opportunity to distribute his attention is more restricted under *concurrent* task conditions. The more temporal overlap between concurrent tasks, the greater the possibility of their mutual interference. The notion of introducing a second concurrent task as a load condition has been adopted by some experimenters as a means of measuring the load imposed by the first task (see Knowles, 1963, and Rolfe, 1971, for a review of such tasks).

Physiological Demands

Load has often been conceptualized in physiological terms, as the amount of work or effort required of the operator to perform a task. From this, inferences can be made about the degree of load on the operator. An example is the effort to determine the physiological "cost" involved in flying a helicopter (Billings et al., 1970).

Response Variables

The nature of the responses required of the operator may also act as a load factor. For most of the input variables described so far, a corresponding output variable exists (number of required responses, response similarity, required response rate, etc.). The mechanisms inducing response load may be presumed to be much the same as those inducing input load. In this section we discuss only the general concept of the *response criterion,* which can serve as an overt or covert input condition.

Naturally, there is no load if the operator lacks an explicit or implicit response criterion. If, for example, he has all the time in the world to respond and is unconcerned about quality of response, load does not exist. One cannot, however, conceive of a system without some explicit or implicit output requirement that is used by the operator as a standard to pace his own output.

The response criterion, as a direct reflection of the system requirement, makes the stimulus possess value or payoff. In complex indeterminate systems, in which the output is often highly dependent on the operator's interpretation of the response criterion, it is probable that input factors interact with motivational variables; thus system output can be either much more or much less than one would otherwise expect with a certain input. If the criterion is not fully understood or accepted by the system's operators, an MMS capable of performing at level *X,* for example, may actually perform at level *X–n.* The problem of partial acceptance or nonacceptance of the criterion is more likely to occur when the system is indeterminate and the operator has greater freedom to select a strategy to deal with or accommodate to the criterion.

User Loads

Social-benefit systems [the sociotechnical systems of DeGreene, 1973], such as transportation or schools, provide a product or service that is received and employed by users. When the number of system users is very large, or their demands are excessive in relation to system capacity, they impose a user load on that system because of their expectations. For example, in a time-sharing computer system wherein users compete for the limited time available from the system, queues and increased waiting time may result. A bus or railway system may be loaded when the number of people to be transported is greater than available system resources can satisfy.

Social benefits and users may be found in all systems, social, industrial or military. Moreover, where the MMS has a hierarchy of subsystems, each higher-order subsystem can be considered as a user of the lower-order "feeder" subsystem and its products and may impose demands on the lower-order subsystems that load the total system.

The previous conditions are more descriptive of the individual operator and his responses than they are of the system. This is because the system is loaded primarily through its operators and users. The ways one can overload equipment mechanically or electrically (e.g., excessive heat or cold, a surge of power) are quite limited. The question is how variations in operator responses to load inputs are transformed into system effects.

A highly idealized version of the relationship between load and performance appears in Figure 2-1, which suggests that initially, as input load increases, performance improves. Theoretically, at some point (the exact point varies with a host of antecedent and interactive conditions), the operator "saturates"—that is, the demands imposed on him by the input equal his capacity to perform. If load continues beyond that point, load demands exceed this capacity, and the operator's performance deteriorates.

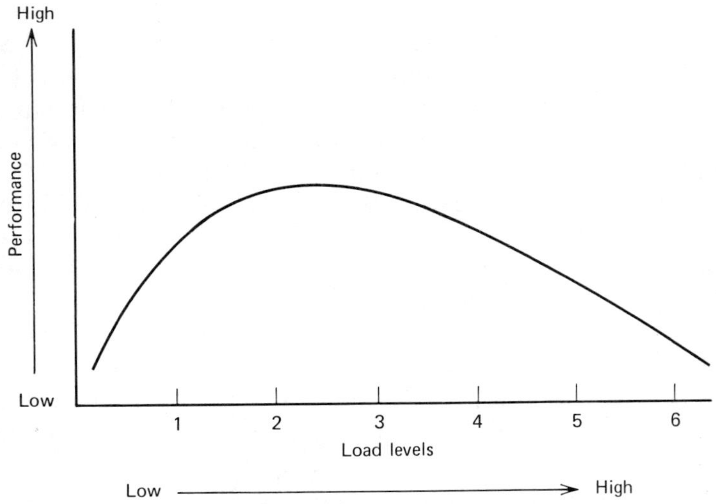

Figure 2-1 Hypothesized relationship between load and performance.

The inverted-U relationship expressed in Figure 2-1 bears a striking similarity to the Yerkes-Dodson law (Yerkes and Dodson, 1908) and to the formulations of arousal theory (e.g., Hebb, 1955; Malmo, 1959), which postulates a general drive state related to performance in much the same way indicated in Figure 2-1, Welford (1973) has also related signal detection theory to arousal and the Yerkes-Dodson relationship.

Stress and arousal theory have therefore been linked to load. Some researchers have hypothesized that low levels of stress (associated with low input load levels) have a stimulating effect that causes the operator to work harder, faster, even perhaps more accurately. In the Siegel and Wolf (1969) model, for example, initial stress buildup is assumed to have an organizing effect on operator performance as long as the value of the stress remains less than *Mj* (stress threshold). When stress exceeds *Mj,* the effect is disorganizing, and *Mj* can be considered to be the operator's breaking point. An *Mj* value of 2 indicates that the operator begins to work more slowly and less accurately at the point at which he has more than twice as much work to do (at average speed) as time in which to do it. Prior to this point, any backlog of essential tasks creates a stress factor that makes the operator's actions faster and more accurate.

In this formulation, since light input loads are correlated with an increase in performance, it may appear as if the light load *causes* an improvement in performance.

If arousal theory is correct, one might hypothesize that at very low input levels the operator is "input starved"; thus the increase in stimulation produced by an increased load activates him to work harder, faster, more efficiently. Some support for this is suggested by the evidence that the operator in the traditional vigilance situation—infrequent stimuli only slightly above the operator's threshold—exhibits a characteristic performance decrement as a function of time on watch.* However, as Adams et al. (1963) have shown, when the stimuli to be monitored are more frequent and well above threshold (hence more stimulating), the classic decrement does not appear. Case 73 also suggests that for easy tasks that are quite boring, the addition of some difficulty or stress factor improves performance. Stress in the form of electric shock has also been found to counter the detrimental effects of sleep deprivation in a radar-watching task (Bergstrom et al., 1973) and to produce more effective auditory vigilance performance (Kennedy and Coulter, 1975).

It is possible, however, that at low input levels the operator's capacity to

*See also Cumming and Croft (1973), who discuss information processing under varying conditions of task demand.

respond is greater than the load imposed, and the operator merely works up to a capacity that is natural and "easy" for him. Or he may view an easy task as a challenge, adjusting his output goals to play a game with himself.

Evidence for the latter alternative is presented by McDonald and Ellis (1974), who found in a simulated driving task that as task difficulty increases, the operator works faster *than required.* However the curvilinear relationship between stress and performance postulated by Siegel and Wolf (1969) was only partially confirmed by McDonald and Ellis (1974, 1975). Other attempts to verify the Siegel-Wolf stress performance relationship have produced variable results (Klein and Cassidy, 1972; Miller et al., 1964). Fineberg (1975) attempted to find the inverted-U relationship between performance and arousal in a perceptual obstacle-avoidance task. Results did not entirely support the relationship.

The relationship in Figure 2-1 presents the following additional difficulties.

1. The problem of attaching absolute quantitative values to points along the load level continuum has forced such theorists as Senders (1970) and Siegel and Wolf (1969) to make highly simplistic assumptions of behavioral functioning. A complex multidimensional concept like load cannot be compared to a psychophysical stimulus like the intensity of a point source of light, which has a threshold of perceptibility (or at least a threshold "range").

2. The figure suggests that the intervals between load levels are equal (i.e., the load continuum is a linear one). Such equivalence defines load as a *physical* input; but subjectively the operator may not perceive, and his performance may not reflect, that the interval (in whatever input units it is expressed) between levels 1 and 2 is equal to the interval between levels 2 and 3, and so on. Of course it could be argued that this is exactly what we want to find out: how performance relates to the physical input. However if the load continuum is not linear, its nonlinear relationship to performance becomes difficult to interpret.

It is also possible that different types of work load are induced by different functions and task situations, each with its own mechanisms, each requiring special modes of operator adaptation.

From a system design standpoint, which demands high precision, the load-performance relationship shown in Figure 2-1 has serious flaws. Singleton (1971) points out that the inverted-U relationship as an explanation of performance differences calls for careful definition, since it can be used to "explain" everything.

GENERAL PERFORMANCE EFFECTS

Conclusions

1. The general effect of load is to degrade performance.
2. Load effects are found in all functions, tasks, and systems.
3. Load may be induced in various ways (e.g., approach rate and number of aircraft under air traffic control, number of targets monitored, and number of simultaneous raids in air defense).
4. The load effect is variable, the operator response to load-inducing conditions often being disproportionate to the input.

Illustrative studies

Air Traffic Control

The most important behavioral studies dealing with air traffic control were performed by Kidd and his colleagues in a brilliant series at the Ohio State University (see Kidd, 1959, for a summary). They used air traffic control simulators consisting of 30 radar target generators and analogue facilities to provide realistic plan-position readouts on cathode ray tube (CRT) displays.

Case 1. In one study in which pairs of operators served as radar controllers in a simulated approach-control task, the approach rate was one aircraft per minute (high load) versus one every 2 minutes (low load). Approach time error (difference between actual and desired average approach times) increased with load, 11.13 seconds for low load, 18.43 seconds for high load.

Case 2. In another study of the Ohio State University (OSU) series, the entry rates per aircraft were 90-, 60-, and 30- second intervals. All measures were affected by increasing load. Table 2-1 presents system as well as operator measures, a characteristic of the OSU studies.

In line with previous studies (Schipper et al., 1956; Versace, 1956), performance decrement was sharply accelerated when the average interval between aircraft dropped below 60 seconds.

Case 3. The results of this study tend to confirm those of the previous one. When the average intervals between aircraft arrivals were 90 and 120 seconds, there was a 12.1% increase in delay under the higher input rate (statistically significant at the .05 level). There were also more separation errors and go-arounds for the higher input rates. As Kidd puts it, "Traffic input rate . . . has had an entirely consistent effect from study to study:

Table 2-1 The Effect of Entry Rate on Air Traffic Control Performance[a]

	Average Interval Between Aircraft (seconds)			
Measure	90	60	30	Significance Level
Percentage delay	40.1	75.6	129.6	<.001
Percentage excess fuel consumed	52.0	78.2	146.8	<.01
Separation errors per Aircraft Processed	.031	.077	.110	
Mean departures permitted per 30-minute problem	.026	.033	.064	

[a] From Kidd, 1961, case 2.

performance falls off as traffic load is increased. When this variable is increased at more than two levels, the performance decline is accelerated" (p. 12).

Case 4. In another experiment designed to test the effect of noise on air traffic control performance, six versus eight aircraft under simultaneous control were compared. There was more than 30% decrement in performance (additional flight time for the larger number of aircraft under control). However there was no interaction between noise and task load.

Case 5. Here load was defined as the level of aircraft monitoring required of the controller: aircraft not equipped with airborne position information (API) displays required the controller to determine speed, altitude, and heading changes; with API-equipped aircraft, he merely had to monitor.

There were four API levels: (1) no API, in which all pilots had no position information and were under positive vector, speed, and altitude control throughout the approach, (2) 37.5% of the pilots had access to API, (3) 62.5% of the pilots had access to API, and (4) all pilots had API. A constant control load of eight aircraft was employed. There was an approximately linear decrease in percentage delay—17%—as positive control requirements decreased. This is comparable to what was discovered by Kinkade and Kidd (1959a).

Air Defense

In the air defense situation, aircraft tracks are monitored by an observer who assigns appropriate weapons (e.g., fighters, missiles) to attack the enemy. This situation, like that of air traffic control, involves detection, monitoring, tracking, and distribution of the operator's attention.

Case 6. The purpose of this study was to investigate two methods of work distribution (organizational structure) as a function of task load. The simulation involved a metal board comprising a 50 × 40 grid having three target areas in the center. Model planes with magnets attached formed "friendly" and "enemy" air forces. The task was to defend the target areas against bomber attack, and planes were moved a given number of grid squares in accordance with rules.

The load variable was number of aircraft employed (15 = high load; 10 = low load). Not surprisingly, load performance scores, based on points added for downing aircraft and subtracted for having one's own aircraft downed or the target bombed, were: 399 (high load) and 598 (low load) ($p = .05$).

However, the organizational structure variable (which will be discussed later), was not significant, nor was the structure × load interaction.

Case 7. In another study a team of five operators was assigned to detect, display, transform, redisplay, and correlate data concerning the temporal position of dot sequences displayed in a rectangular grid. Load was defined as the number (10, 20, 40, 60) of task units to be processed per unit time. The study dealt with the relation between experience and performance as a function of load, where experience was defined as the number of problem situations in which subjects participated. Mean total error was an almost linear function of load. However, this relation was a function of the performance measure—total error—selected. Experience will be discussed later.

Case 8. A series of studies by Adams and his colleagues (e.g., Adams and Humes, 1963; Adams et al., 1962) has explored visual load in the context of a simulated air defense monitoring system. The experimental task involved multiple stimulus sources, symbolic stimuli, and persistent critical stimuli, features common to monitoring tasks in modern MMS. The task simulated a semiautomatic air defense system in which a digital computer periodically displays information about an aircraft to an operator. An animated 35mm strip film presented data on a screen for a 1000-mile surveillance area, and aircraft had speeds of 450, 900, or 1800 knots. Each aircraft symbol had its own three-digit identification number and letter (G). The subject's task was to detect the change of the G to F and report as fast as possible, for example, a symbol marked G521 that changed to F521. Only one signal changed at a time, and the critical F signal persisted for 20 seconds before reverting to G.

Mean detection latencies of 1 to 2 seconds were found with the moderate load condition (6 signals); heavy load (15 signals) increased detection latencies by a factor of 3.

Case 9. In a similar study utilizing the same experimental situation as in Case 8, load was contrasted between 36 and 6 signals. Load was also varied in terms of response complexity: detection *(D)* alone (light load), versus detection plus evaluation *(E)* (heavy load). Again, no appreciable decrement in response latency over time for persistent stimuli was found, but performance under heavy load was significantly less efficient than under moderate load (Figure 2-2). Of even greater importance, performance with the detection response was also superior to that of the detection evaluation response (confirming that response complexity can act as a load condition).

Case 10. The load condition here was the number of enemy tracks displayed in any one session (60–96). Upon appearance of the first incoming targets, the operator identified them, determined their speed, altitude, and raid status, selected appropriate counterweapons, and assigned the weapons against the threats.

The five highly trained subjects exhibited a generally rising performance trend with load (number of "kills" as in Figure 2-3) that tended to converge on an asymptote, followed by some leveling off in number of tracks "killed" at higher levels (see session II). Along with increasing load there was a rising system cost in terms of weapons expended and a somewhat less evident "damage" trend (i.e., number of aircraft lost).

Subjects were obviously able to handle the increased track load. The increased performance under higher load levels is in marked contrast to previous cases in which increased load resulted in reduced performance. In previous cases the comparison of load levels was on *discrete* loads (e.g., 10 vs. 5 aircraft). In Case 10, performance with increasing load was measured *continuously.* The discrete situation may exaggerate negative load effects because it pictures extremes and subjects are not permitted to work up to their capacity.

For the first time also, we see an operator strategy involving selective attention based on stimulus value and payoff (interactive variables). Subjects tended to concentrate their attention on tracks involving higher level threats (e.g., bombers) rather than on those with lower level threats (e.g., fighters).

Case 11. In a similar situation, nine experienced commanders in two groups (varying in experience) were presented over 120 sessions with 60 to 96 tracks and a variety of aircraft and missile weapons. Two levels of threat were presented.

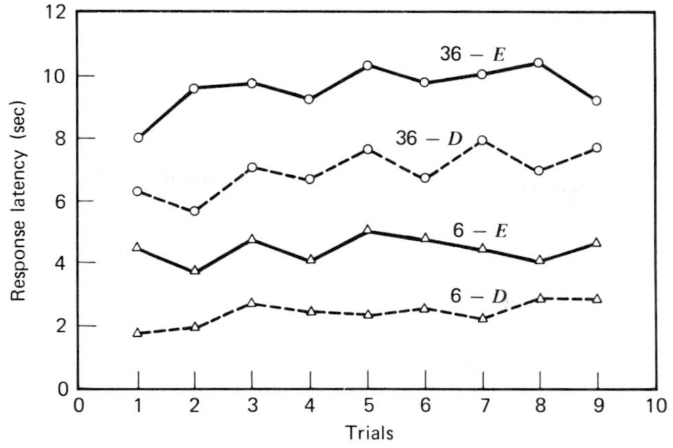

Figure 2-2 Response latency as a function of number of signals (6, 36) and response complexity *(D, E)*. With permission, from Adams et al., *Hum. Factors,* 1961, Case 9.

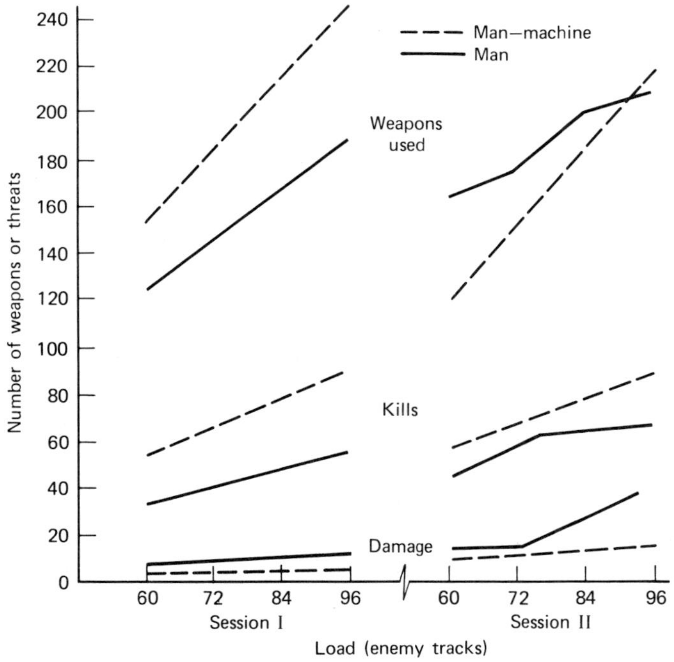

Figure 2-3 Number of kills and weapons expended as a function of track load. With permission, from Connolly et al., 1961, Case 10.

As in Case 10, the number of kills increased as the load increased (Figure 2-4). A definite pacing effect was observed; as load increased, commanders increased the rate at which they reacted to the situation. Frequent exposure to near-saturation task loads tended to ensure maximum performance. Nor did track load have a significant effect on percentage of weapons lost and cost.

However, load effects were by no means absent. For example, the backlog of tracks unassigned to weapons increased with load. Group I commanders (more experimental experience) generally sustained less damage than did those of group II, indicating the effect of experience on performance. Group II, however, showed considerable improvement with limited practice except at the highest loads. Group II demonstrated

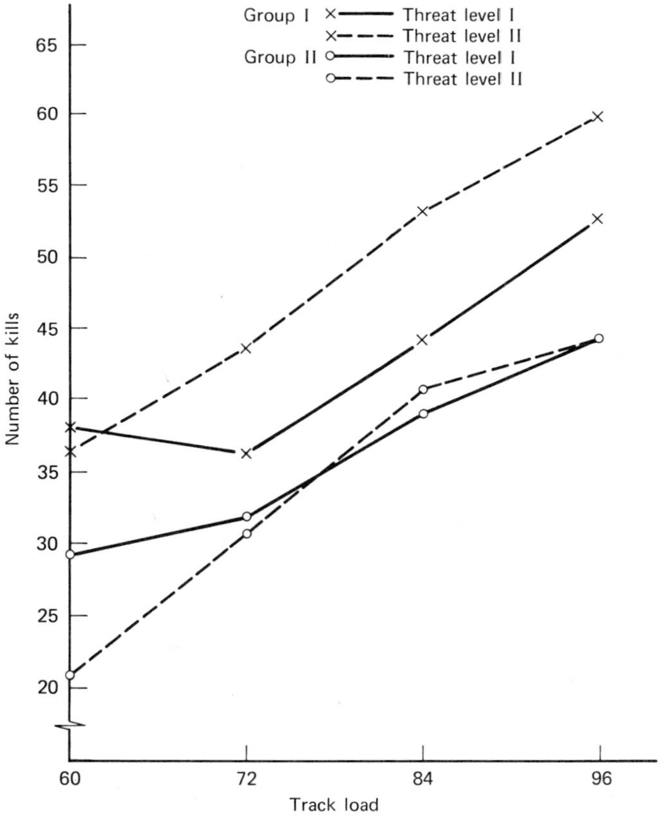

Figure 2-4 Number of kills as a function of threat level and experience level. With permission, from Fox and Vance, 1961, Case 11.

ability to "keep up" with the load but tended to become more saturated at the highest loads than the more experienced group. Effects comparable to those reported in Cases 10 and 11 were found in a subsequent study involving the same situation (Connolly et al., 1961).

Case 12. Another series of early studies on air defense dealt with what was termed "raid load," the number of simultaneous raids with which the Air Control Officer (ACO) had to deal. The ACO's performance is dependent in part on the amount of time available to intercept an enemy raid after he has taken control and before the raid has reached some protected line. Raids in Case 12 overlapped in time but were not simultaneous. Hence the amount of temporal overlap could be considered to be another measure of load. Controller performance in a mockup of an airborne CIC was studied at four levels of raid overlap: 40, 55, 70, and 85%. Mean percentage of penetration distance (by enemy aircraft) significantly decreased as time available for intercept increased from 2 to 6 minutes. This relationship interacted with raid load. Successful interceptions of enemy aircraft significantly increased as a function of increasing time to intercept. No significant difference in mean percentage of penetration distance was found as a function of variation in *percentage* of raid overlap.

Case 13. In another study in the same series, mean percentage of penetration distance increased as raid load increased but declined sharply when time to intercept increased from 2 to 6 minutes (as we saw previously). Of greater significance, order transmissions (communications) per raid increased (an indication of adaptation). However, no relationship between controller performance and number of orders transmitted per raid was found.

Case 14. In a third study using the same experimental situation, it was found that all intercept time/load fractions within a single ratio of minutes per raid were of equal difficulty, affecting ACO performance equally, irrespective of the absolute values of the time and load components. Thus a situation of two raids with 2 minutes of time available was neither more nor less difficult than a situation of six raids with 6 minutes to intercept. Number of raids and time to intercept apparently cancel each other out, the former being load- inducing, the latter load- counteractive.

The concept of load-counteractive variables (and the possibility of these existing in other situations) is intriguing.

Case 15. Target load in a radar detection situation may be broadly defined in terms of space and time density, the first referring to the distribution of targets on the scope face, the second involving the total number of occurrences per unit time, including new and old targets, target course changes, and so on. Detection time was a positive exponential function of the number of targets, and there was an almost linear relation between time between successive reports and number of targets on the scope (Figure 2-5). However the number of plots per unit time remained constant under varying load (mean, 5.26 per minute). Increasing the number of targets had no negative effect on accuracy of range and bearing estimation. In fact, there was some improvement, suggesting that load may affect dissimilar system functions differentially.

Case 16. Similar results were reported in another early radar detection study in which six experienced officers in a Navy aircraft CIC were faced with five different time densities. As one would expect, probability of detection for new and old targets decreased markedly as time density increased (Figure 2-6). The functional relationship between the probability of detecting discrete events and time density was demonstrated to be a probability function taking the form $p = ae^{bx}$ where x represents time density in other than linear form.

More recent studies give comparable results. In research by Bell et al. (1974) subjects performed a 90-minute watch-keeping task requiring

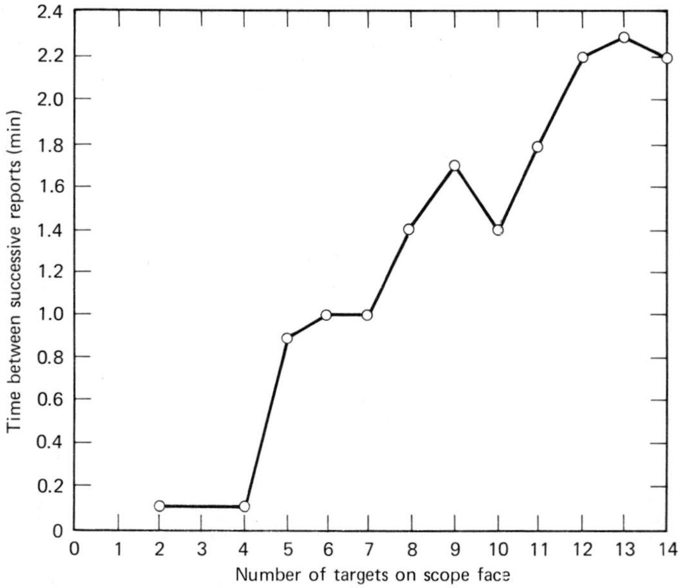

Figure 2-5 Time between successive target reports as a function of number of targets. With permission, from Lefford and Taubman, 1950, Case 15.

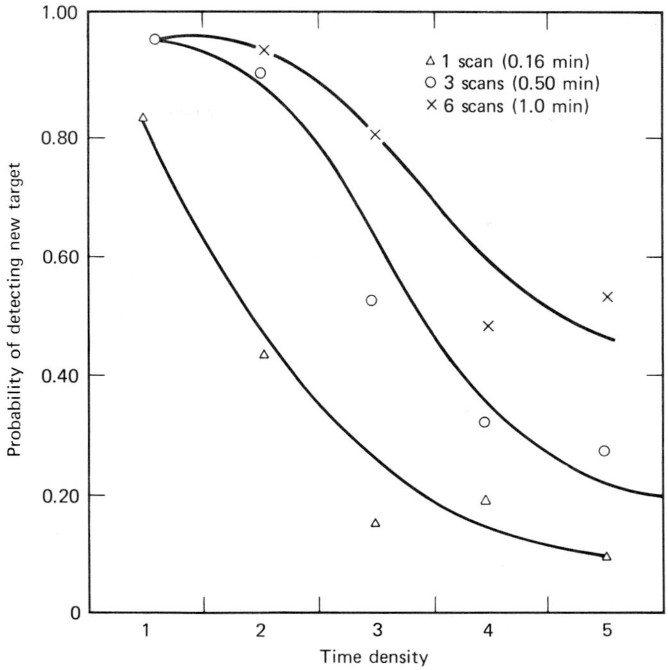

Figure 2-6 Probability of target detection as a function of number of targets per unit time. With permission, from Sinaiko et al., 1951, Case 16.

detection of a plus sign that appeared occasionally in a matrix of solid circles. Stimulus density (among other variables) varied between 4 and 12. Although detection times for the two densities were roughly the same, accuracy was significantly affected: on the average, subjects made 39% more errors with the higher density.

Case 17. In a semiautomatic air traffic control and airborne warning/surveillance system, track initiation required the operator to identify a radar return as belonging to a suspected aircraft. The operator designated which one of five returns was to be tracked by computer by light-penning the return on his CRT and depressing a key. Track maintenance was performed in the same manner except that a different response key was depressed. The load variables were as follows: (1) target introduction rate (two, three, and four targets per minute), (2) clutter density (90, 180, and 360 blips per scan), and (3) target velocity (500 and 900 knots per hour). Velocity had a significant effect on three of the performance measures utilized. Target introduction rate and clutter both exerted significant influence on track initiation and maintenance, degrading performance as rate and clutter density increased. The major impact of these variables however, was, at their extreme levels (four targets per minute and 360 pieces of clutter) (Figure 2-7). There was a significant clutter × rate interaction, as one would expect. Most performance errors in track initiation and maintenance were affected significantly by clutter and rate, but not by velocity.

Although clutter density significantly affected performance, this effect occurred only at levels that were unlikely, the investigators felt, to occur in operational situations. The estimate of the maximum load that could be handled by a single operator was seven simultaneous tracks (up to three new targets per track). (Metters, 1974, found a mean saturation level of 10-12 targets in his study.) There was a gain in performance through additional experience (subjects for this study had previously served in a preliminary experiment).

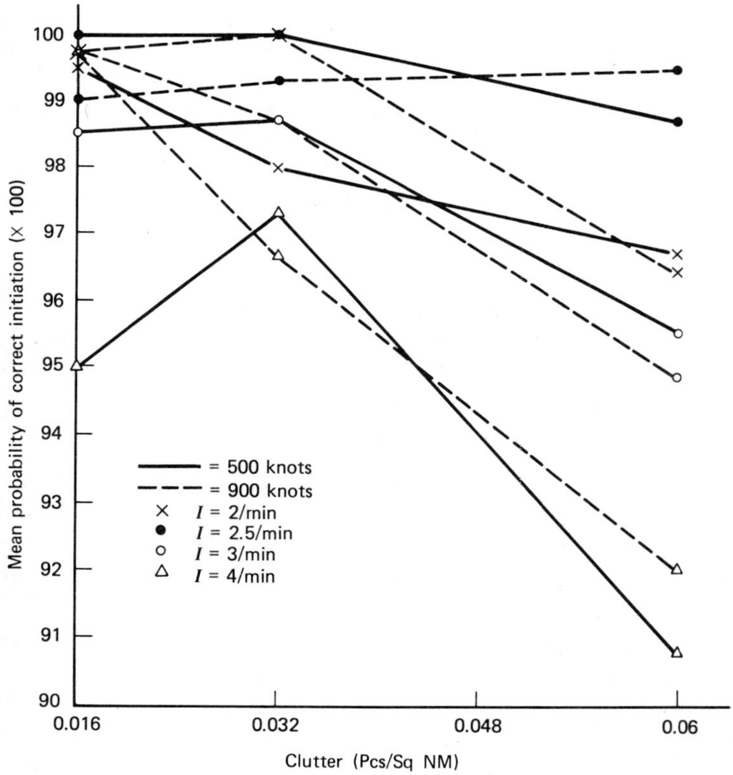

Figure 2-7 Probability of correct track initiation as a function of number of targets, clutter, target introduction rate *(I)*, and velocity. With permission, from Mills and Bauer, 1971, Case 17.

Visual Functions

Load effects have also been studied in visual tasks calling on the operator to perform functions other than detection. In these tasks the load variable is usually the number of items displayed (display density), which the operator must either identify, locate, recognize, count, or recall.

Case 18. In a study that demanded the counting of the number of symbols displayed on slides, the load was 20, 60, and 100 items per slide. Average counting time and counting errors increased almost linearly as a function of number of items displayed.

Case 19. Similar load effects were also found in recognition and recall of letters projected on slides: 79.4, 49.5, and 36.3% of the stimuli were transmitted correctly with 4, 8, and 12 letters per slide. Number of responses increased directly with load: 3.52, 4.49, and 4.81 responses made for 4, 8, and 12 letters per trial.

Case 20. In this case accuracy of report was found to be an almost linear inverse function of the number of letters per slide presented. Load rate (i.e., the rate at which load was increased or decreased over time) was varied experimentally. This variable is hypothetically important because, as in an air traffic control situation, load might increase as a function of nearness to some critical time and then decrease. Differences in performance among the various rates had negligible effects.

Case 21. The effect of load also appeared in a visual search task in which the stimuli were coded. There were 40, 80, and 120 symbols presented per display. The operator tasks were to identify, locate, count, compare, and verify. The differences in the number of correct responses made per minute for the three load levels—11.36, 8.45, and 7.60—were highly significant statistically.

Conclusions. Many of the load situations described previously involved only a single function or task. Multiple-task situations present a somewhat different type of load. A single type of input (e.g., aircraft tracks) may impose a load by increasing its frequency of occurrence or the number of instances of the same type of input. More commonly in the real world, however, one finds that different *types* of inputs requiring different behavioral functions (e.g., monitoring and communicating) are presented concurrently. Multiple-task situations are *inherently load inducing* because the tasks compete with each other for the operator's attention.

The system context therefore affects the performance of the individual task. For example, McKenzie (1969) recorded data on five monitoring/tracking tasks, individually and as part of total system operation. Performance of four of these tasks during system operation degraded slightly or significantly; one task improved.

The multiple-task situation has been examined most intensively by Chiles, Alluisi, and Adams (see case 22 following) in their studies of the effects of extended operations on performance. Their task battery consisted of two active tasks (compensatory tracking and mental arithmetic) together with three passive tasks (monitoring a pair of meters, responding—choice RT—to a pair of red/green lights, and responding to a peripherally located amber light). The workload effect can be measured by comparing performance of a single task, such as monitoring, against performance with the full task battery.

Case 22. The effects characteristically found are revealed in this study in which two groups of four subjects each were tested over six sessions. Each session consisted of four performance hours. During two of the four hours, the performance of the full battery was required; during the other half of the session the monitoring tasks were performed alone. Data were recorded for each 15-minute period of performance.

The percentages of correct auditory vigilance detections for the two conditions are plotted in Figure 2-8. It is immediately apparent that the difference between the monitoring-only and full battery conditions is highly significant. Similar effects were found for the other types of tasks.

The authors of the study speculated that within the context of this task battery, "workload" probably represented the amount of "free" time the operator had to scan or concentrate on the monitoring tasks. Thus a

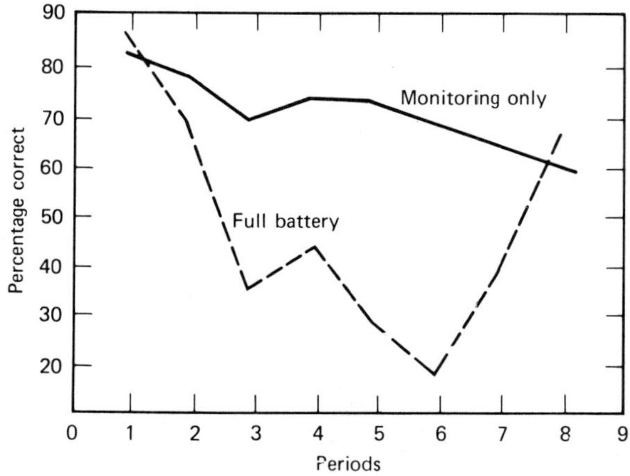

Figure 2-8 Effect of multiple-task loading on accuracy of auditory monitoring. With permission, from Chiles et al., *Hum. Factors,* 1968, Case 22.

complex of tasks obliges subjects to distribute their attention over a variety of displays if they are to maximize their performance. In such tasks subjects must be flexible in their information handling behavior.

Case 23. Effects similar to those of Case 22 were demonstrated in a multiple-task situation utilizing the same inputs. The work load variable was the combination of either two or three tasks in comparison with the baseline task (single-task performance).

The results of the study indicated that load had an effect on certain tasks but not on others, and that the load effect (for monitoring performance) was somewhat inconsistent. Although performance in monitoring and tracking was affected by the total battery, arithmetic performance was unaffected. Adding a third task to the monitoring requirement produced an improvement in performance with certain measures, although those effects were not statistically significant and the cause of the improvement is obscure.

The investigators pointed out that the reasons for the differential effect of load on the tasks affected must lie partly in factors specific to the tasks themselves and their organization within the overall tasks, rather than in a general load condition. For example, the effect of work load on movement time in responding to the choice reaction time task in Case 23 was probably the result of a perceptual-motor conflict due to the occasional illumination of a red or green light during the arithmetic problem.

Work load resulting from the addition of tasks is an excellent example of what can be termed "intrinsic" load because the load results from the competition between the tasks themselves, not from inputs arising outside the system.

Thus load may be designed into task structure, either deliberately (as a consequence of mission requirements) or inadvertently (as a result of design error). Load may also arise from either temporary or continuous extrasystem (environmental) contingencies. Intrinsic load should be of particular interest to system developers because it is potentially more controllable, through appropriate design, than is load resulting from extrasystem inputs. Whether the effect on operator/system performance of these different load conditions is the same is not known.

Case 24. This multiple-task situation involved both monitoring and tracking. The load variables were (1) speed of information presentation (measured in terms of number of deviations from a null point—24, 36, and 48—programmed through a number of dials in a given unit of time), and (2) number of dials through which information was presented (6, 12, and 18). In terms of reaction time and time off target measures, both load variables were significant at the .05 level, as Table 2-2 indicates.

Thus the operator should be able to handle a greater inflow of informa-

Table 2-2 Load Effects in a Multiple-Task Situation[a]

Presentation Rate	Mean Transformed Value	Number of Dials	Mean Transformed Value
48 units	10.10	18	10.24
36 units	9.28	12	9.70
24 units	8.98	6	8.42
	Time Off Target (seconds)		
48 units	96.18	18	87.08
36 units	79.30	12	85.11
24 units	70.32	6	73.61

[a] From Olson, 1963, case 24.

tion if it is channeled through fewer displays. Sending the same amount of information through a greater number of channels would presumably reduce performance.

Case 25. The variability of load effects is revealed in a study in which there were two five-man crews and three signal rates (9.4, 15.5, and 20.6 signals per hour). The critical signal to be detected was a significant shift in any of four meter pointers. When detection time served as the performance measure, input rate did not significantly affect the tasks, either singly or in combination with the others.

Case 26. What differences in performance result when the operator must process two sources of information simultaneously, as compared with alternating between the two information sources? Subjects performed a compensatory tracking task while listening to and memorizing a list of eight letters. Not surprisingly, time-shared performance in simultaneous processing was reliably worse than control performance; equally understandable, the decrement was greater for the more complex signal (Figure 2-9). This decrement was reduced by practice, and again, the reduction was greater with the complex signal. Recall became worse with time-shared and better with the less complex signal.

Similar results were found by Wiener (1975) in a study of simultaneous monitoring and tracking. In his situation, however, inclusion of knowledge of results for certain groups almost completely eliminated the negative effects of time-sharing.

The results just presented suggest that operators can reallocate their processing capacity across input channels in proportion to the probability that each channel will carry information relevant to the response required

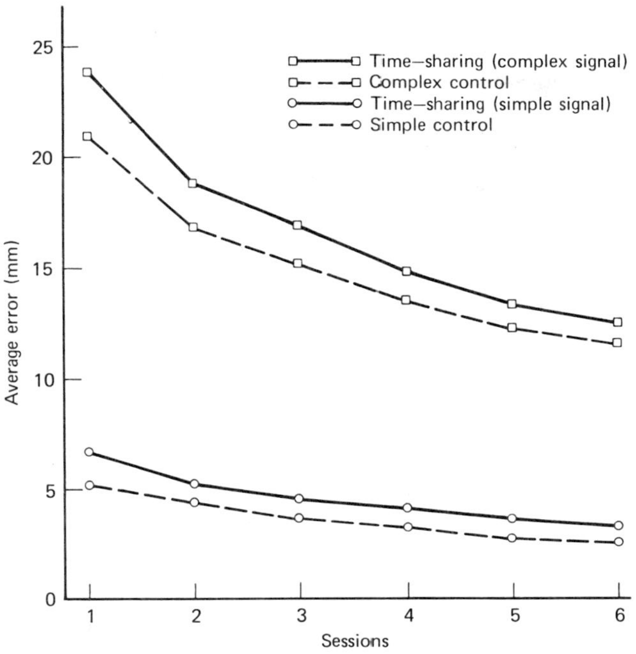

Figure 2-9 Tracking error as a function of time-sharing and signal complexity. With permission, from Shulman and Briggs, 1971, Case 26.

on each trial. In fact, there is some evidence that time-sharing performance can be improved by training (Gabriel and Burrows, 1968).

In such multiple-task situations it is necessary for the operator to develop a strategy for the allocation of his attention. At times this strategy is maladaptive. For example, an operator may take relatively large losses of time on high-probability/high-frequency information channels to ensure response to another channel in which signals are very infrequent. This is almost certain to result in performance decrement. Decrement in Case 26 also was attributed to the sequential processing situation, even though information sources did not overlap. These results apply also to measures of speed. Decrement occurred primarily when there was heavy load.

Conclusions. The effects of various types of load inputs may be additive, manifesting themselves in the following ways: (1) two input variables individually have an effect on performance, but their combined effect is much greater than their individual effects; (2) alternatively, neither of two input variables has a loading effect on performance, but when combined they exert a significant effect.

Illustrative Studies

Case 27. Load in one information processing study, was defined in two ways: input rate (17, 20, 24, 30, and 40 messages per minute presented visually), and number of stimuli (2, 3, and 4 alphanumerics per word). Both variables exhibited significant load effects. At given input rates, increases in number of stimuli produced large decrements in performance; with number of stimuli constant, input rate increases also adversely affected performance. The important point is that interaction effects were compounded (Figure 2-10). The interaction effect yielded the largest single component of the statistical variance. Performance as measured by percentage of errors fell off very rapidly when input rate and number of symbols were combined.

Case 28. A study of performance in air traffic control illustrates situation 2 in the foregoing Conclusions section. The controllers (nine subjects) used a simulated surveillance radar scope to vector aircraft. The load variable was number of aircraft (three, six, eight) being vectored. The additional variable contributing to load was the speed of the aircraft (all at 145 mph; at three different speeds: 130, 145, and 160 mph; at six different speeds: 120–190 mph).

When all aircraft were flying at the same speed, variations in performance measured in terms of constant error were insignificant. Moreover, the controller could handle eight aircraft as well as he managed three. When the simulated aircraft were flying at different speeds, however, there was a marked difference in quality of vectoring as the number of aircraft to be controlled increased. Probably the irregular pattern of aircraft speeds added to the task a load dimension (complexity) that did not exist when varying numbers of aircraft were flying at the same speed. With the combination of speed differences and increased number of aircraft, the number of collisions increased markedly (Figure 2-11). The wider the range of speeds, the greater the decrement produced by the load condition.

Case 29. The same additive effect appears in an entirely different context—a radar detection task in which the load conditions were (1) target densities (i.e., number of echoes: 5, 10, and 15) and (2) blip scan ratio (the probability of a contact "painting" on the display: .5, .75, and 1.0). A trend was found for increased target densities and lower blip scan ratios to degrade performance, but the individual load conditions were not statistically significant. The *interaction* of the two variables was, how-

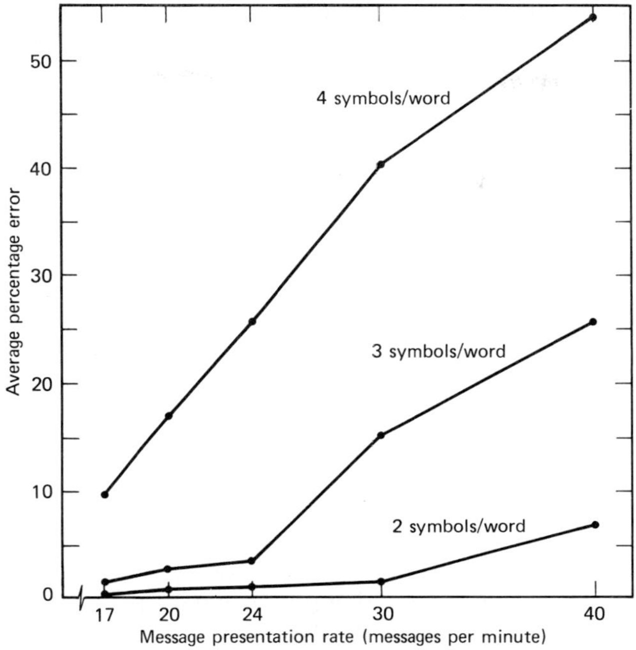

Figure 2-10 Error as a function of input rate and number of stimuli. With permission, from Jeantheau, 1959, Case 27.

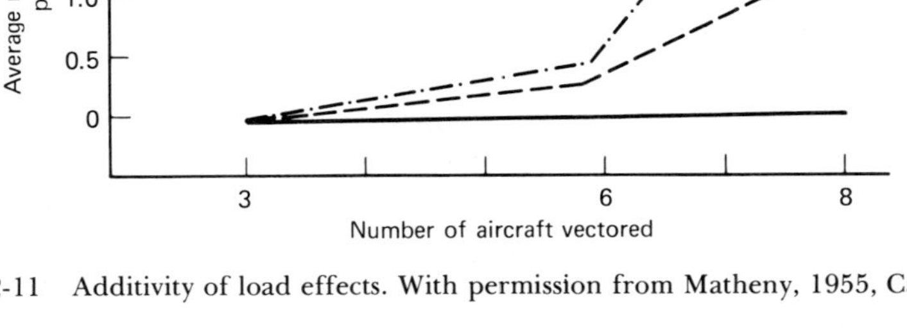

Figure 2-11 Additivity of load effects. With permission from Matheny, 1955, Case 28.

ever, significant; there was a sharp decline in performance when target density was more than 10 and the blip/scan ratio was less than .75. There were also significant individual differences among operators, an effect one must expect. The investigators could not attribute marked variations displayed even among trained and experienced operators to differences in visual acuity or general intelligence.

Conclusions. In multiman systems consisting of several subsystems, load effects may be variously distributed, perhaps because of individual differences in system structure or in the functions performed by subsystems. The effects of load on the operator and the system may not be equal.

Illustrative Studies

If, as we assume, the MMS is an entity distinct from the operator, differences in load effects should appear if the load effect can be partitioned between the operator subsystem and the total system. Unfortunately, very few studies address the problem, because many studies are not performed in a system context and, even if they are, there is so much identity between the system and operator output that it is difficult to differentiate between the two. To permit this differentiation, system output measures must be distinctly different from those of the operator. For example, the operator response measure might be number of errors or detection time; the system output might be fuel consumption or number of flights launched.

Case 30. One illustrative study was performed in a full-scale mockup of an airborne CIC, with three teams of from two to four search operators and two evaluators. The load variable was defined in terms of target minutes on the scope, a target minute being defined as the movement of a target on the scope for one minute. Target minutes were additive. Load was also defined in terms of events (the appearance and disappearance of the target on the scope). (See also Case 15.) The four levels of load were 40, 80, 120, and 160 target minutes, together with corresponding 6, 12, 18, and 24 events. Six days of practice were given.

System saturation occurred during the heaviest load density, although the individual search operators were *not* saturated, as evidenced by the finding that two operators reported information as effectively as four. Saturation of the system must have occurred at a location other than the search operator stations, but exactly where is not clear, since load density made no difference to evaluator performance. Other variable load effects noted in previous cases were also found in this study. For example, there were nonlinear effects: performance showed a steady increase through load level 3, after which additional load produced no increase in responses. However, rate of reporting increased with increasing load. As the load density increased, course and speed report production remained proportionately constant; other amplifying information proportionately decreased, however, and this may be interpreted as an operator filtering strategy. Percentage of detections decreased as load increased, but detection of new targets remained constant for load levels 2, 3, and 4. The initial light load showed superior detection time when compared to these three loads. Position reports were more accurate under intermediate loads than under either the light or heavy load conditions, which supports the nonlinear relationship between load and performance suggested in Figure 2.1.

OPERATOR FACTORS

Conclusions. Significant individual differences among operators in response to load are often observed. Much of this variability results from the operator's adoption of a "strategy" or method of adapting to increased stress.

Illustrative Studies

There are two types of operator variability: interoperator and intraoperator. It is well known that different operators often respond differently to the same load. The same operator may also respond differentially to the same load condition at different times as a consequence of temporary changes in such factors as alertness and motivation.

That operators respond differentially to the same load may suggest that the load taps an individual aptitude to resist the load effect. Indeed, personality variables may substantially mediate the effects of load variables (Sales, 1969). This simply confirms what novelists and other observers of life have often noted, that individuals differ in the way in which they handle inputs. The effect of long-term experience is to reduce this variability; yet even so one still finds significant differences among skilled operators (see Case 29).

From a system design standpoint therefore, it seems wise to select operators based on their "load resistance" capacity. Unfortunately, we have almost no data relating this capacity to enduring aptitudes or traits.

Variability in response to load may result from the operator's tendency to adopt (deliberately or unconsciously) a "strategy" or a method of adapting to the increased stress of system inputs (Morgan and Alluisi, 1969). There is no necessary inconsistency between hypothesizing an inherent individual ability to resist load and the ability to adopt a more efficient strategy. Those who are inherently resistant to load may display this trait because they are more efficient in selecting adaptive strategies.

One of the most significant findings from studies of load is that the system and the operator are not completely captives of the input load condition. Just as the operator increases his efficiency at low load levels, he adapts once more when load becomes very stressful, by applying one or more of the following mechanisms:

Response Selection. The operator continues to make the responses he values and reduces his output of less valued responses. Nonessential behaviors are eliminated.

Response Omission. In tasks that permit this modification, responses may be omitted.

Response Queueing. The operator may delay responding during periods when inputs overlap, in the hope that he can catch up during a lull.

Filtering. The operator may neglect to process certain information categories. Filtering differs from response omission, however; in response omission the operator fails to respond at all, whereas in filtering he responds to the input but in terms of fewer dimensions. For filtering to occur, the input must possess more than one dimension that can be evaluated in terms of importance.

Criteria Modification. A variation of the filtering mechanism is to lower discrimination criteria: the operator reduces his categories of discrimination, responding to inputs with less precision (e.g., chunking). In a communications situation, for example, the operator may reduce message duration by eliminating secondary or unnecessary words.

Load Balancing or Time Sharing. The operator may employ multiple channels in parallel. He may partition his attention between input sources.

For the operator to use the preceding mechanisms to adapt, the task situation must permit him to modify his responses. In highly proceduralized (determinate) systems, in which inputs are unequivocal and demand an invariable response, these mechanisms cannot be easily employed. Freed (1962) has pointed out that human interactions play a vital role in the reliability of the MMS.

One might assume that adaptation occurs primarily in tasks characterized by considerable cognitive loading. However, operator adaptation can also occur in tracking tasks. Ince and Williges (1974) found that the operator lags in adaptation to changing system dynamics but does adapt when given sufficient time. As the rate of change increases, a larger change must occur for the operator to be able to detect it, but he needs less time to do the detecting.

Case 31. In an empirical study of air traffic control at Orly (Paris) Airport, the load variables were number of aircraft (four to eight) and number of aircraft already under control (zero, two, and four). The dependent measure was the agreement between routing solutions chosen by controllers. This was at a maximum when traffic load was very low (four aircraft) or very high (eight aircraft).

In low traffic aircraft were processed individually; as traffic increased, the number of direct (shortest) routes decreased and a greater amount of data per aircraft was processed. Most significantly, when the traffic level increased, the controller reduced the number of variables he processed. In low traffic he took into account performance data, flight levels, and geographic data. In high traffic he considered only flight levels, since aircraft fly out of the stacking area from the lowest flight levels. Here strategy took advantage of environmental constraints.

Apparently the controller's freedom to select solutions is greater when there is middling traffic. At low traffic density there is little constraint on the controller, but the number of possible solutions is limited. The number of possible solutions increases when traffic increases, but constraints also increase because of air space saturation or the operator's limitations. Message duration decreases slightly but significantly; secondary or unnecessary words and messages disappear (filtering, chunking). The controller becomes less passive *vis avis* the pilot, and he initiates more messages. Thus when work load increases with traffic, the increase is modified by changes in operator strategy.

Case 32. In other air traffic control studies the strategy adopted by controllers was that of maintaining the maximum degrees of freedom as long as possible. When aircraft were far from the center of the control zone, no attempt was made to set up precise landing sequences or to predict possible conflict between aircraft. Instead, the controller made judgments about entire groups of aircraft (e.g., that he would be able to accommodate a given number of aircraft). As aircraft approached the GCA gate, successive decisions were made that gradually affected the required temporal or spatial separations establishing the landing sequence. This type of strategy requires minimum communications and changes in flight paths and speeds and makes it easier to accommodate chance effects up to the last few minutes of the approach path. Controllers demonstrated the capacity for tolerating irregularity (a load-inducing factor?) but also introduced it where it had not existed initially. A moderate degree of irregularity may facilitate human performance. It is possible that the impact of input predictability on performance increases as the complexity and duration of the task increases and as the nature of the task precludes use of a predetermined or fixed-pattern solution.

Case 33. Similar results were found in an air defense (SAGE) situation. As task load (number of tracks) climbed to three times initial load, the crews organized themselves to cope more effectively. They invented and tried out new procedures to handle the load, and they simplified the task by learning to discriminate between more and less important tracks and by paying more attention to important information. Crews did not work appreciably harder as the load increased, casting some doubt on the concept of load as work. Rather, they adapted by changing their strategy, reducing the number of tracks carried by attending more to potentially hostile ones. (Note the relationship of operator strategy to stimulus value/payoff: obviously hostile tracks possess greater value and payoff. See also Case 10.)

Case 34. In the same situation outlined in Case 33, when traffic load was increased by a factor of 5 there was no change in the proportion of critical tracks carried, in the number of simultaneous tracks carried, and in rate of verbal communications. (The operator worked "up to" some maximum load level.) To accept this load, however, operators decreased the average time spent carrying each critical track and decreased the proportion of noncritical tracks carried.

The effects described in Cases 33 and 34 are graphically illustrated in Figure 2-12 taken from Case 34. As the authors wrote-

"The proportion of penetrating tracks from dangerous directions (and, therefore, potentially hostile) remained essentially the same (90 to 95 percent) regardless of the load increase. But there was a marked decrease in the proportion of unimportant tracks carried (interaction of class and load). The increase in load provoked the elimination of nonessential behavior. Load increases finally caused a pruning of almost all behavior not critical to defending the area . . . The crews also (as load went up) tended to carry tracks for shorter and shorter times, and also with fewer and fewer reports" . . . "At the beginning of each problem, the crew . . . carried everything until it was identified (or recognized as being nonthreatening). Once they had the continuity completed, the 'plot everything' behavior disappeared . . . " (p. 109).

It was noted that

"Whenever a novel element (was) introduced into the system . . . there was an 'initial burst' activity associated with making an analysis of the new phenomenon. The characteristics of this activity are different from the

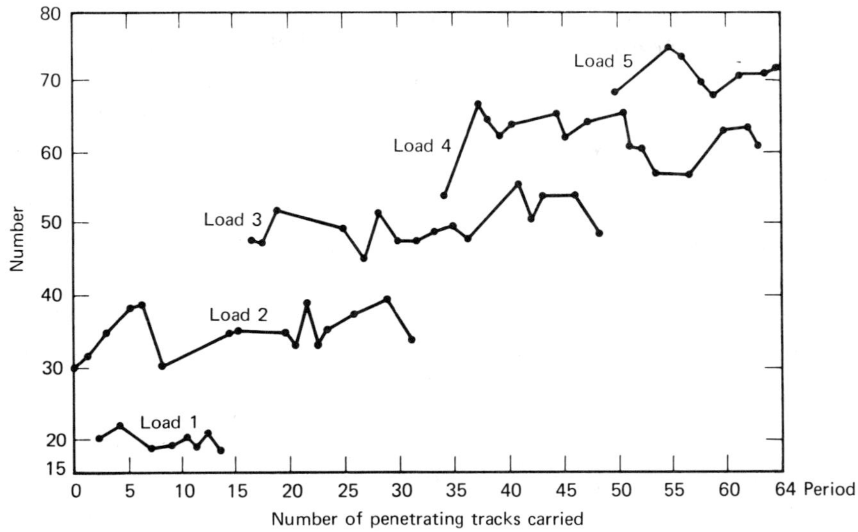

Figure 2-12 Number of enemy tracks carried as a function of traffic load and practice. With permission, from Sweetland and Haythorn, *Behav. Sci.,* 1961, Case 34.

processing activities: the initial burst activity is essentially diagnostic. The new phenomenon is analyzed and the results of this analysis determine the treatment (processing) to be used" (p. 111).

Conclusions. Load effects are modified by incentives, by stimulus value, and by expected payoff. As part of the operator's strategy of dealing with increasing load, he responds selectively to the inputs he considers or has been instructed to consider to have most importance in accomplishing his task.

Illustrative Studies

The motivated operator enters system operation by initially diagnosing the situation (consciously or unconsciously) to determine an appropriate strategy for dealing with inputs (even if he decides after this initial scrutiny that no strategy is necessary or possible). For such a strategy to succeed, of course, the situation must contain multiple inputs, it must be possible to differentiate among them in terms of value, and it must be possible to respond to these inputs differentially.

Although all this is entirely logical, the results of the experimental studies in this area are somewhat inconsistent, for several reasons.

1. Strategies based on subjective stimulus value judgments may be

canceled out by more structural variables such as stimulus exposure duration. For example, if duration is too short for the operator to process data effectively, he may not be able to apply value criteria to stimuli.

2. A distinction must be made between an incentive like a monetary reward, which presumably motivates the operator to work harder, and stimulus value, which serves primarily as an attention director. Cases 33 and 34 clearly indicate that operators respond differentially to differentially valued stimuli. In the absence of other guidance, however, incentives may not direct the operator to perform more effectively.

The data relating to incentives come mostly from vigilance phenomena. Bergum and Lehr (1963), using military subjects, contrasted monitoring without interruption with monitoring interrupted by a superior officer who pointed out errors of omission. The latter group achieved a significantly higher rate of detections. Similar effects were reported by Ware et al. (1964) for groups treated "autocratically" or "democratically." Bergum and Lehr (1964) found monetary incentives to have a facilitating effect in a vigilance task, but the effect was short-lived. Levine (1966) viewed the effect of target value and cost in a decision framework; that is, these variables are assumed to determine the decision criteria the observer sets for himself. Studying the effect of value and cost on auditory detection and identification, Levine found that increasing costs for misses and false detections led to poorer detection, but value had no effect. Increasing load affected only identification. Similarly, Williges (1971) found that differential payoff had little effect in simple visual monitoring. The reader is referred to Davies and Tune (1969), who reported a series of vigilance studies in which incentives of various kinds did improve monitoring behavior.

Taub (1965) investigated the effects of differential value on the detection and recall of alphabetic stimuli in a visual search task and reported that subjects made more correct identifications, initial responses, and false reports of high value stimuli than of low value stimuli. The percentage of initial responses of high value symbols, as well as the difference in the total percentage of high and low value symbols correctly transmitted, increased directly with increases in *ratio of value.* Karsh et al. (1970) discovered in a sequential keeping-track task that information categories having the highest value (in terms of reward) led to the best performance, but performance with lesser valued categories deteriorated. Christ (1969) found similar although smaller interactions between ratio and value. However, contrary to Taub's findings, differential ratio of value had no effect on the frequency with which subjects gave a high value letter as the initial report. In an effort to resolve this discrepancy, Hearns and Moss (1968) varied the method of payoff using a similar visual search task, observing that while subjects made more high and low value responses in two of the payoff conditions, ratio of value had no effect on performance.

Taub and Abrams (1971) (Case 19) found that differential ratio of value had an effect on performance, but there was a significant interaction with exposure duration. Differences in performance as a function of value ratio increased with exposure duration; failure to exhibit the value effect may have resulted from overly short exposure intervals and the response criterion employed. (Although ratio affected initial responses, transmissions, and number of responses, it had little consistent effect on accuracy.) The operator's ability to successfully employ a selective strategy in a search task almost certainly depends on time to view and search the display.

Case 35. The effect of target value was studied in five experiments in which subjects had to recall English letters projected by slides. Accuracy with high- value targets was slightly superior to performance with lower-value targets, although overall performance was not generally affected by the ratio of high to low value. The most significant outcome of these studies was the finding that the ratio of value between two categories of targets was a less important determinant of level of performance than was number and range of different values. As number of differential values increased, accuracy decreased compared to the situation in which all targets are considered equally valuable.

Although these discordant results do not disprove the general principle of differential operator responses based on stimulus value, they suggest that additional variables complicate the situation.

Case 36. The load variable was the number of vehicle targets shown in each slide for 0.6 second, and the task was to identify each target; the performance measure was number of targets correctly identified. As one would expect, load effects were significant. Of greater importance, the interaction of load with target value was also significant, and mean percentage of correct responses was (surprisingly) *reduced* with increasing load for valued targets, although there was no reduction for no-value targets. Subjects presented with no-value stimuli performed at a much higher level of accuracy (for higher loads) than the two groups given payoff instructions. A possible explanation is that the greater the opportunity for processing the information in the display, the greater is the opportunity for error in selecting the items of information to be processed. Since subjects in the no-value group had no reason to scan displays selectively or to report selectively, they tended to report everything, whereas subjects with high value targets tended to concentrate on certain targets and, therefore, reported fewer. Selective response to inputs may produce more or less effective system performance, depending on operator instructions and system output criteria.

Case 37. In a study aimed at determining the parameters influencing the monitoring of multichannel displays, the experimenters investigated the function of payoff ratio, display format, stimulus on-time, load, and channel payoff ratio. Each of 32 subjects was instructed to keep track of the status of a 6- or 10-channel display rack of digital readouts. At intervals the operator was presented with a new item of information or was interrogated regarding one of the readouts.

All the variables studied had significant effects on performance. Significantly more errors were made with the 10-channel task than with the 6- channel one and with the low value channel interrogations than with the high value queries, which suggests some selectivity in processing information. Unsurprisingly, significantly fewer errors were made with the 8-second on-time than with the 5-second period. Postexperimental interviews revealed that subjects often used a spatial encoding strategy for keeping track. A significant payoff effect was found: the percentage of total errors made with high value channel interrogations decreased as the payoff ratio increased. This suggests that the greater the disparity between the value or significance of different information channels, the greater the degree of filtering.

Conclusions. Load effects may also be modified by previous training and experience. It seems reasonable that better trained and more experienced operators should be better able to cope with greater loads. Oddly enough, the amount of evidence on this point, while supporting real world observation to some extent, is not great.

Illustrative Studies

Experience as a factor affecting response to load has not been extensively investigated. Two studies previously cited (Cases 7 and 11) included an experience factor.

In Case 7 the task was modeled after an air defense situation. Experience (defined as number of problems experienced prior to testing) was found to be a statistically significant variable (although small) (Figure 2-13). The task involved five operators in a series relationship. The only operator position at which improvement in performance as a function of experience occurred was the first (i.e., the input position).

Case 11 was an air surveillance/weapon assignment situation. Experienced operators were generally superior to those less experienced. The backlog of unassigned tracks built up faster for less experienced

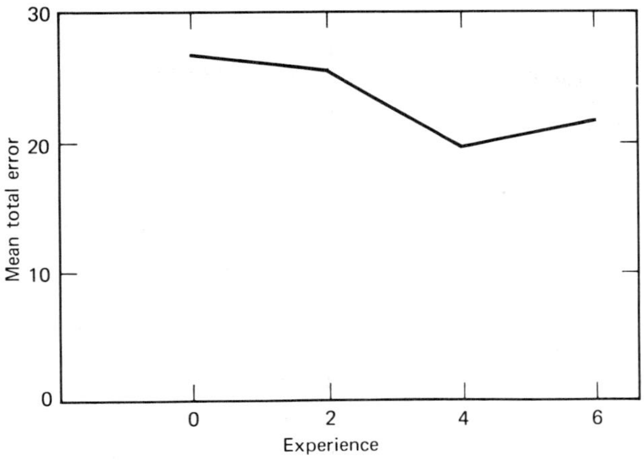

Figure 2-13 Error as a function of experience level. With permission, from Behan, *Hum. Factors,* 1961, Case 7.

operators and remained longer. A more severe threat situation affected the less experienced group more significantly. Less experienced personnel tended to saturate more at the highest load levels than did more experienced operators.

On the other hand, Wright et al. (1964) compared the performance of experienced and inexperienced radar observers in detecting targets on a PPI display. In three tests they found no significant differences between experienced and inexperienced operators, although the usual significant differences between individuals appeared. One test revealed reliable differences between simple and complex conditions.

Case 38. The question then arises: how should crews be trained to function under high load conditions? Under a high load or a graduated (increasing) load? The experimental results reported in Case 38 are equivocal on this point. A graduated load pattern does not seem to be a necessary condition for the development of more efficient procedures; teams modify their behavior at a rate that is idiosyncratic and relatively independent of task load.

Kidd (1959) also investigated the effect of graduated load on learning. Air traffic controllers were trained under constant high input load or under a graduated load. Test performance of the controllers trained under constant high load was superior to that of controllers trained under increasing load. It is possible that under high load operators received more knowledge of results at a higher rate than they did under the alternative condition, simply because more inputs were presented to them.

Case 39. One ordinarily thinks of experience in long-range terms (months or years). In any task, however, events occurring at times *T*-3, *T*-2, *T*-1 and so on serve as experiential inputs (history) for the operator's performance at time *T*. What then is the effect of a prior consistent base work rate on such periods of high operator load, such as emergencies? In this case 29 subtasks were performed. Most involved simple signals such as a light flashing and discrete responses such as operating a toggle switch that turned the light off. Once the signal appeared at the console, it remained on until the operator operated the correct control.

Base rates were 400, 200, and 100 signals per hour (well below the subject's capacity); bursts (high load) were 4000 signals per hour, lasting 40 seconds, followed by 10 seconds of no signals, after which signals returned to base level. Periods between high load bursts were 5, 15, and 25 minutes.

Analysis of response time during high load periods indicated no significant effects of prior base work rate on performance. Response time was an almost linear function of number of overlapping signals that caused the operator to utilize certain adaptive strategies (e.g., queueing) to maintain effective performance. Speed stress effects appeared to be a function of the immediate operator work load independent of preceding load levels. This ties in with our previous statements about load being more a response to immediate events than to long-term system requirements.

ORGANIZATIONAL FACTORS

Conclusions. Organizational factors modify load effects, but less strongly than variables inherent in the input condition.

Illustrative Studies

Since the system is often a fairly complex organization, one would expect to find load effects modified by organizational factors. Among the factors that have been investigated are the following.

1. Crew organization (e.g., task or personnel dependency, amount/type of cooperation between operators).
2. Crew size.
3. Organizational flexibility (e.g., number of contingent responses permitted).
4. Division of responsibility among operators in responding to inputs.

5. Distribution of load among personnel (i.e., load weighing more heavily on certain subsystems than on others).
6. Communication factors.

The data do not suggest a very clear-cut or strong relationship between these factors and load. Perhaps we should not expect a close relationship, because system or crew organization is not an input per se but serves as the matrix within which inputs function.

Crew Organization

In Case 1, pairs of operators simulated an approach/control task. This study contrasted two types of organization, one involving *non-compensatory,* the other *compensatory* functions. The former situation was one in which a team member had to complete an aircraft approach precisely on the one-minute (high load) or two-minute (low load) schedule, regardless of any delays his partner experienced. In the latter situation a team member compensated for his partner's delay in making the approach by speeding up his own approach to maintain an average one-or two-minute schedule.

It was hypothesized that approach times should be closer to criterion in compensatory teams and this is what happened, but primarily under low load. Moreover, flight errors (misses and safety infractions) were expected to be greater in the non-compensatory condition, and this was what was found, but only during high load.

Despite the statistically significant differences found, however, the difference between types of team organization appeared to be minimal at best (at least at the conclusion of training). Performance by the different organizations tended to converge as testing proceeded, perhaps reflecting a tendency for learning to cancel out the effects of arrangement differences. Alternatively, as errors reduce toward zero, the amount of variability that can occur is so highly restricted that it may prevent real differences in organizations from showing their effects. In any event, the major differences illustrated in this study are differences between load conditions.

Crew Size

Some system managers attempt to deal with fluctuating work loads by augmenting the crew under peak load conditions. Studies of the performance of small teams (e.g., Kidd, 1958; Kinkade and Kidd. 1959b) have cast some doubt on the utility of crew augmentation. If a unitary task (i.e., one normally suitable to single-man operation) is distributed among more than one operator, the gain in performance is slight. If a single operator could handle the task under moderate input load conditions, adding one or two helpers does far less than double or triple load capacity. This principle seems to be most pronounced for decision making as opposed to more routine tasks.

In a previously reported study (Case 2) of air traffic control using simulated aircraft targets presented on a CRT, two independent variables were tested: size of the crew (one-, two-, and three-man operation) and load (aircraft arrival at 90-, 60-, and 30- second [average] intervals). Since the input load per controller decreases proportionately as crew size increases, a progressive improvement in performance with the larger teams would be expected.

The results in Table 2-3 (comparing teams of various sizes, where the input to the total system was held constant) do not fully verify that expectation. There was some small improvement as crew size increased and individual controller load decreased, but the change was not statistically significant. When load per controller was held constant (resulting in an increasing total system load when control unit size is increased), performance was subject to progressive decline with increasing team size. There was no observable interaction between crew size and input load.

In an airborne CIC situation (Case 30) in which number of search operators was varied, no differences in performance were noted as a function of this variable, regardless of the load imposed. Rate of reporting (elapsed time between successive position reports) found two- and three-operator systems equal in speed, and both were superior to the four-operator system.

Obviously increasing the size of a crew to deal with increased work load is not a very efficient solution to the problem, unless the additional operators must handle new tasks arising out of the increased work load.

Table 2-3 Effect of Crew Size on System Performance[a]

	Crew Size		
Criteria	One Man	Two Men	Three Men
Mean percentage delay	129.6	122.6	122.5
Mean percentage excess fuel	146.8	134.4	126.8
Separation errors per aircraft:			
Airborne	.110	.092	.077
Runway	.064	.052	.051
Missed approaches	.038	.033	.043
Mean departures	5.2	5.3	5.4

[a] From Kidd, 1961, case 2.

Yet new tasks are unlikely to arise if the system has been properly designed. Moreover, the whole thrust of modern job design* is to reduce the number of personnel by combining functions.

Organizational Flexibility

Case 40. A study was performed to test the effect of strict versus flexible operating rules on a simulated terminal air traffic control system under a variety of traffic conditions with constant load. Twelve college students were trained in control procedures, were assigned to two crews, and operated the system in a sequence of 24 problem sessions in which operating rules were systematically varied. In the strict rule condition the crew had to control the aircraft through all points along the complete route specified in the flight plan; no deviation was permitted. In the flexible condition flight plans contained only entry and exit points, and each crew was free to organize traffic within the terminal complex. Flexibility in aircraft takeoff and landing procedures was permitted by a reduction in aircraft separation standards.

Figure 2-14 suggests that a change in criteria from strict to flexible rules tended to improve system performance. The improvement was noted in nine out of ten indices for both crews. Many of these changes were statistically significant. See also Cases 31 and 32.

Division of Responsibility

Case 41. In this study three assignment structures were tested, and there were three different vigilance tasks, each one containing four critical signal categories. The three structures were (1) homogeneous—each of four operators responsible for all signals, (2) paired—two signals assigned to one pair of operators, two to the other pair, and (3) individual—each of the four signals assigned to a different operator. In one of the three tasks the critical signals were certain code patterns in any auditory message. In the second task the critical signals were the appearance of nonsense words, scrambled in a letter matrix. The critical signals in the third task were certain types of information concerning a briefly presented street scene.

The results of this study indicated that vigilance varied with the different team structures, the paired structure being generally superior. Unfortunately the raw data that would permit an evaluation of the significance of these results are not available. There was apparently some interaction between the effectiveness of the team structures and the different tasks. (Chapter 5 also deals with this topic.)

*At least as one sees it in ship manning (Gaites, 1974).

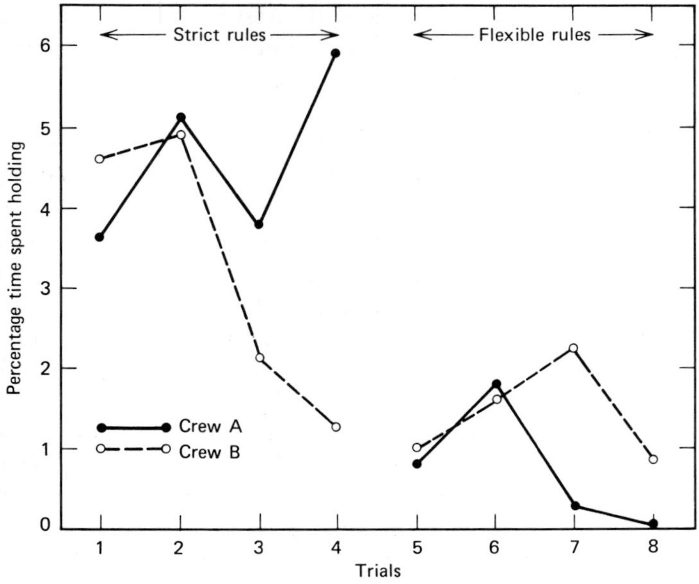

Figure 2-14 Aircraft hold time as a function of rule flexibility. With permission, from Alexander and Cooperband, *Hum. Factors,* 1964, Case 40.

Distribution of Load

Case 42. This study was aimed at identifying the adaptive mechanisms utilized by a multiman system to adjust to unequally distributed loads. The system involved a team of five operators (commanding officer, operations officer, and three reconnaissance [search] officers) whose task was to direct reconnaissance flights by vectoring aircraft and evaluating the resultant information. Stimuli were presented on a CRT, and responses were observed by closed-circuit TV and audiomonitoring.

The primary variable was the distribution of load (nine aircraft) under two conditions: equal (3-3-3) and unequal (1-3-5, 1-1-7, 1-4-4) distribution among the reconnaissancce officers. Total system output, the number of enemy installations detected, was not significantly affected by load distribution. System output increased systematically over time even though level of load distribution was changed regularly. Nor did cost (in terms of total number of flights, total fuel used, and total aircraft use time) vary significantly as load distribution changed.

In contrast to system performance, subsystem (operator) performance varied significantly as load increased or decreased. Analysis of information gathered for reconnaissance officer (RO) loads shows that RO's performed differently with different loads. Load was not excessive, since it did not produce breakdown. There were significant differences among crews.

Operator and system load effects apparently differ, as hypothesized, and internal system adjustments seem to dampen out individual operator responses to load. Internal compensations (i.e., changes in operating procedures) were managed easily and did not alter task requirements appreciably. Among the compensations possible were (1) *load leveling*—operators with smaller loads could receive and evaluate information relayed by other operators with larger inputs, (2) *centralization of control*—the operator with the largest input could receive and evaluate filtered information relayed by other operators, and (3) *alliance*—the operators with smaller loads could exchange raw or filtered information to increase their joint effectiveness. Operators made no attempt to equalize load among themselves, however. Adjustments occuring at the subsystem level were made on an individual basis; each operator affected by load changes exerted the effort he considered necessary to deal with the load.

Communication Factors

The available evidence suggests that crew output is an inverse function of the extent to which operators are required or permitted to coordinate or communicate with one another (Briggs and Naylor, 1965; Johnston, 1966; Williges et al., 1966).

Communications as a Load-Inducing Condition. When communications functions as a load condition, we would expect this variable to affect system performance like any other load. For example, in a simulated air traffic situation involving heavy verbal communications (Case 4), load was defined as percentage of interruption of communications. A near-linear decline in performance resulted as a function of channel intermittency.

Communications as affected by Load. In the study already described as Case 1, a pair of operators served as radar controllers in a simulated approach-control task. The hypothesis was that the greater the load, the less a crew can afford the luxury of communication, and hence the higher the load, the greater the disruptive effect of communication.

Contrary to expectation, crew communication was uninfluenced by all levels of load. It was, however, influenced by type of crew organization—compensatory versus noncompensatory. The correlation between communication frequency and team performance was significant only for the noncompensatory high load condition, and here crew communication did not augment system output but depressed it.

The hypothesis is that communication is most disruptive in the organization that least requires communication (noncompensatory). It appears that when there is least need for communication and least freedom to achieve it, communication hinders crew performance. When there is enough time and reason to communicate, communication neither retards nor enhances crew performance.

Case 43. This study correlated communications with information load defined in terms of preprogrammed messages input to the crew (2, 6, 10, 14, 18, and 22 messages during each 30-minute period). Four-man teams participated in a complex game in which subjects had to make geopolitical decisions in a simulated small-scale conflict having Vietnam-like characteristics. The four-man teams were divided into two two-man decision-making and acquisition units: decision makers functioned on the basis of information gathered by the acquisition teams, who could communicate by telephone only. Information acquisition teams could initiate any number of calls to secure additional information to be passed to decision makers.

The purpose of this study was to determine the effects of suboptimal, optimal, and superoptimal information loads on information transmission characteristics. An optimal load was defined on the basis of previous research as 10 messages per 30-minute period.

Table 2-4 indicates that when information load was low, the number of calls exceeded the number of messages received by information acquisition teams but did not reach the level of optimality. The number of calls slightly exceeded the optimal number under high load level but remained well below the number of messages received. All calls made by all teams were relevant. However some information passed by the information acquisition teams to their decision makers was imaginary when information load was low (64% during load level 2, 17% during load level 6).

The results of this study are of interest with respect to an adaptive concept of system functioning. Even though no instructions were given to adjust the number of calls asking for information, the information acquisition teams did increase the calls beyond those required when information

Table 2-4 Mean Number of Calls as a Function of Message Input Level[a]

Level of Message Input					
2	6	10	14	18	22
5.24	6.71	9.24	10.18	11.12	12.88

[a] From Streufert, et al., 1972, case 43.

was scarce and suboptimal and decreased their calls (relative to load) when information was superoptimal. Apparently people communicate when they need information, a hardly surprising conclusion.

Another experiment, described in Case 41, was designed to investigate the amount of communication required and problem-solving performance as a function of the complexity of relationships between information and response action. Problems of nine types were given to four-person groups, each problem requiring the use of several items of information that were distributed across group members. The nine problems were graded in terms of the amount of information that had to be transmitted between group members for a correct solution to occur. At one extreme, each group member had all the information he needed for making action decisions, and at the other extreme, each group member had to obtain all relevant information from his co-workers. The time taken to solve the problem was found to vary directly with the total amount of information to be transmitted; an analysis of the communications between loci showed a very close correspondence between the required information transmission and the number of messages that occurred.

INPUT VARIABLES

Conclusions. The characteristics of the inputs serving as load variables clearly determine load effects. The major input characteristics that have been considered in the literature are as follows.

1. Number of channels through which inputs are presented.
2. Input rate (i.e., rate at which inputs enter the system and are presented to operators).
3. Stimulus-response relationships, particularly the complexity of these relationships.
4. Amount and type of information presented by the inputs.
5. The physical display that presents the input.

Illustrative Studies

Number of Channels

One condition that can load the system is the number of separate channels through which inputs are presented to the operator. (See the definition of load by Conrad, 1951.)

The question has importance in the field of system design, since most multiple-task situations involve multiple channels. For example, the duties of air traffic controllers and tactical commanders often require personnel to keep track of and recall continuously changing information presented over a number of channels. The practical question is: what is the maximum/minimum number of channels as a function of the information to be presented? It is desirable to know this not only because too many or too few channels may degrade operator performance but also because the system developer would prefer to minimize the need for additional complex equipment. However, one should not expect to determine a single generalizable number of channels for input presentation. That number will vary as a function of many other variables, such as type of input, number of inputs, frequency of input change, and operator response requirements.

In the classic studies by Yntema (Yntema, 1963; Yntema and Schulman, 1967), the subject was read a set of messages that informed him of the status of a number of information channels. At intervals he either was interrogated about the current status of one of the channels or was given a message to change the present status of a channel. Yntema's results suggest that the operator's capacity to track a random situation (in which any information channel may be in any given state with equal probability) is fairly low—on the order of only two or three channels at once. The reason, of course, is that there is some limit on the operator's ability to divide his attention. (For a systematic treatment of psychological research on attention, see Kahneman, 1973.) In Case 37 subjects made significantly more errors with 10 channels than they did with 6 channels.

Case 41 used a code-detection task in which one of three possible four-place dot-dash critical signals was embedded in one of three carrier signals of six places each. The three carrier signals were presented simultaneously, and the subject was responsible for detecting one, two, or three of the critical signals in one, two, or three of the carrier channels. In accordance with expectations, the results showed that the probability of detection for a particular critical signal was decreased when an increase was caused in either the number of signals or the number of channels for which a subject was responsible.

Gould and Schaffer (1966, 1967) studied the effects of divided attention (produced by a secondary task) on visual monitoring of multichannel displays. Additional variables included number of channels monitored (4, 8, 12, and 16), rate of display change (6 or 12 times per minute), number of signals simultaneously watched for (8 or 24), and number of levels within channels (2 or 8). Although the divided attention did not degrade performance significantly, all the other variables had significant effects, the rate of display change having the greatest effect.

Case 44. The same effect also appears when more complex stimuli are used. In this study observers watched 16 or 24 TV screens for an hour. The screens showed still scenes with occasional movements lasting 2 to 40 seconds. The observer's response was to press an alarm, and the performance measure was time to press. The times increased from 2.29 seconds (16 TVs) to 3.10 seconds (24 TVs). The percentage of missed incidents also increased from 3.3 to 12.5%. As discussed previously, increased response requirements apparently add to load. When observers had to answer questions every 30 seconds, errors increased from 3.3% in the standard condition to 15%. Similarly, when the observer was required to report *all* movements observed rather than merely the sighting of an incident, missed incidents increased from 3.3 to 8.3% (effect of response complexity). These results were replicated and expanded in a subsequent study (Tickner and Poulton, 1973).

Input Rate

The fact that high input rates will load the operator and degrade his performance is well attested, at least for psychomotor tasks.

Many researchers have found performance decrement with increased speed stress (Alluisi et al., 1957; Conrad, 1951; Dorfman and Goldstein, 1971; Gould and Schaffer, 1967; Wagner et al., 1954; and Cases 24, 27 and 50). Monty (1973), summarizing the results of a series of studies on keeping track of sequential events, indicates that "at slow rates of presentation (i.e., one event per second or less), individual events falling into two categories can be discriminated one at a time and mentally tallied with considerable accuracy. Thus, if accuracy is the sole criterion of performance, the number of categories presented should be extremely small." As the number of categories or the trial length increases, errors per trial increase rapidly. For a given trial length it is better to present fewer categories with many possible occurrences of each than a larger number of categories with fewer occurrences. In general, the slower the rate of presentation, the better the performance.

However, the effect is at least partially task determined. For example, Johnston et al. (1966) found that monitoring of alphanumeric stimulus displays was poorest with lower input frequency and higher stimulus density. Mackworth (1968) found that signal rate had no impact on the classic vigilance task.

The examples below illustrate this point. It is of greater interest to ascertain whether at higher input rates the operator's behavior changes qualitatively as well as quantitatively. For this point, much less information is available.

Case 45. This study utilized a symbolic discrimination task analogous to a pilot maneuvering his aircraft while attending to cockpit displays. The load variables were input rate (32, 53, and 77 slide presentations per minute) and stimulus complexity (4-, 9-, and 12-celled matrices in each slide). Fifty-eight students were tested over three 90-minute sessions.

As expected, input speed had highly significant performance effects. The rate of transferring information fell steadily as input speed was increased. However the total variance associated with this factor went from 78% on the first day to 69% on the third day (which may indicate a certain operator adaptation to input conditions). Stimulus complexity accounted for only 2% of the variance on day one but went up to 12% on the third day.

The way the operator responded to overload was by reducing his minimal stimulus analysis time. With practice, this period became shorter within a reasonably small range. When overload did occur it was not accompanied by breakdown, at least not after extensive practice.

Obviously increased input rates reduce performance accuracy and rate of information processing. Minimal time required to handle a stimulus accurately also increases with stimulus complexity. Hence an interaction exists between input rate and complexity; the more complex the stimuli, the lower the input rate threshold for stressing the operator.

The impact of speed stress in tasks such as those employed in Case 45 is similar to that found in simple discrimination learning tasks. Once the operator's channel capacity is matched, performance falls off steadily. It is possible that when subjects are overloaded, they tend to recirculate data in immediate memory, which therefore increases average response time and reduces response rates.

Adaptation can occur. Practice helps: time to respond adequately decreases with practice, and the operator can do even better if he is trained to use a more effective filtering strategy.

Case 46. This study illustrates the importance of task characteristics in affecting responses to input rate. Subjects had to detect a deflection of a meter point from its null position, a type of task characteristic of classic vigilance studies in which stimuli are at or slightly above threshold and do not possess a complex structure.

When signals were presented at two rates (60 vs. 12), the presentation rate proved to be nonsignificant for the initial half of the sessions. As monitoring time increased, performance degraded (as one usually finds in the typical vigilance study), and the input rate factor became significant.

The major purpose of the study was to determine whether response complexity could aid performance, as was suggested by Adams and Boulter (1960) who found that response complexity reduced the customary vigilance decrement. The experimenter in Case 46 increased response complexity over the simple detection response (control) condition by requiring subjects to operate a patch panel after detection without an indicator light (condition A), with a light (condition B), and without patch cords and jacks (condition C). Subjects performing under condition A made significantly fewer errors (18.44%) than control subjects (24.22%). Comparisons between the control condition and conditions B and C were, however, insignificant.

Parenthetically, however, it should be noted that the situation with regard to response complexity in vigilance tasks is muddied. Montague et al. (1965) utilized two response conditions (simple and complex) in a complex monitoring task lasting 18 hours. No significant difference in performance resulted from increased response complexity. On the other hand, Childs and Halcomb (1972) found that increased response complexity produced superior performance in a monitoring task. They attributed this effect to increased arousal of the subject. Clearly the final word on response complexity has not been said, but apparently under certain conditions (e.g., when the task is very monotonous), response complexity can stimulate the operator.

Case 47. In a monitoring task in which the critical stimuli to be monitored were changes in voltmeter readings on a simulated instrument panel, the three input rates were changes in dial readings, each 6, 10, and 15 seconds. This load condition was significant: as the input rate increased, the number of errors increased significantly (Figure 2-15).

Case 48. Case 48 utilized the same experimental situation described for Case 47, except that the load values in the voltmeter readings occurred every 10, 15, and 20 seconds. As would be expected, the effect of input rate as a load-inducing variable was highly significant.

Case 49. Stimuli were presented on a control display panel consisting of nine voltmeters; when activated, each meter deflected its pointer to its full excursion and remained there for 2 seconds unless a correct response was made, in which case the pointer returned to zero. The stimulus-response conditions were systematically complicated by increasing the number of conditions the subject had to keep in mind while responding. In condition I the status of two indicator lights determined a correct response. In condition II the indicator lights and the color of a rectangle painted on the meter determined the correct response. In condition III, in addition to the lights and rectangle, the presence or absence of a red frame bordering the meter was critical.

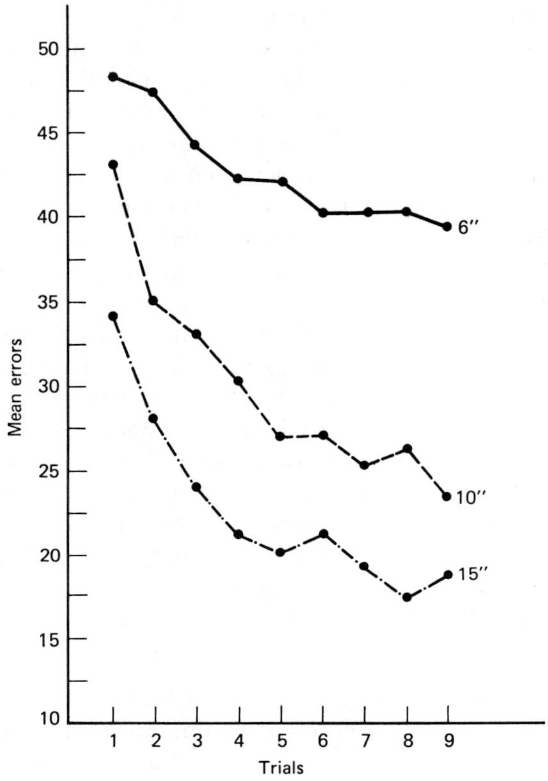

Figure 2-15 Error as a function of input rate. With permission, from Lanzetta and Roby, *J. Abnom. Soc. Psychol.,* 1957, Case 47.

Significant differences were found between each stimulus condition (Figure 2-16). As input rate increased, subjects increased their performance up to some point (response maximum) at which load became excessive and performance decreased (the inverted-U relationship). The important point is that the amplitude of the initial maxima was an inverse function of the number of relevant stimulus dimensions, as was also the time of occurrence of these maxima. The amount of performance decrement following the maxima was an inverse function of the number of relevant stimulus dimensions. (That task difficulty is directly related to the different number of relevant stimulus dimensions and response alternatives has been shown by Posner, 1964.) Increasing complexity tended to flatten out the curves in Figure 2-16, perhaps because complexity produced performance decrement more quickly.

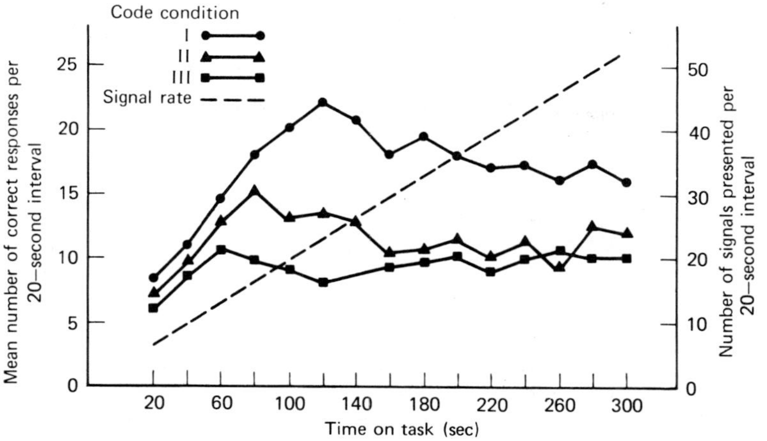

Figure 2-16 Accuracy as a function of varying input conditions. With permission, from Huntley, *Hum. Factors,* 1972, Case 49.

As we have come to expect, individual subject variance was highly significant. All subjects attended to some meters more than others. In the less complex stimulus dimension task, however, subjects did not differ systematically in this. With the most complex situation, subjects developed distinct strategies.

"Rational filterers" established meter priorities on the basis of the minimum amount of attention required by meter codes. "Arbitrary filterers" established their priorities on an arbitrary basis or because meters were located next to one another. Differences in performance resulting from the two strategies (Figure 2-17) reveal an interesting reversal of strategy effects as time on task increased for "arbitrary filterers."

The results of Case 49 are similar to those found elsewhere in the literature. Conrad (1951, 1954), Mackworth and Mackworth (1956), Miller (1964), and Klemmer and Muller (1969) discovered that omissions and errors became greater with increases in input rate. Significant positive correlations between response rates and errors were found also by Fitts (1966) and Pachella and Pew (1968), who showed that information processing performance demonstrated a speed-accuracy tradeoff in which increases in response rates are made at the expense of error rate increases.

Operators apparently perform differently under high information input conditions as compared with low information input conditions. At high input rates, operators tend to pay proportionately more attention to

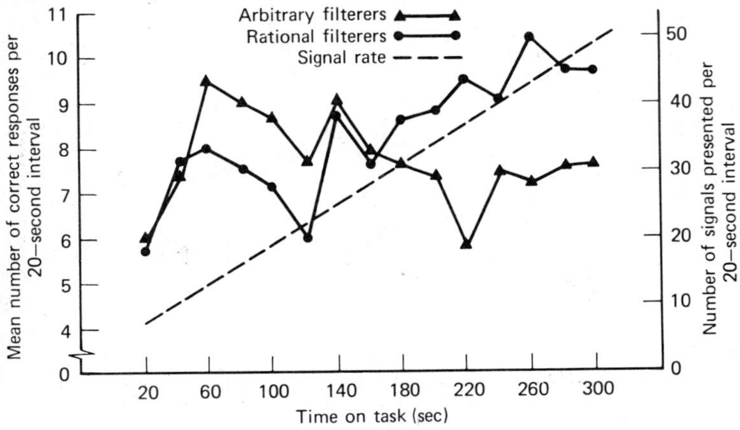

Figure 2-17 Accuracy as a function of a different operator strategies. With permission, from Huntley, *Hum Factors,* 1972, Case 49.

high value information sources at the expense of monitoring low value sources. Similarly, Miller (1964) showed that at high imput rates operators may respond selectively to groups of similar stimuli (development of a strategy), whereas at low rates information is more likely to be processed on a first come, first served basis.

Case 50. In a study designed to determine information handling rates for sequentially presented stimuli, the relationship between the rates, size of display surface, and the effect of exposure time variations, the subject was presented with a 5 × 5 matrix of 25 lights. All lights illuminated presented one of four colors. Either 1, 9, or 25 lights were illuminated on any single trial, and subjects reported the name of the color presented. Two different exposure times were used: 250 and 330 milliseconds. Speed of light presentation was systematically increased or decreased in successive trials until subject responses were 50% incorrect.

As speed of presentation increased, there was a monotonic decrease in response correctness. Of greater significance, increasing speed of presentation resulted in greater differences between experimental conditions. The interaction between input rate and matrix size was highly significant. Exposure time was not significant, which is somewhat surprising, although the times used may not have been sufficiently different from each other to have an effect.

Case 51. A number of experiments (e.g., Jenkins, 1958; Kappauf and Powe, 1959) indicate that probability of detection in a classic vigilance task is directly related to critical signal rate. However the vigilance task also involves other variables, such as the noncritical signal and total stimulus rates, which may be of equal or even greater importance in determining vigilance performance. Colquhoun (1961) observed that many of the studies reporting a direct relationship between performance and frequency of critical signal presentation had used tasks in which an increase in the number of critical signals was accompanied by a decrease in normal or noncritical signal rates. Thus the investigator concluded that detection efficiency was determined by the probability that a stimulus will be a critical signal (i.e., number of critical signals divided by total number of stimuli presented). Subsequent studies (Jerison and Pickett, 1964; Jerison et al., 1965; Johnson et al., 1966) have tended to verify this conclusion. Yet other work (Jerison, 1965) casts doubt on the idea, and Case 51 was intended to clarify the question.

To compare the relative effects of critical signal rate, normal signal rate, and critical signal probability, a Mackworth-type vigilance task was utilized (a clock in which the normal signal was 9 degrees of movement and the critical signal to be detected had a sweep-hand movement of 20 degrees). No consistent differences were attributable to critical signal rate or to critical signal probability, although performance was poorer with the faster rate of stimulus presentation than with the slower rate. The increase in total number of stimuli did not affect rate of vigilance decrement but lowered performance in all time periods. Performance was inversely related to the rate at which noncritical (normal) signals were presented.

Two explanations can be advanced. First the normal signal rate may determine performance at low critical signal probabilities, whereas at high probabilities the probabilities themselves may determine performance; alternatively, when detection performance reaches asymptote, it is no longer influenced by further changes in rate or probability.

Case 52. In one of the more interesting attempts to simulate a civilian system (taxi dispatching), the effects of load and information accessibility on two-man team performance was studied. The primary task of the subject in position 1 on the team was to assign cabs to pick up passengers. This involved choosing cabs to minimize both time and miles traveled by the vehicle. For example, three passengers might have to be picked up when five cabs were available, resulting in 60 selection possibilities. The primary task of the subject in position 2 was to monitor the positions of the cabs, start them on their runs, and keep records.

In the low load condition passengers entered the system at 2-minute intervals; in the high load condition they entered at 1-minute intervals. Another variable was access to information. The full access condition allowed communication between team partners, and access to each other's displays was possible; in the restricted access condition this was not permitted. Performance was measured by earnings according to a specific rate per mile, and there were penalties for delays.

The load factor was not a significant determinant of earnings, although teams working with high load conditions did somewhat better than those with low load (Figure 2-18). The effect of high load may have been to cause teams to work harder, a hypothesis that agrees with findings of Lanzetta and Roby (1956). The effect of access to information was much more significant, as was the interaction between information access and load. The effect of free access to information was to improve crew performance more under high (where need for information is greater) than under low load conditions. In terms of the overall criteria, performance was best for high load and high access to information. Greater input load

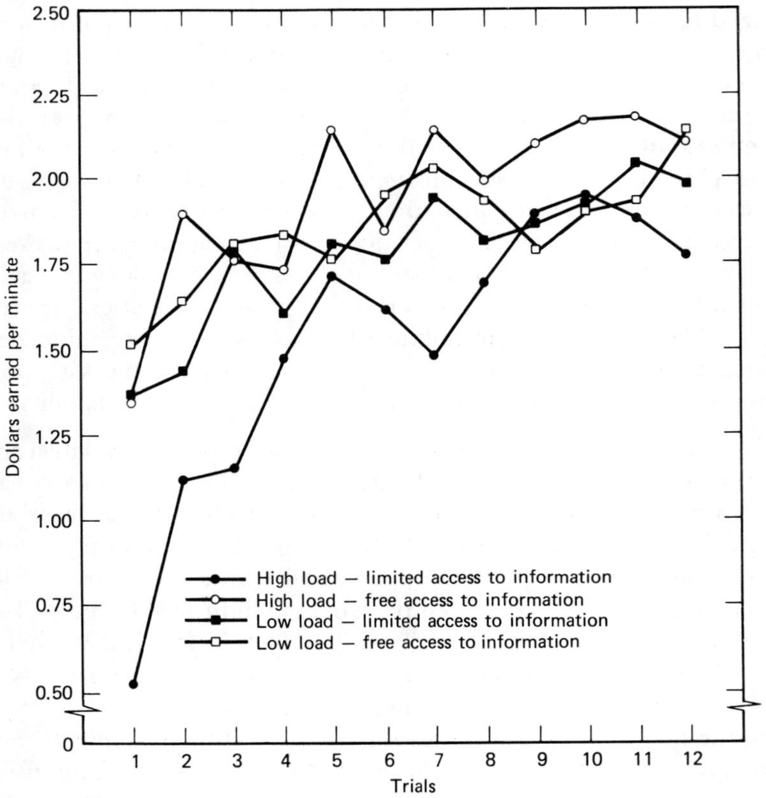

Figure 2-18 Earnings as a function of load and access to information. With permission, from Moore, 1961, Case 52.

did not lead to performance degradation, probably because operators were not overly stressed by the load. Free access to information reduced passenger waiting times under both load conditions, possibly because team members could share the load, thus reducing it (load balancing). None of the experimental conditions led to significant differences in bookkeeping errors.

Stimulus-Response Complexity

Complicating the stimulus, the response, or both, these variables can act as a load condition. (See also cases 46 and 49.)

Case 53. In this study the two operators had to depress a telegraph key as quickly as possible to various combinations of critical indicator lights. Stimuli were sometimes redundant (both subjects respond to all critical stimuli) and sometimes nonredundant (each subject responsible for half the stimuli). Stimulus complexity was defined as the difference between four and eight critical combinations. This definition of complexity emphasizes the number of stimuli as the complexity element, but this is obviously only one (and the simplest) dimension of complexity.

Unsurprisingly, the number of positive errors increased as the subject proceeded from four to eight stimuli. Reaction time (RT) was faster for lesser stimulus complexity, as Brown (1963) had also found. RT for the redundant condition was higher than for the nonredundant condition, a finding attributable to the necessity for integrating subject operations. Redundant operation resulted in fewer failures to detect a critical stimulus than did nonredundant operation (a consequence of compensatory team actions). Nonredundant operation produced fewer false alarms.

In Case 9 operators either had to perform a detection response alone or both detect and evaluate (classify) the stimuli. As one would expect, response latency was increased during the more complex response condition. However the typical decrement found in monitoring performance over time was not influenced by the increased complexity.

Case 54. This study was an attempt to determine whether the performance effects of speed stress vary as a function of differences in the type and complexity of information processing activities included in the loaded task. The multitask situation involved a primary monitoring or decision task and simultaneous secondary discrete tasks. The basic activity was identifying variations in four geometric forms (rectangle, trapezoid, triangle, parallelogram) presented in the four quadrants of a CRT. On each presentation a form appeared in one of seven variations differing in base-altitude ratio. On each presentation or frame of the display, the subject searched the display for the occurrence of five different classes of events.

The MEAN task required the subject to estimate the population mean for the rectangles in the upper right quadrant, keeping track of the form values on successive frames, to discriminate among five possible distributions. The FORM task required him to identify the form value for the trapezoid appearing at the lower right of each frame. The PAIR task required detection of pairs and triples (formed of identical base-altitude dimensions) appearing in a single frame. In the RUN task the subject was supposed to remember the location of pairs and triples and check for the appearance of another pair or triple in the same quadrant location in the next frame. The LINE task presented the relevant information within only a single quadrant location, a much simpler task than the others.

Frame times (display times) were successively reduced from 10 to 7 and 4.5 seconds, which resulted in highly significant increases in percentage of error, particularly at the faster exposures. However the change in percentage of error was different for each type of task at each display time (Figure 2-19). The overall error rate of 18.4% was largely attributable to the MEAN task, which accounted for 73.5% of the errors. In other words, the complexity of the task performed was significant for performance.

Subjects used a priority strategy in dealing with the various tasks based on the relative frequency of response required for the tasks, their display location, and the relative contribution of individual tasks to error scores.

There was no indication that attempts to speed up performance under time stress resulted in loss of accuracy on high priority tasks not requiring search of the display. The major effects of stress appeared in the performance of the tasks involving spatial and temporal uncertainty and calling for search of the display to identify significant events. High frequency predictable activities can be expected to be least sensitive to stress in complex tasks, showing little decrement until time is no longer adequate. In multivariate complex tasks, activities requiring search and memory for low probability events "break" first under stress.

Amount and Type of Information

If we assume that the system needs a certain amount and type of information to perform its mission, we can conceive of information as an input load condition. Thus a system could be exposed to: (1) too much information, (2) too little information, or (3) inaccurate, distorted, or irrelevant information that does not correspond to system requirements.

This is easy to understand conceptually until we examine the variable in greater detail. To make much sense of information load, it is necessary to

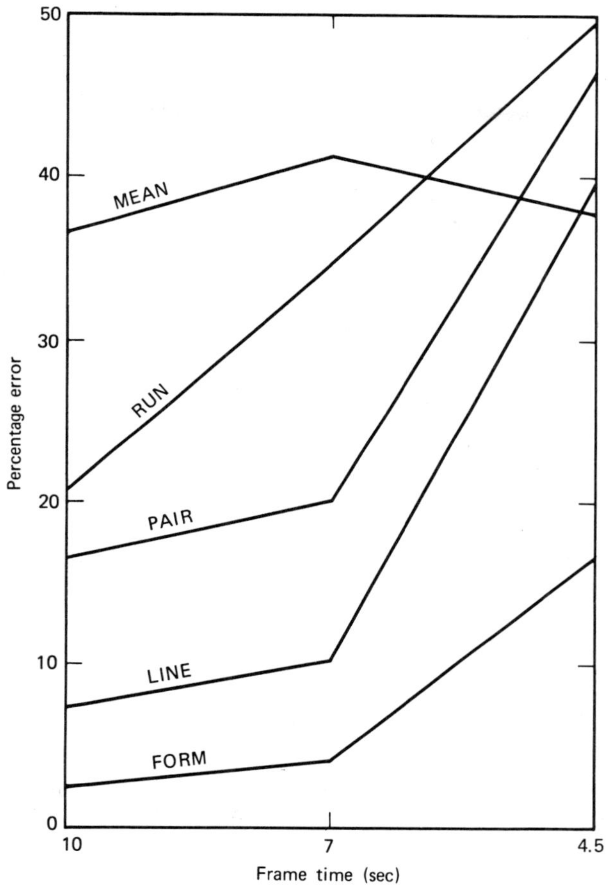

Figure 2-19 Error as a function of differences in task types and reduced exposure time. With permission, from Weisz and McElroy, 1964, Case 54.

specify system requirements in information terms. This is difficult because information as a variable appears to apply only to the operator, not to the system, which may explain why system requirements are almost never described in information terms.

The definition of information poses another problem. It is easier to specify number of messages than to indicate the information these messages contain.

The studies described in this section do not deal with information in the sense used by Shannon (1948), although we conceptualize information in terms of reduction of uncertainty. At the moment, the measurement of information in terms of "bits" can be made only for highly discrete behaviors, such as the identification of tone frequencies or light intensities, or the counting of dot matrices. For such behaviors, it is possible to estimate operator information handling capacities. Such estimates of capacity range from 3 bits per stimulus event or dimension (Miller, 1956) to 10.4 bits (Klemmer, 1957).

Unfortunately the MMS behaviors we are concerned with are far more complex, thus not easily translatable into "bit" measures [although see McKendry et al. (1971) for an attempt to measure information utility quantitatively]. The measurement of complex system information handling operations also involves the problem that such tasks entail the element of "meaning," and in the presence of "meaning," precision exits, since the criterion of what is meaningful to the operator is likely to be highly idiosyncratic. However despite our inability to quantify information in anything like "bits," it is necessary to deal with information load because many complex MMS (e.g., air traffic control) are specifically information processing systems.

Studies such as Case 43 have suggested that the range of information used and the flexibility of the ways in which that information is organized in decision making are highest at some optimal level of environmental load and diversity. As environmental load and diversity increase or decrease from the hypothesized optimal level, decision-making processes are characterized by reduction in both the amount (range of dimensions) of information considered and the number and range of alternatives generated. The hypothesized relationship depicted in Figure 2-20 is very close to the relationship between load and performance postulated in Figure 2-1.

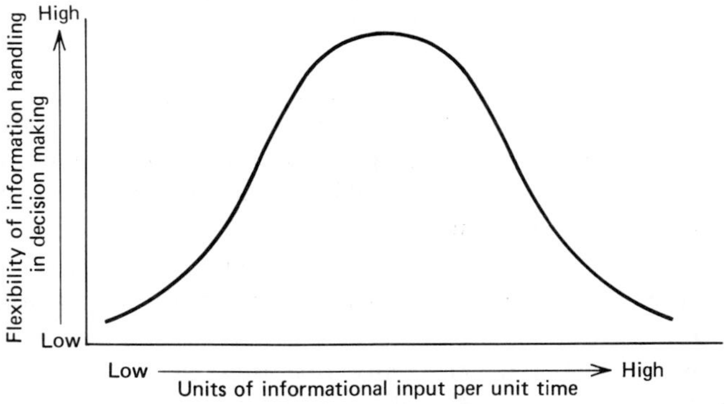

Figure 2-20 Flexibility in handling information as a function of amount of information input per unit time.

Case 55 This study had several purposes: to ascertain whether operators are aware of the interaction between environmental factors and the level of decision-making performance, to see whether they adjust to optimize their behavior, and to observe the effects of individual differences in conceptual orientation on performance (i.e., between subjects characterized as "concrete" vs. "abstract"). Teams of four subjects played the same tactical game described in Case 43 over seven half-hour periods during which the informational input was varied during each period from 2 to 5, 8, 10, 12, 15, and 25 input units. Each input unit described one event in the experimental environment and represented either feedback concerning the subject's previous decisions or information concerning enemy actions. To keep the success-failure level of each team constant, each team unwittingly played against a prearranged program.

A number of previous studies had indicated that the optimal information load for generating the highest level decision-making processes occurs during the 8, 10, and 12 input unit periods. In this study the subject was asked to indicate at the end of each playing period whether he wished more or less information in the next period. The amount of change in information requested could be randomly distributed across loads, or, as case 43 suggested, less would be requested after very high input loads and more after very low periods. If the latter were true, it would appear that personnel tend to gravitate toward an optimal level environment.

As Figure 2-21 indicates, significantly more information was requested in the extremely suboptimal period (2 inputs) than in the extremely superoptimal period (25 inputs). Individuals characterized as structurally concrete or abstract did not, however, differ significantly in terms of their information requests. Concrete subjects requested relatively more information than did abstract subjects, except for periods of 8 and 10 inputs. Attempts to differentiate cognitive and personality styles in information-handling have not been overly successful.

Case 56. A later study utilized the same tactical game and the same information load described for Case 55. It was desired to investigate the effect of input information on the number of risky decisions made, that is, whether maximum risk taking occurred during optimal or during super- or suboptimal information conditions. Decisions were categorized as risky (aggressive rather than defensive), preparatory (in anticipation of later aggressive/defense moves), retaliatory (defensive moves), and withdrawal (withdrawing men and equipment from endangered positions). Significant effects as a function of information load were obtained (see Figure 2-22). Most risky decisions were made in periods of optimal information load, but risky decisions as a function of load tended to flatten out at the high information load levels. Risk taking as a function of time spent

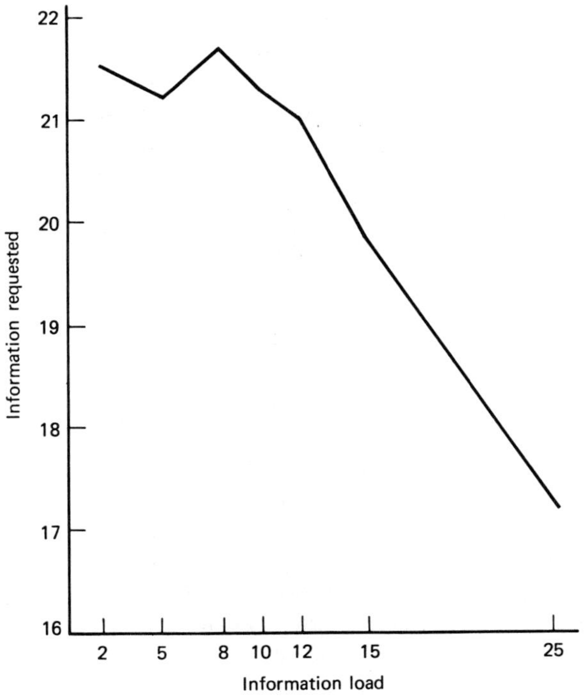

Figure 2-21 Amount of information requested as a function of amount of information input. With permission, from Streufert et al., 1964, Case 55.

in playing the game also tended to rise. [The same phenomenon was noted also in Streufert and Streufert (1970).] The results of the time analysis showed a striking similarity to the work of Davis et al. (1968), obtained in a much simpler, gambling risk experiment.

The presence of irrelevant information also appears to add to the information processing task, at least in complex displays. However the results of the relatively few studies on this variable have produced somewhat contradictory results. Archer (1954) found no effects, and Monty (1973) indicates that "irrelevant information presented along with relevant information in a sequential manner will not degrade performance provided the irrelevant information is introduced in a manner that will enable the operator to pace his rehearsal of the current state of the information." Other investigators (Crawford, 1962; Baker et al., 1960; Eriksen, 1955; Rabbitt, 1964) consider that irrelevant information has negative effects. Most recently Goldstein and Allen (1971) and Goldstein (1972) found that irrelevant stimuli increased response time and errors.

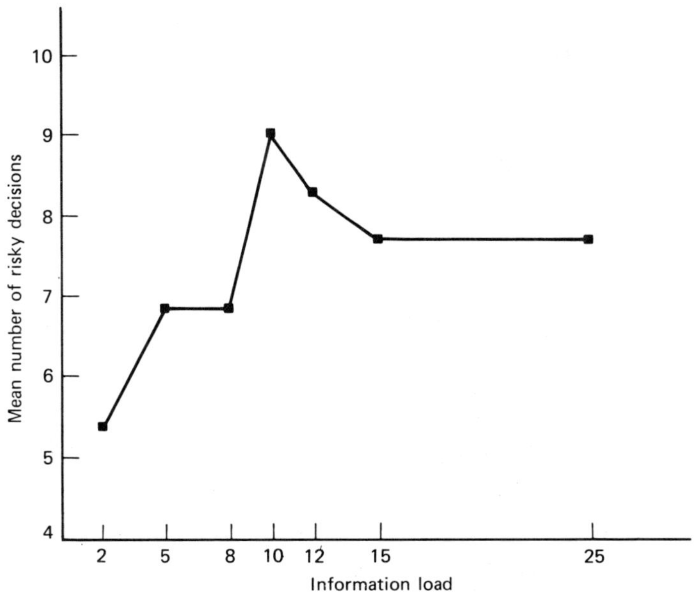

Figure 2-22 Effect of information load on risk taking. With permission, from Streufert and Streufert, *Psychonomic Sci.,* 1968, Case 56.

The effect of irrelevant information may ultimately be a function of the complexity of the information-processing involved, irrelevance being critical for complex behaviors, of little significance for simple ones (Streufert, 1972).

Display Factors

Since most of the inputs that load the system are visual, it is reasonable to suppose an interaction between load conditions and display characteristics. (And in fact that is what we find. Thus Poulton (1968) discovered that the probability of missing at least one target symbol in a simulated electronic display increased approximately linearly with number of targets and number of symbols present. Coffey (1961) found that performance was superior with a lower density of alphanumeric material. Howell and Tate (1966) had much the same result. Monty et al. (1967), determined that performance in a keeping-track task was superior with numbers and letters to performance with symbols.)

The studies in this area have been largely in the context of military information processing systems and describe load in terms of number of inputs.

Case 57. The following series of information assimilation studies was performed over several years at the U.S. Army Research Institute for the Behavioral and Social Sciences (ARI):

1. Information assimilation from alphanumeric displays.
2. Coding updated alphanumeric data in individual and group displays.
3. Accuracy of information assimilation from updated alphanumeric displays.
4. Information assimilation from symbolic displays.
5. Confidence in information assimilation from symbolic displays.

We discuss first the study that dealt with the effects of *amount and density of alphanumeric information* in an automated information processing system. Subjects were required to extract specific items from slide-projected displays containing varying amounts and densities of data. The data were embedded in a columnar format and were located in five different positions. Figure 2-23 is an example of such an alphanumeric display.

In general, as amount of data presented increased, mean search time increased substantially. The slight differences in search time that resulted

Figure 2-23. Example of coded updated alphanumeric data display.

Status of friendly tactical units[a]					
Unit	Activity	Effective Strength	Terrain	Armor Status	Weather
23	**LANDING**	77	FARMLAND	92	DAMP
72	REBUILDING	96	LOWLAND	85	**OVERCAST**
57	ASSEMBLING	87	RIVERS	91	SNOW
82	WITHDRAWING	**76**	MEADOWLAND	**82**	HUMID
34	FLANKING	80	MARSHLAND	76	RAIN
13	SUPPORTING	**71**	DESERT	96	HURRICANE
26	SURROUNDING	85	**FLATLAND**	**86**	SUNNY
99	SCREENING	90	SWAMP	87	WINDY
64	REGROUPING	78	**JUNGLE**	83	HAIL
41	PLANNING	83	LAKES	89	DRY
28	TRAINING	79	VALLEY	80	FREEZING
37	HOLDING	75	**CLIFFS**	78	**STORM**
18	PENETRATING	72	FOREST	**75**	HOT
53	**ASSAULTING**	94	HILLS	90	FOG
31	DEFENDING	78	MOUNTAINS	94	CLEAR

[a] With permission, from Ringel et al., 1966, case 57.

from changes in density suggest that packing data fairly tightly into an alphanumeric display does not impair efficiency of data extraction. Searches for pertinent data near the bottom of displays lasted approximately 22% longer than searches for information near the top. When the number of columns to be searched was increased from two to four, search time increased 100%.

Investigation of the time taken to locate updated alphanumeric elements of information on a CRT revealed that the mean time to locate *coded* updates was approximately 65% shorter than mean time required to locate uncoded updates. This finding lends support to the incorporation and use of coding capabilities in alphanumeric information processing.

More important, performance time was significantly shorter with individual displays than with a group display, raising some doubts about the merits of group displays, at least for certain kinds of tasks. For example, group displays permit the viewers less freedom than do individual displays to touch, trace, and mark on them—activities that may help in the performance of the task.

In another study subjects were presented with successive pairs of slides. The first slide of a pair contained 12, 18, or 24 military symbols positioned on a map background. The second slide of the pair was identical to the first except that 2, 4, or 6 symbols were removed, added, or repositioned. Three levels of enhancement to make the updated changes conspicuous were achieved by three different techniques: (1) hard copy of previous slide to compare with uncoded updated slide, (2) single-cue coding, and (3) double-cue coding. These were contrasted with a condition of no enhancement.

While viewing the second slide, subjects engaged in information extraction (defined as counting and identifying). After the second slide was removed, subjects were given an information-assimilation task in which they indicated the updates they had noted on the second slide.

In general, increasing either the amount of data or the amount of updating resulted in degraded extraction and assimilation performance. In the assimilation task, however, double-cue coding completely prevented the degrading effect of increased data. Double-cue coding improved extraction 97% and assimilation 57% over unaided performance. Single-cue coding improved extraction and assimilation 68 and 47%, respectively. Hard copy history failed to improve extraction and improved assimilation only slightly, chiefly when repositioned updates were presented. These findings clearly indicate the desirability of incorporating conspicuity coding capability in information processing systems. This permits major performance improvement without using relatively costly color techniques.

In still another study subjects were given one minute in which to locate alphanumeric entries that had been updated on the CRT and to cross out the entries in a test booklet. As the number of data elements increased, *accuracy* declined from 86 to 62% for coded updates and from 38 to 11% for uncoded updates. As the number of data elements updated increased, accuracy declined from 88 to 56% for coded updates but remained approximately stable at 22% for uncoded updates. The percentage of errors of omission was higher than that of errors of commission for both coded and uncoded updates. These findings emphasize the importance of establishing and using only essential data.

The end products of automated military information processing systems are maps, "totes," and overlays in which situation information, intelligence data, and administrative data are presented and updated. A study was conducted to determine how information assimilation is affected by amount of data presented and extent of data change introduced by updating *symbolic displays.*

Slides containing military unit flag symbols such as might be depicted on maps and overlays were presented to 32 subjects. Four levels of amount of data were produced by varying the number of symbols (12, 16, 20, and 24) per slide. Following the presentation of a slide, a second slide was shown—basically the same as the first except that 2, 4, 6, or 8 symbols had been removed. Measurement of accuracy of information assimilation was based on the subject's ability to identify the locations from which symbols had been removed.

The results showed a rather large decrease in accuracy of assimilation, from 85% with 12 symbols per slide to approximately 48% correct with 24 symbols. Doubling the number of symbols resulted in halving—approximately—the accuracy. As the amount of change in a single update was increased, accuracy remained the same but more errors of omission than of commission were made.

Accuracy of data assimilation may be only one means of evaluating the effectiveness of displays used in decision making. Feelings of *certainty* about that accuracy may ultimately play an important role in the formulation of decisions; and it was desired to determine the interrelationships among amount of data presented, extent of changes introduced in updating, accuracy of information assimilated, and confidence in that accuracy.

The method was the same as in the previous study, but each subject also had to provide an estimate of his confidence in the correctness of his answers. As the number of symbols in a single display increased, mean accuracy and confidence tended to decrease together in approximately straight-line fashion. As the number of symbols removed increased and as the display was updated, confidence tended to decrease, but accuracy remained the same. Apparently confidence can be affected by a display variable that has no effect on accuracy of information assimilation.

Emergencies may arise when personnel must rapidly screen large numbers of displays for estimates of threat or enemy buildup. A study of such quick estimates was conducted to determine how accuracy was affected by varying brief exposure times, visual angle, and numerosity (amount of data) in a display. After practice, half the subjects received a sequence of slides (from 4 to 22 symbols per slide) at 500 milliseconds followed by the same slides, reoriented, at 200 milliseconds; half received the reverse exposure order. After making a number estimate, each subject indicated his confidence in the correctness of his estimate.

Results indicated that the use of rapid exposure may be feasible for scanning large numbers of displays. However, depending on the requirement of the situation, tradeoffs between accuracy and duration of exposure will be necessary. Confidence statements tend to be conservative, and shorter time exposures have more effect on a man's *confidence* in his performance than on his *actual* performance. Reduced confidence might cause a subject to withhold action beyond the time at which he is able to make a competent decision.

Case 58. Type of data displayed is one of the factors detrmining load and confidence effects. In a COMCON situation (Case 42) an eight-man team evaluated aggressor threats. Load was the number of enemy concentrations (1, 2, 4). Confidence, the subject's probabilistic estimate that his hypotheses would explain the threat, varied with load, but only when individual data classes (main attack, combat support, logistics, order of battle) were differentiated.

Tactical information may be displayed either alphanumerically or in graphic (e.g., map) form. There is great practical value in knowing whether one medium or the other has an advantage. In particular, does the form selected affect the operator's ability to withstand such load conditions as number of inputs? The results of the comparisons so far performed are somewhat equivocal.

In a nonmilitary task, Howell and Tate (1966) found graphic displays to be generally superior to alphanumeric if subjects were required to recall the displayed material. Indirect support for graphic superiority came from Newman and Davis (1962), who determined spatial coded material (graphic) to be superior to alphabetic coded material in both speed and accuracy of recall. On the other hand, Silver (1965) reported that decision accuracy in a nonmilitary task was unrelated to display mode.

Case 59. In these two experiments the two display modes were compared in a military threat-evaluation task. Subjects observed a series of displays describing three enemy forces. The composition and position of units within the three forces were updated at each presentation. Subjects were required to determine which of the three forces was most likely to attack. Although speed and accuracy of decision were related to rate of updating (experiment 1) and to time stress (experiment 2), in neither experiment were performance differences significantly dependent on mode of display.

Two further experiments involving a tactical classification task were designed to provide a more definitive answer to the question. These studies compared the alphanumeric and graphic displays under two conditions of memory and no-memory requirement and under four levels of complexity.

No meaningful differences between the two media appeared, but the complexity factor was quite significant, as were the memory variable and the interaction between the two factors. As we would expect, operators performed much more poorly when memory was required. Problem solution time also increased with level of complexity, although only the first three levels of complexity were significant. Apparently the advantages or disadvantages of the display formats under investigation are determined by functional (e.g., task) requirements. The type of display to be employed, therefore, may be decided on the basis of relative cost.

The reader may wish to consult the following additional studies dealing with display factors.

1. Siegel and Fischl (1971) performed a multidimensional scaling analysis of subject performance on 12 air-defense-oriented visual information displays and isolated seven factors, including stimulus numerosity and coding.

2. Coffey (1961) found that the vertical versus horizontal arrangement of alphanumeric material produced no differences in performance.

3. In a comparison of different types of graphic displays, Schutz (1961) discovered that preference should be given to line-type graphs, followed closely by vertical bar graphs. Irrelevant points and missing data degraded performance.

4. Smith et al (1965) reported that color coding was superior to superscript or underline coding in formatted displays. Color coding was found effective by Munns (1968) in making the observer's task easier on simulated radar displays, but it did not significantly increase effectiveness. Wedell and Alden (1973) learned that color was not significantly more effective than numeric coding in a keeping-track task (based on the air traffic control situation), although it was useful in carrying information concerning the number of items per stimulus category. They feel that color should be used as an aid to search or to fulfill an alerting function but not as a primary information source.

5. Kanarick and Petersen (1971) also found that redundant color coding (used as background) failed to aid performance in keeping-track performance with 10 digital readouts, although it increased the number of strategies subjects employed. On the other hand, color coding for information location (as in checkpoint location on maps) was most effective when many categories of information must be coded and the number of objects per category is small (Shontz et al., 1971). Christ (1975) has presented design guidelines for color coding.

6. Baker and Goldstein (1966) contrasted sequential and "batch" displays in problem solving. In the former situation only the response alternatives permissible at any given time were displayed at that time; in the latter, all response alternatives, permissible or not, were displayed. Significantly more time was required to learn solutions using the "batch" display, because of the longer search time required in this condition.

7. Fell and Laughery (1969) found that the mode in which alphanumeric information is presented (visual and auditory) is unimportant, provided the modes are not mixed. For sonar detection, however, the auditory mode is superior and a combination of auditory-visual is superior to auditory alone (Colquhoun, 1975).

8. Sonn and Rizy (1974), like others before them, reported that the B-scan sonar display format was superior to the standard PPI display.

9. The organization of displayed stimulus material is important, as indicated by Monty (1968). He presented categories of sequential stimuli always on the same display (ordered organization) or on random displays (scrambled organization). The former was significantly superior to the latter.

10. As far as accuracy is concerned, there is little or no difference in group performance in scanning large visual displays versus use of small individual displays. Performance, however, is almost 15% faster with the large group display (Smith and Duggan, 1965). (But see case 57.)

For further information on the effect of display variables on operator performance, the reader is referred to Meister and Sullivan (1969) and Heglin (1973).

SUMMARY OF CONCLUSIONS

1. In general, excessive inputs and stringent output requirements degrade operator performance, a conclusion that replicates real world experience.

2. The load effect can be observed in all functions and tasks tested. Although we know a good deal about the effect of load on the individual operator, the effect of load on multiman systems is less clear, since such

systems have been rarely tested. However, load effects are often highly variable. An input level one would assume to be load inducing sometimes has no discernible effect at all, or, when several system functions and tasks are involved, affects only some of them. An alternative result is that performance degradation due to load can be observed only when particular measures are employed, other measures showing no effect.

3. Some types and frequencies of input (i.e., a low load level) appear to stimulate the operator to improve his performance.

4. Often the operator/system response to a load variable is not proportionate to the amount of load input, that is, is less than one would expect (nonlinearity characteristic).

5. The effects of combining two inputs, either one of them not load inducing, may be to create a load and to degrade operator performance (additivity characteristic). In the same vein, the effect of load on performance may be modified by interaction with coexistent variables (modifiability characteristic).

6. In relatively simple operator systems the effect of the load on the operator and the system is often the same; in multiman systems the load effect on the operator may differ from that on the system.

7. When the opportunity permits, the operator can and will reduce load effects by adopting various strategies of handling inputs.

8. Operator responses to load variables are highly individual and may be modified by learned stimulus values and expected response payoff. They can also be affected by the operator's previous training and experience.

9. Load effects may be influenced by organizational variables such as system and crew organization, task distribution, team size, and load distribution. However the relationship is by no means as clear-cut or as strong as would be expected.

10. In tasks requiring the use of displays (usually visual), coding tends to reduce the negative effect of load variables.

DEVELOPMENTAL IMPLICATIONS

This is the first question to answer. Is the system susceptible to overloading and in what way? The potential for load is highest for systems in which *(a)* inputs are complex, numerous, occur with very high or low frequency, and are relatively weak (close to the operator's threshold) and unpatterned, or *(b)* responses are contingent on multiple factors. It is unlikely that all these factors will be present in the same system; the more there are, however, the more likely it is that a load-inducing situation exists.

The essence of the necessary analysis is an examination of the characteristics of the input (and to a lesser degree, output), based on a review of system requirements. Theoretically the demands imposed on the operator must be matched against what is known about the operator's capabilities. We say "theoretically" because in practice this is a very difficult task to perform except on a highly intuitive basis (which is still better than making no comparison). Where a major discrepancy exists, a load-inducing situation exists.

The literature provides general guidance and helps to pinpoint solutions; but it cannot supply specific design values (e.g., the number of inputs that load the operator in situation X). For specific situations it may be necessary for the HF specialist to perform small-scale ("quick and dirty") experiments to determine the values needed to optimize a particular task.

The most efficient design solution to load is to simplify inputs and outputs and to eliminate time-sharing and task competition as much as possible. However, most systems are the way they are because of unmodifiable mission requirements. The following suggestions, which are more appropriate to indeterminate than to highly programmed systems, may be helpful.

1. It is desirable to develop the system in such a way that its operating procedures are highly preprogrammed and relatively invariant (except under emergency conditions). When the type of system operations makes this impossible (e.g., environmental inputs are erratic), operating procedures should be maximally flexible.
2. Train operators deliberately to develop an adaptive strategy, and perform training under full load conditions.
3. Because multiple-task situations often lead to performance degradation, reduce the number of tasks that must be performed concurrently.
4. When tasks or individual inputs and outputs vary in importance, indicate their relative priority to the operator.
5. Since short-term memory requirements amplify load effects, reduce the operator's memory requirements as much as possible by providing mnemonic aids or by automating the memory function.
6. Give the operator maximum access to all the information he needs. That information should be provided by display media, not through direct verbal communication. Reduce to a minimum the need for operator interaction (particularly communication). Reduce to as few as possible the number of channels through which information is presented to enable personnel to perform their tasks.
7. If a large mass of structured data (e.g., alphanumerics, symbols) must be presented, code the data by color, shape, and so on. Reduce input variability as much as possible, to permit the operator to develop patterns of expectancy.

8. When the task is simple, highly repetitive, or dull, deliberately add a little load (e.g., complexity) to stimulate operators.

GENERAL REFERENCES

Adams, J. A. and Boulter, L. R. *Monitoring of complex visual displays: I. Effects of response complexity and intersignal interval on vigilant behavior when visual load is moderate.* Report AFCCCDD-TN-60-63, University of Illinois Aviation Psychology Laboratory, 1960.

Adams, J. A. and Humes, J. M. Monitoring of complex visual displays: IV. Training for vigilance. *Hum. Factors,* 1963, **5,** 147–153.

Adams, J. A. et al. Monitoring of complex visual displays: III. Effects of repeated sessions on human vigilance. *Hum. Factors,* 1962, **4,** 149–157.

Adams, J. A. et al. Monitoring of complex visual displays: V. Effects of repeated sessions and heavy visual load on human vigilance. *Hum. Factors,* 1963, **5,** 385–389.

Alluisi, E. A. et al. An information analysis of verbal and motor responses in a forced-paced serial task. *J. Exp. Psychol.,* 1957, **53,** 153–158.

Archer, E. J. Identification of visual patterns as a function of information load. *J. Exp. Psychol.,* 1954, **48,** 313–317.

Baker, C. A. et. al. Target recognition on complex displays. *Hum. Factors,* 1960, **2,** 51–61.

Baker, J. D. and Goldstein, I. Batch vs. sequential displays: Effects on human problem solving. *Hum. Factors,* 1966, **8,** 225–235.

Bell, R. A. et al. Watchkeeping performance as a function of certain properties of the viewing situation. *Hum. Factors,* 1974, **16,** 65–69.

Bergstrom, B. et al. Effects of sleep loss and stress upon radar watching. *J. Appl. Psychol.,* 1973, **58,** 158–162.

Bergum, B. O. and Lehr, D. J. Effects of authoritarianism on vigilance performance. *J. Appl. Psychol.,* 1963, **47,** 75–77.

Bergum, B. O. and Lehr, D. J. Monetary incentives and vigilance. *J. Exp. Psychol.,* 1964, **67,** 197–198.

Billings, C. E. et al. Physiological cost of piloting rotary wing aircraft. *Aerosp. Med.,* 1970, **41,** 256–258.

Briggs, G. E. and Naylor, J. C. Team versus individual training, training task fidelity, and task organization effects on transfer performance by three-man teams. *J. Appl. Psychol.,* 1965, **49,** 387–392.

Brown, D. W. The effect of observer redundancy on display monitoring efficiency. *J. Psychol.,* 1963, **56,** 413–419.

Childs, J. M. and Halcomb, C. G. Effects of noise and response complexity upon vigilance performance, *Percept. Mot. Skills,* 1972, **35,** 735–741.

Christ, R. E. Effects of payoff and probability on recall of multisymbol displays. *J. Gen. Psychol.,* 1969, **80,** 81–92.

Christ, R.E. Review and analysis of color coding research for visual displays. *Hum. Factors,* 1975, **17,** 542–570.

Colquhoun, W. P. The effect of "unwanted" signals on performance in a vigilance task. *Ergonomics,* 1961, **4,** 41–51.

Colquhoun, W. P. Evaluation of auditory, visual, and dual-mode displays for prolonged sonar monitoring in repeated sessions. *Hum. Factors,* 1975, **17,** 425–437.

Coffey, J. L. A comparison of vertical and horizontal arrangements of alphanumeric material—Experiment I. *Hum. Factors,* 1961, **3,** 93–98.

Connolly, D. W. et al. *Tactical decision making: II. The effects of threatening weapon performance and uncertainty of information displayed to the decision-maker on threat evaluation and action selection.* Report ESD-TR-61-45 and AFCRL 1100. Operational Applications Laboratory, Bedford, Mass. December 1961. (AD 288-878.)

Conrad, R. Speed and load stress in a sensor-motor skill. *Brit. J. Indust. Med.,* 1951, **8,** 1–7.

Conrad, R. Missed signals in a sensor-motor skill. *J. Exp. Psychol.,* 1954, **48,** 1–9.

Conrad, R. Some effects on performance of changes in perceptual load. *J. Exp. Psychol.,* 1955, **49,** 313–322.

Crawford, A. The perception of light signals: The effect of the number of irrelevant lights. *Ergonomics,* 1962, **5,** 417–428.

Cumming, R. W. and Croft, P. G. Human information processing under varying task demand. *Ergonomics,* 1973, **16,** 581–586.

Davies, D. R. and Tune, G. S. *Human Vigilance Performance.* New York: American Elsevier, 1969.

Davis, J. H. et al. Risk taking; task, response pattern, and grouping. *Organ. Behav. Hum. Performance,* 1968, **3,** 124–142.

DeGreene, K. B. *Sociotechnical Systems; Factors in Analysis, Design and Management.* Englewood Cliffs, N.J.; Prentice-Hall, 1973.

Dorfman, P. W. and Goldstein, I. L. Spatial and temporal information as cues in a time-sharing task. *J. Appl. Psychol.,* 1971, **55,** 554–558.

Eriksen, C. W. Partitioning and saturation of visual displays and efficiency of visual search. *J. Appl. Psychol.,* 1955, **39,** 73–77.

Fell, J. C. and Laughery, K. R. Short-term memory: Mode of presentation for alphanumeric information. *Hum. Factors,* 1969, **11,** 401–406.

Fineberg, M. L. *Effects of object distance, shape, closing velocity and aviator experience in obstacle avoidance behavior.* Dissertation, Catholic University, Washington, D.C., May 1975.

Fitts, P. M. Cognitive aspects of information processing: III. Set for speed versus accuracy. *J. Exp. Psychol.,* 1966, **71,** 849–857.

Freed, A. M. Human interactions in man-machine systems. *Hum. Factors,* 1962, **4,** 389–396.

Gabriel, R. F. and Burrows, A. A. Improving time-sharing performance of pilots through training. *Hum. Factors,* 1968, **10,** 33–40.

Gaites, R. A. The Navy and reduced shipboard manning. *Nav. Eng. J.,* December 1974, 73–80.

Goldstein, I. L. Irrelevant information: Is there any? *Proceedings,* American Psychological Association Convention, 1972, pp. 677–678.

Goldstein, I. L. and Allen, J. C. Effects of irrelevant stimuli on the processing of information in complex displays. *J. Appl. Psychol.,* 1971, **55,** 110–113.

Gould, J. D. and Schaffer, A. Visual monitoring of multi-channel displays. *IEEE Trans. Hum Factors,* 1966, **HF-7,** 69–76.

Gould, J. D. and Schaffer, A. The effects of divided attention on visual monitoring of multi-channel displays. *Hum Factors,* 1967, **9,** 191–202.

Hearns, J. F. and Moss, S. M. Effect of method of payoff on the detection of targets in a visual search task. *J. Exp. Psychol.,* 1968, **78,** 569–573.

Hebb, D. O. Drives and the C.N.S. (conceptual nervous system). *Psychol. Rev.,* 1955, **62,** 243–254.

Heglin, H. J. *NAVSHIPS-display illumination-design-guide. Section II, Human Factors.* Report NEL/TD 223, Naval Electronics Laboratory Center, San Diego, Calif., January 1973. (AD 770-478/6 GI.)

Howell, W. C. and Tate, J. D. Influence of display, response, and response set factors upon the storage of spatial information in complex displays. *J. Appl. Psychol.,* 1966, **50,** 73–80.

Ince, F. and Williges, R. C. Detecting slow changes in system dynamics. *Hum. Factors,* 1974, **16,** 278–285.

Jenkins, H. M. The effect of signal rate on performance in visual monitoring. *Am. J. Psychol.,* 1958, **71,** 647–661.

Jerison, H. J. Human and animal vigilance, *Percept. Mot. Skills,* 1965, **21,** 580–582.

Jerison, H. J. and Pickett, R. M. Vigilance: The importance of the elicited observing rate. *Science,* 1964, **143,** 970–971.

Jerison, H. J. et al. The elicited observing rate and decision processes in vigilance. *Hum. Factors,* 1965, **7,** 107–128.

Johnston, W. A. Transfer of team skills as a function of type of training. *J. Appl. Psychol.,* 1966, **50,** 102–108.

Johnston, W. A. et al. Human vigilance as a function of signal frequency and stimulus density. *J. Exp. Psychol.,* 1966, **72,** 736–743.

Kahneman, D. *Attention and Effort.* Englewood Cliffs, N.J.: Prentice-Hall, 1973.

Kanarick, A. F. and Petersen, R. C. Redundant color coding and keeping-track performance. *Hum Factors,* 1971, **13,** 183–188.

Kappauf, W. E. and Powe, W. E. Performance decrement at an audio-visual checking task. *J. Exp. Psychol.,* 1959, **57,** 49–56.

Karsh, R. et al. Effects of knowledge of results and method of payoff on keeping-track performance. *J. Psychol.,* 1970, **75,** 73–79.

Kennedy, R. S. and Coulter, X. B. Research note; The interactions among stress, vigilance, and task complexity. *Hum. Factors,* 1975, **17,** 106–109.

Kidd, J. S. et al. *The effect of enroute flow control on terminal system performance; A study in human engineering aspects of radar air traffic control.* Report 57–663, Wright Air Development Center, Wright-Patterson AFB, Ohio, April 1958. (AD 142–096.)

Kidd, J. S. *A summary of research methods, operator characteristics, and system design specifications based on the study of a simulated radar air traffic control system.* Technical report 59–236, Wright Air Development Center, Wright-Patterson AFB, Ohio, July 1959. (AD 235-126.)

Kinkade, R. G. and Kidd, J. S. *The effectiveness of procedural variations in the use of target identification and airborne position information in the performance of a simulated radar approach control system.* Technical report 58–624, Wright Air Development Center, Wright-Patterson AFB, Ohio, March 1959a. (AD 216-361.)

Kinkade, R. G. and Kidd, J. S. *The effect of team size and intermember communication on decision-making performance.* Report 58–474, Wright Air Development Center, Wright-Patterson AFB, Ohio, April 1959b. (AD 245-621.)

Klein, T. J. and Cassidy, W. B. Relating operator capabilities to system demands. *Proceedings,* Human Factors Society 16th Annual Meeting, October 1972.

Klemmer, E. T. A further study of information transmission with matrix patterns. *USAD CRC Res. Rep.* 57–1, September 1957.

Klemmer, E. T. and Muller, P. F. The rate of handling information; Key pressing responses to light patterns. *J. Mot. Behav.,* 1969, **1,** 135–147.

Knowles, W. B. Operator loading tasks. *Hum. Factors,* 1963, **5,** 155–161.

Lanzetta, J. T. and Roby, T. B. Group performance as a function of work-distribution patterns and task load. *Sociometry,* 1956, **19,** 95–104.

Levine, J. M. The effects of values and costs on the detection and identification of signals in auditory vigilance. *Hum Factors,* 1966, **8,** 525–537.

Mackworth, J. F. The effect of signal rate on performance in two kinds of vigilance task. *Hum. Factors,* 1968, **10,** 11–18.

Mackworth, J. F. and Mackworth, N. H. The overlapping of signals for decisions. *Am. J. Psychol.,* 1956, **69,** 26–47.

Malmo, R. B. Activation. A neuropsychological dimension. *Psychol. Rev.,* 1959, **66,** 367–386.

McDonald, L. B. and Ellis, N. C. Relationship between predicted stress and measured attentional demand in a simulated driving task. *Proceedings,* Human Factors Society, 18th Annual Meeting, October 1974, 117–122.

McDonald, L. B. and Ellis, N. C. Stress threshold for drivers under various combinations of discrete and tracking workload. *Proceedings,* Human Factors Society, 19th Annual Meeting, October 1975, 488–493.

McKendry, J. M. et al. A subjective value approach to information utility. *Hum. Factors,* 1971, **13,** 503–509.

McKenzie, R. E. et al. *Neptune; A multielement task system for evaluating human performance.* Report SAM-TR-69-25, USAF School of Aerospace Medicine, Brooks AFB, Texas. October, 1969. (AD 700-313.)

Meister, D. and Sullivan, D. J. *Guide to human engineering design of visual displays.* Final report, contract N00014-68-C-0278. Bunker-Ramo Corp., Canoga Park, Calif., August 1969. (AD 693–237.)

Metters, D. B. Modeling the saturation level of a human radar operator. Report GE/EE/74-73, USAF Inst. Technology, Wright-Patterson AFB, Ohio, December 1974. (AD A009-203.)

Miller, G. A. The magical number seven, plus or minus two; Some limits on our capacity for processing information. *Psychol. Rev.,* 1956, **63,** 81–97.

Miller, G. E. et al. *Human factors aspects of reliability.* Report U2296, final report, contract DA 36-039 SC-90877, U. S. Army Electronics Research and Development Laboratories, Fort Monmouth, N.J., January 1964. (AD 431-611.)

Miller, J. G. Psychological aspects of communication overloads. In R. W. Waggoner and D. J. Carels (Eds.), *International Psychiatry Clinics: Communication in Clinical Practice.* Boston, Little, Brown, 1964, pp. 201–224.

Montague, W. E. et al. The effects of signal rate and response complexity on eighteen hours of visual monitoring. *Hum Factors,* 1965, **7,** 163–172.

Monty, R. A. Spatial encoding strategies in sequential short-term memory. *J. Exp. Psychol.,* 1968, **77,** 506–508.

Monty, R. A. Keeping track of sequential events: Implications for the design of displays. *Ergonomics,* 1973, **16,** 443–454.

Monty, R. A. et al. Stimulus characteristics and spatial encoding in sequential short-term memory. *J. Psychol.,* 1967, **65,** 109–116.

Moore, H. G. *The effects of load and accessibility of information upon performance of small teams.* Dissertation, University of Michigan, Ann Arbor, October 1961. (AD 268–462.)

Morgan, B. B. and Alluisi, E. A. Effects of work-load stress on skilled performance with a code transformation task. *Proceedings,* American Psychological Association Convention, 1969, pp. 753–754.

Munns, M. Some effects of display symbol variation upon operator performance in aircraft interception. *Percept. Mot. Skills,* 1968, **26,** 1215–1221.

Newman, K. M. and Davis, A. R. Relative merits of spatial and alphabetic encoding of information for a visual display. *J. Eng. Psychol.,* 1962, **1,** 102–126.

Pachella, R. G. and Pew, R. W. Speed-accuracy tradeoff in reaction time; Effect of discrete criterion times. *J. Exp. Psychol.,* 1968, **76** (1), 19–24.

Posner, M. I. Information reduction in the analysis of sequential tasks. *Psychol. Rev.,* 1964, **71** (4), 491–504.

Poulton, E. C. Searching for letters or closed shapes in simulated electronic displays. *J. Appl. Psychol.,* 1968, **52,** 348–356.

Rabbitt, P. M. Ignoring irrelevant information. *Brit. J. Psychol.,* 1964, **55,** 403–414.

Rolfe, J. M. The secondary task as a measure of mental load. In W. T. Singleton et al. (Eds.), *Measurement of Man at Work.* London, Taylor & Francis, Ltd. 1971, pp. 135–148.

Sales, S. M. Effects of variations in workload. *Proceedings,* American Psychological Association Convention, 1969, pp. 333–334.

Schipper, L. M. et al., *Human engineering aspects of radar air traffic control: IV. A comparison of sector and in-line control procedures.* Technical report 56–69, Wright Air Development Center, Wright-Patterson AFB, Ohio, September 1956. (AD 110–528.)

Schutz, H. G. An evaluation of formats for graphic trend displays—Experiment II. *Hum. Factors,* 1961, **3,** 99–107.

Senders, J. W. The estimation of operator workload in complex systems. In K. B. DeGreene, (Ed.), *Systems Psychology.* New York: McGraw-Hill, 1970, pp. 207–216.

Shannon, C. E. A mathematical theory of communication. *Bell Syst. Tech. J.,* 1948, **27,** 379–423, 623–656.

Shontz, W. D. et al. Color coding for information location. *Hum. Factors,* 1971, **13,** 237–246.

Siegel, A. I. and Fischl, M. A. Dimensions of visual information displays. *J. Appl. Psychol.,* 1971, **55,** 470–476.

Siegel, A. I. and Wolf, J. J. *Man-Machine Simulation Models.* New York: Wiley, 1969.

Silver, C. *Development of criteria for evaluation of large-screen displays.* August 1965. (AD 621–231.)

Singleton, W. T. The measurement of man at work with particular reference to arousal. In W. T. Singleton et al. (Eds.), *Measurement of Man at Work.* London: Taylor & Francis, Ltd, 1971, pp. 17–25.

Smith, S. L. and Duggan, B. C. Do large shared displays facilitate group effort? *Hum. Factors,* 1965, **7,** 237–244.

Smith, S. L. et al. Color coding in formatted displays. *J. Appl. Psychol.,* 1965, **49,** 393–398.

Sonn, M. and Rizy, E. F. The effect of display format on sonar detection and localization. *Hum. Factors,* 1974, **16,** 368–372.

Streufert, S. C. *Relevance and Load: effects on simple and complex decision making.* Technical Report 3, Contract N00014-67-A-0226-0030, Purdue Univ., November 1972.

Streufert, S. and Streufert, S. C. Effects of increasing failure and success on military and economic risk taking. *J. Appl. Psychol.,* 1970, **54,** 393–400.

Taub, H. A. Effects of differential value on recall of visual symbols. *J. Exp. Psychol.,* 1965, **69,** 135–143.

Tickner, A. H. and Poulton, E. C. Monitoring up to 16 synthetic television pictures showing a great deal of movement. *Ergonomics,* 1973, **16,** 381–401.

Versace, J. *The effect of emergencies and communications availability with differing entry rates. A study in human engineering aspects of radar air traffic control.* Technical Report 56–70, Wright Air Development Center, Wright-Patterson AFB, Ohio, December 1956. (AD 118-320.)

Wagner, R. C. et al. *Preliminary investigations of speed and load as dimensions of psycho-motor tasks.* TR-54-45, Air Force Personnel and Training Research Center, Lackland, AFB, Texas, 1954.

Ware, J. R. et al. The role of experimenter attitude and contingent reinforcement in a vigilance task. *Hum. Factors,* 1964, **6,** 111–115.

Wedell, J. and Alden, D. G. Color versus numeric coding in a keeping-track task: Performance under varying load conditions. *J. Appl. Psychol.,* 1973, **57,** 154–159.

Welford, A. T. Stress and performance. *Ergonomics,* 1973, **16,** 567–580.

Wiener, E. L. On simultaneous monitoring and tracking. *J. Appl. Psychol.,* 1975, **60,** 100–105.

Williges, R. C. The role of payoffs and signal ratios in criterion changes during a monitoring task. *Hum. Factors,* 1971, **13,** 261–267.

Williges, R. C. et al. Role of verbal communication in teamwork. *J. Appl. Psychol.,* 1966, **50,** 473–478.

Wright, A. D. et al. *Radar target detection as influenced by experience and training. Research Memorandum,* U.S. Army Air Defense Human Research Unit, (HUMRRO), Fort Bliss, Texas, October 20, 1964. (AD 455–767.)

Yntema, D. B. Keeping track of several things at once. *Hum. Factors,* 1963, **5,** 7–17.

Yntema, D. B. and Schulman, G. M. Response selection in keeping track of several things at once. *Acta Psychol.,* 1967, **27,** 325–332.

Yerkes, R. M. and Dodson, J. D. The relation of strength of stimulus to rapidity of habit-formation. *J. Comp. Neurol. of Psychol.,* 1908, **18,** 459–482.

CASE REFERENCES

Case 1. Johnston, W. A. and Briggs, G. E. Team performance as a function of team arrangement and workload. *J. Appl. Psychol.,* 1968, **52** (2), 89–94.

Case 2. Kidd, J. S. A comparison of one-, two- and three- man work units under various conditions of work load. *J. Appl. Psychol.,* 1961, **45** (3), 195–200.

Case 3. Kidd, J. S. *Some sources of load and constraints on operator performance in a simulated radar air traffic control task.* Technical Report 60-612 Wright Air Development Center, Wright-Patterson AFB, Ohio, March 1961. (AD 255-232.)

Case 4. Kidd, J. S. Work team effectiveness as a function of mechanical degradation of the intra-team communication system. *J. Eng. Psychol.,* 1963, **2,** 1–14.

Case 5. Kinkade, R. G. and Kidd, J. S. *The effect of different proportions of monitored elements on operator performance in a simulated radar air traffic control system.* Technical report 59–169, Wright Air Development Center, Wright-Patterson AFB, Ohio, June 1959. (AD 227–420.)

Case 6. Lanzetta, J. T. and Roby, T. B. Group performance as a function of work distribution patterns and task load. *Sociometry,* 1956, **19,** 95–104.

Case 7. Behan, R. A. An experimental investigation of the interaction between problem load and level of training. *Hum. Factors,* 1961, **3,** 53–59.

Case 8. Adams, J. A. et al. Monitoring of complex visual displays: V. Effects of repeated sessions and heavy visual load on human vigilance. *Hum. Factors,* 1963, **5,** 385–389.

Case 9. Adams, J. A. et al. Monitoring of complex visual displays: II. Effects of visual load and response complexity on human vigilance. *Hum. Factors,* 1961, **3,** 213–221.

Case 10. Connolly, D. W. et al. *Preliminary summary report, tactical decision making: II. The effects of track load on damage, cost and kills.* Report ESD-TR-61-43, Operational Applications Laboratory, Air Force Electronics Systems Division, Bedford, Mass., November 1961. (AD 285–819).

Case 11. Fox, W. R. and Vance, W. H., Jr. *Tactical decision making: I. Action selection as a function of track load, threat complexity, reliable data presentation and weapon uncertainty.* Report ESD-TDR-61-42, Operational Applications Laboratory, Air Force Electronic Systems Division, Bedford, Mass., November 1961. (AD 404–275.)

Case 12. Fox, G. J. et al. *The effect of overlapping raid situations on ACO intercept performance.* Report SPECDEVCEN 279-3-22, Special Devices Center, Port Washington, N.Y., July 7, 1954 (AD 65247.)

Case 13. Fox, G. J. et al. *ACO performance as a function of number of simultaneous raids and time available for interception.* Report SPECDEVCEN 279-3-15, Special Devices Center, Port Washington, N.Y. November 20, 1953. (AD 43900.)

Case 14. Connolly, D. W. et al. *An experimental investigation of the validity of an empirical measure of threat of enemy raids using ACO performance as a criterion.* Report SPECDEVCEN 279-3-17, Special Devices Center, Port Washington, N.Y., November 20, 1953 (AD 43981.)

Case 15. Lefford, A. and Taubman, R. E. *Preliminary studies of detection time and other factors involved in AEW performance.* Technical report SDC 279-3-4, Special Devices Center, Port Washington, N.Y., August 1950. (AD 641–446.)

Case 16. Sinaiko, H. W. et al. *Functional relationship of time density to the detection of discrete radar target events.* Report SDC 279-3-6, Special Devices Center, Port Washington, N.Y., April 1951. (AD 641–452.)

Case 17. Mills, R. G. and Bauer, M. A. *Aircraft track initiation and maintenance in a single-operator simulated surveillance system. Technical Report II.* Report AMRL-TR-71–76, Aerospace Medical Research Laboratories, Wright-Patterson AFB, Ohio, October 1971. (AD 741-902.)

Case 18. Smith, S. L. and Thomas, D. W. *Display color coding compared with three shape codes for a class counting task.* Report MTS-12, MITRE Corp., Bedford, Mass., December 1963.

Case 19. Taub, H. A. and Abrams, C. Effects of target value and exposure duration on recall in a visual search task. *J. Appl. Psychol.,* 1971, **55** (4), 393–398.

Case 20. Teichner, W. H. et al. *Experiments on the effects of input variables on multi-target alphabetic displays.* Report ESD-TDR-63-546, Decision Sciences Laboratory, Hanscom Field, Mass., July 1963.

Case 21. Hitt, W. D. An evaluation of five different abstract coding methods—Experiment IV. *Hum. Factors,* 1961, **3,** 120–130.

Case 22. Chiles, W. D. et al. Work schedules and performance during confinement. *Hum. Factors,* 1968, **10,** 143–196. Consult Case 22 for comprehensive bibliography.

Case 23. Chiles, W. D. and Jennings, A. E. Effects of alcohol on complex performance. *Hum. Factors,* 1970, **12,** 605–612.

Case 24. Olson, P. L. Variables influencing operator information processing. *Hum. Factors,* 1963, **5,** 109–116.

Case 25. Chiles, W. D. et al. *Methodology in the assessment of complex human performance: The effects of signal rate on monitoring a dynamic process.* Report AM 69-6, Federal Aviation Administration, Office of Aviation Medicine, April 1969. (AD 697–943.)

Case 26. Shulman, H. G. and Briggs, G. E. *Studies of performance in complex aircrew tasks.* Report Ohio State University, Research Foundation (Air Force Office of Scientific Research, TR-72-0156), December 1971. (AD 735–762).

Case 27. Jeantheau, G. *The differential effects of speed and load stress on task performance.* Report WADC-TR-59-7, Aeromedical Research Laboratories, Wright-Patterson AFB, Ohio, July 1959. (AD 233-460.)

Case 28. Matheny, B. J. *A study of the effects of varying loads of aircraft pips and pip speeds upon vectoring performance in air traffic control.* Report SPECDEVCEN 71-16-14, Special Devices Center, Port Washington, N.Y., March 28, 1955. (AD 90614.)

Case 29. Wallis, D. and Samuel, J. A. Some experimental studies of radar operating. *Ergonomics,* 1961, **4,** (2), 155–168.

Case 30. Cusack, B. L. et al. *The effect of number of search operators, reporting procedures and load on the performance of an airborne CIC system.* Report SPECDEVCEN 279-3-13, Special Devices Center, Port Washington, N.Y., October 30, 1953. (AD 43899.)

Case 31. Sperandio, J. C. Variation of operator's strategies and regulating effects on work load. *Ergonomics,* 1971, **14** (5), 571–577.

Case 32. Kidd, J. S. et al. *The effect of enroute flow control on terminal system performance: A study in human engineering aspects of radar air traffic control.* Report 57-663, Wright Air Development Center, Wright-Patterson AFB, Ohio, April 1958. (AD 142–096.)

Case 33. Kennedy, J. L. The system approach: Organizational development. *Hum. Factors,* 1962, **4,** 25–52.

Case 34. Sweetland, A. and Haythorn, W. W. An analysis of the decision-making functions of a simulated air defense direction center. *Behav. Sci.,* 1961, **6,** 105–116.

Case 35. Teichner, W. H. et al. *Effects of target value and the interaction of short and long term memories on human information processing.* Report 1, Contract Nonr 3357 (06), Office of Naval Research, September 1965.

Case 36. Christ, R. E. and Teichner, W. H. The effects of differential value on the recall of realistic targets. *Hum. Factors,* 1967, **9,** 273–276.

Case 37. Kanarick, A. F. and Petersen, R. C. Effects of value on the monitoring of multi-channel displays. *Hum. Factors,* 1969, **11,** 313–320.

Case 38. Alexander, L. T. and Cooperband, A. S. *System training and research in team behavior.* Technical memorandum 2581, System Development Corporation, Santa Monica, Calif., August 16, 1965. (AD 620–606.)

Case 39. Hartman, B. O. and McKenzie, R. E. *Systems operator proficiency. Effect of speed stress on overload performance.* Report 61–40. School of Aerospace Medicine, Brooks AFB, Texas, June 1961.

Case 40. Alexander, L. T. and Cooperband, A. S. The effect of rule flexibility on system adaptation. *Hum. Factors,* 1964, **6,** 209–220.

Case 41. Roby, T. B. *Research involving communication processes in task oriented groups.* Final report, Tufts University Institute for Psychological Research, ONR contract Nonr 494 (15), May 16, 1967.

Case 42. Feallock, J. B. and Briggs, G. E. *A multiman-machine system simulation facility and related research on information-processing and decision-making tasks.* Report

AMRL-TDR-63-48, Aerospace Medical Research Laboratories, Wright-Patterson AFB, Ohio, June 1963. (Study 1, pp. 64–83.) (AD 411-552.)

Case 43. Streufert, S. et al. Information load, group organization and communication frequency. *Psychonomic Sci.,* 1972, **27** (6), 348–350.

Case 44. Tickner, A. H. et al. Monitoring 16 television screens showing little movement. *Ergonomics,* 1972, **15** (3), 279–291.

Case 45. McKendry, J. M. and Hurst, P. M. Adaptation to speed stress in an immediate memory task. *Hum. Factors,* 1971, **13,** 543–552.

Case 46. Monty, R. A. Effects of post-detection response complexity on subsequent monitoring behavior. *Hum. Factors,* 1962, **4,** 201–207.

Case 47. Lanzetta, J. T. and Roby, T. B. Group learning and communication as a function of task and structure "demands." *J. Abnorm. Soc. Psychol.,* 1957, **55,** 121–131.

Case 48. Roby, T. B. and Lanzetta, J. T. Conflicting principles in man-machine system design. *J. Appl. Psychol.,* 1957, **41** (3), 170–178.

Case 49. Huntley, M. S. Task complexity and serial performance under steadily increasing input rates. *Hum. Factors,* 1972, **14,** 65–75.

Case 50. Nicholson, R. M. Maximum information-handling rates for sequentially presented visual stimuli. *Hum. Factors,* 1962, **4,** 367–373.

Case 51. Taub, H. A. and Osborne, F. H. Effects of signal and stimulus rates on vigilance performance. *J. Appl. Psychol.,* 1968, **52,** 133–138.

Case 52. Moore, H. G. *The effects of load and accessibility of information upon performance of small teams.* Dissertation, University of Michigan, Ann Arbor, October 1961. AFSOR contract AF 49(638)-449. (AD 268–462.)

Case 53. Brown, D. W. and Fox, G. H. The effect of observer redundancy and task difficulty on display monitoring efficiency. *J. Psychol.,* 1965, **59,** 267–274.

Case 54. Weisz, A. Z. and McElroy, L. S. *Information processing in a complex task under speed stress,* Report ESD-TDR-64-391. Decision Sciences Laboratory, Electronics Systems Division, Hanscom Field, Bedford, Mass., May 1964.

Case 55. Streufert, S. et al. *Some effects of discrepancy from optimal information load and diversity.* Princeton University, N.J., Report 2, ONR contract Nonr 1858(42), June 1964. (AD 603–178.)

Case 56. Streufert, S. and Streufert, S. C. Information load, time spent and risk taking in complex decision making. *Psychonomic Sci.,* 1968, **13,** (6), 327–328.

Case 57. Ringel, S. et al. *Human factors research in a command information processing system.* Technical report 1145, Army Personnel Research Office, Washington, D.C., March 1966. (AD 634–313.)

Case 58. Southard, J. F. et al. *An application of Bayes theorem as a hypothesis selection aid in a complex information-processing system.* Report AMRL-TDR-64-51. Aerospace Medical Research Laboratories, Wright-Patterson AFB, Ohio, August 1964. (AD 607-256.)

Case 59. Nawrocki, L. H. *Alphanumeric versus graphic displays in a problem solving task.* Technical research note 227, U. S. Army Behavior and Systems Research Laboratory, Arlington, Va., September 1972. (AD 748–799.)

III
TASK CHARACTERISTICS

SUMMARY. Whereas in Chapter 2 we considered the input elements of tasks, in this chapter we examine task characteristics as these relate to the task as a whole and to task interrelationships. Special attention is given to difficulty, patterning, complexity, organization, and automation, which cut across task content. The common elements of task taxonomies and the features of the wide variety of tasks used in behavioral research are summarized. Studies to develop quantitative predictive indices of equipment-task relationships are reviewed.

INTRODUCTION

The task is critical to the operator and the system because it activates, organizes, and directs their operations. The task as performed is a snapshot of the system in operation. System development is in large part task construction. It is likely therefore that task characteristics significantly influence operator performance; they may account for a major part of the variance and variability in that performance.

Unfortunately we lack a substantial base of data on task effects. In most experiments the task is simply given, and because its characteristics are held constant between experimental groups, it is implicitly assumed that task effects are canceled out. What is overlooked is that even when the task is supposedly held constant, it interacts with experimental variables, affecting their operation in a (usually) unknown manner.

Task variables were implicit in Chapters 1 and 2. Chapter 2 dealt with the stimulus (a critical task element) and stimulus characteristics (e.g., number, rate, value). In this chapter, however, we treat task characteristics in relation to the task *as a whole,* to the *interrelationships* among tasks, and to relationships *between tasks and equipment characteristics.*

It is necessary to make a distinction between *task content* and *task characteristics.* When we talk about different types of tasks (e.g., monitoring a stock quotation display or troubleshooting an automobile engine), we refer to the content of these tasks. Task characteristics, such as difficulty or complexity, cut across task content. Two auditory vigilance tasks, although differing in specific content (one monitoring voice radio, the other Morse code), could be equally complex or difficult. Or the same task could possess different characteristics if it had different speed or accuracy requirements. Task characteristics are, therefore, *general properties* of tasks rather than of any specific type of task.

Task content is more fundamental than characteristics because content stems in the first instance from system requirements and always constrains characteristics. Yet at a later stage of system development (e.g., when one wishes to specify training requirements) task characteristics become important in their own right.

The classic definition provided by Miller (1953) is still valid: a task is "a group of discriminations, decisions and effector activities related to each other by temporal proximity, immediate purpose and a common man-machine output." The elements of a task are, therefore, the stimulus to the operator, which triggers performance of the task, the required response to that stimulus (i.e., the performance criterion), a procedure for performing this response (which includes the equipment to be utilized for performing the task), and a goal or purpose that organizes the whole. The task is obviously more than the inputs of Chapter 2, but less than the subsystem and system.

There are other ways of defining the task. Roby and Lanzetta (1956) suggest that a task can be represented by a space of multiple dimensions, each dimension describing some condition relevant to productivity. Teichner and Olson (1971) define a task in terms of information flow between man and machine or of the functional activity describing the manner in which information is transferred. However these other definitions do not conflict with the Miller definition and are less useful.

Despite a tendency to think of the task as if it were a physical entity, the task as we observe someone performing it is quite different from the task in its abstract (verbal-descriptive) form (e.g., as a checklist). The latter has no physical existence and is purely and simply a conceptual *construct.* It is similar to the distinction between a physical male adult and the word "man." The distinction is important because the conceptual form of the task is a very imperfect representation of its actual performance, and this is where our difficulties begin.

For example, it is commonly assumed that something about the task should suggest *what* the operator should be trained to do and *how,* so that he can perform the physical task well. What is that something? Is it observable in the task as physically performed? Where in the task are the trainable skills and knowledges that presumably can be extracted from it?

Obviously these are deductions or inferences, not observables, hence likely to be somewhat erroneous.

Nonetheless although task characteristics do not exist in a physical form, it is useful to apply them because we can derive meaningful products from that application, personnel selection criteria, training requirements, and man-machine interface characteristics.

Almost all task applications in system development (task construction, task description and analysis, task classification, and some aspects of performance measurement) are conceptual rather than physical and make use of conceptualized task characteristics.

It is therefore impossible to understand operator or system behavior unless one first understands the task to be performed, including its characteristics.

Definitions

In the preceding subsection, the task was defined in terms of its elements (e.g., stimulus, procedure, purpose). Such a definition is useful in constructing a new task and describing it in written form. This definition does not aid us at all in defining properties that differentiate one task from another, however. If, for example, we wish to say that task X will require more training time than task Y, it is necessary to refer to a property (characteristic) of the task that is related to training duration. The major properties of tasks are as follows.

Temporal Relations

Tasks function over time. An individual task has an anticipated or required *duration* that may affect performance through such variables as operator fatigue and motivation (the longer the task, the more likely the operator is to be fatigued). The context in which an individual task is performed (e.g., when it is combined with other tasks) may, however, change its duration (Mills and Hatfield, 1974).

When several tasks must be performed over time, their *sequence* may vary, although not randomly, since sequence is largely determined by system requirements. The order in which they must be performed will influence task relationships. The sequence may be more or less repetitive (e.g., 1, 2, 3, 4, 1, 2, 3, 4 vs. 4, 3, 2, 2, 1, 3, 4, etc.). When tasks are performed in a regular (patterned) sequence, the operator should find it easier to make necessary responses. For example, Monty (1973) indicates that irregular rates of stimulus presentation are disruptive because the operator cannot anticipate the occurrence of the next stimulus. Concurrent performance of tasks often results, as we saw in Chapter 2, in degraded performance of one or more of these tasks.

Functioning

To perform a task, the operator must call on certain psychological processes (functions) that are required by (inherent in) the task but cannot be observed. Functions serve as intervening variables that translate task requirements into performance.

Task content usually determines function. However, one effect of task characteristics may be to modify functional requirements. For example, if a task is organized with divided responsibilities, communication between operators may be necessary.

Because of the centrality of functions, dissimilar tasks are easily categorized in terms of their underlying common functions (e.g., monitoring, communicating, controlling). Most task taxonomies are based on functional groupings.

Dependence

Because a task is ordinarily performed with other tasks, either sequentially or concurrently, relationships among tasks exist. We pointed out in Chapter 1 that a task may be wholly or partially dependent on or independent of one previously occurring. The prediction of task success requires consideration of such dependency relationships.

Complexity

It is difficult to define "complexity" unequivocally because of the many conditions that may produce it. A task may possess a varying number of subtasks and elements, and the greater the number of these, the more complex the task. A task is complex when it can be performed in various ways depending on the input configuration. The larger the number of such contingent responses (options), the greater the number of stimulus-response connections, the more complex the task. Thus task complexity is tied directly to system indeterminacy.

Organization

Organization is an essential characteristic of the multiman system, since the manner in which operators interact defines task organization. The effects of this organization can be either to simplify or to complicate task performance. For example, coordination needs increase task complexity.

Divisibility

As Chapter 2 indicated, tasks may be divisible into subtasks. For example, in early sonar equipment bearing to and range of the target required two equipments and two operators who had to coordinate their responses.

Newer sonar devices incorporate both functions in the same piece of equipment, run by a single operator.

Difficulty

The most common characterization of tasks is that they are *easy* or *difficult*. Task difficulty is a special construct related in a complex way to the preceding characteristics, and because of this it presents special problems that are discussed in greater detail later.

Criticality

Although not a property inherent in the task, criticality is a judgment of importance assigned to it by the operator on the basis of system requirements and values. The significance of criticality is that the operator's criterion of importance determines the amount of attention he gives any single task in a multitask situation. Because criticality is not a characteristic of the task itself, it is not discussed in this chapter, but we see its effects in terms of stimulus value and payoff.

Automation

The procedure for performing a task may require varying degrees of manual or equipment implementation. The less manual intervention required (and conversely, the more equipment devices required to perform the task), the more automated the task. From a system developer's standpoint automation is probably *the* critical task characteristic because his design decisions deal largely with automation. Many of the preceding task characteristics are affected by degree of task automation.

Task characteristics are highly interrelated. For example, dependency partially determines complexity; and complexity in turn leads to difficulty. Such interrelationships make definition of individual task characteristics more difficult, especially their interpretation in terms of design implications.

Interactive Variables

Variables that are not part of the task itself nevertheless affect task performance.

System Context

It is possible that the nature and size of the system, the number of man-man and man-machine interactions, and the hierarchical level in the system at which tasks are performed all have some effect on performance of the individual task. This effect may not be very great, since these variables merely provide the setting in which tasks are performed.

System Requirements

One would also expect the system (mission) requirement to influence the individual task by establishing the *criteria* (e.g., accuracy, speed) according to which the task should be performed. For example, emphasis on either speed or accuracy in output will result in more but less accurate or fewer but more accurate responses.

Training and Experience

It is self-evident that training and experience determine how effectively the task will be performed.

Research Considerations

What questions should research on task characteristics address? The following items may permit us to evaluate the adequacy of the task research literature.

1. How important are task characteristics like complexity or dependency to operator/system performance, as contrasted with specific task content? For any single task it would be desirable to say something like: function (e.g., visual monitoring of PPI displays) accounts for 30% of performance variance, other task characteristics (e.g., complexity) account for a further 20 to 30%, specific task content (e.g., display parameters) accounts for 20%, and 20% is unaccounted for. Although we cannot presently make such statements, that ought to be one goal of our research.

Any generalizations and predictions that can be made about operator performance depend on the task and its characteristics. If task performance is largely determined by specific content, it becomes more difficult to extrapolate research conclusions.

2. What factors determine individual task characteristics (e.g., complexity, patterning, difficulty) and how? This requires exploration of the relationship between individual characteristics.

3. What is the effect of individual task characteristics on operator and system performance? Are these effects stable? Do they change as a function of training or practice as, for example, the combination of abilities contributing to performance apparently changes with practice (Fleishman, 1967)?

4. Do certain task characteristics have an effect on the operator's motivation and attitude toward his job? For example, is there a certain level of task difficulty at which a decrease in the operator's motivation to perform (e.g., his level of aspiration) could be predicted?

5. How does one derive from task characteristics guidelines for operator selection and training and for prediction of operational performance? This question is critical because unless the HF specialist can derive

these procedures, his contribution to the design of the personnel subsystem will be severely limited. Despite written guides, the derivation of selection and training requirements is now largely an intuitive process.

TASK TAXONOMIES

A taxonomy is a classification system; a task taxonomy is a classification of behaviors involved in task performance. The development of a universally accepted classification system is considered to be fundamental to science. Without one it is difficult for behavioral scientists to communicate with one another. Although a taxonomy does not suggest hypotheses to be tested, it indicates the variables to be investigated. The commonality of task behaviors described determines in part the extent to which data can be generalized from one task to another. Hence the continuing calls (e.g., Melton and Briggs, 1960; Fleishman, 1967) for the development of an acceptable taxonomy.

Granted the "scientific" need for a task taxonomy, one may still ask whether it is necessary for "practical" (developmental) purposes. The answer is yes; in this context too the HF specialist needs a conceptual frame of reference because the taxonomy will influence his approach to the analysis of system requirements and the derivation of tasks. The subject of task taxonomies therefore warrants discussion, even though taxonomic efforts are primarily analytic rather than empirical.

Any task taxonomy is based on a concept of the behaviors inherent in the task and how they function; thus it is limited by our conceptual rigidities. Since Western behavioral concepts are couched almost exclusively in verbal terms, the task taxonomies developed have been equally verbal. Consequently differences among taxonomies often turn out to be largely semantic—for example, the contrast between "scan" and "survey." Although there may well be subtle differences involved in the actual behaviors represented by these two terms, the differences are extremely difficult to define either observationally or descriptively.

Another manifestation of the verbal influence is the way HF specialists both create and analyze a task. The way in which we develop task instructions bears a not so remarkable resemblance to the way in which we parse a sentence. For example, the task *throw switch to ON position* breaks down to: the *operator* (noun-subject of the sentence, usually implied and not stated) *throws* (verb) a *switch* (noun-object) to *ON position* (adverb modifier of the verb). We tend to analyze the task in the same way:

Stimulus	Response	Object
Operator/input	Verb	Equipment operated

To develop a more effective taxonomy, a radical departure from traditional taxonomic methods may be required; but such a departure would necessitate overcoming our verbal limitations.

Implicit in the concept of taxonomy, as derived from the physical and biological sciences, is the notion of universal acceptance; hence the popular belief that only one taxonomy is possible. For behavior, at any rate, this is quite untrue. Alternative taxonomies are possible, based on the *purpose* for which they were assigned: training (Cotterman, 1959; Gagne, 1965), selection (Fleishman, 1967), and human engineering of equipment (Colson et al., 1974) are examples. Classifications have also been proposed in terms of (1) the behaviors *observed* during task performance, (2) the behaviors, functions, or processes presumably *required* during task performance, (3) the *abilities* the operator must possess to perform the task, and (4) the *characteristics* of the task in terms of its stimulus and response properties (e.g., complexity, dependence).

Although it is desirable to secure agreement among HF specialists that a particular taxonomy will be generally applied for a given purpose, this agreement is a convention rather than a specification of absolute truth. No one behavioral taxonomy is inherently truthful, although one taxonomy may be more *useful* than another. This may be why no single taxonomy has so far been generally accepted by HF specialists, even though one or two are most commonly referred to in the literature.

Taxonomies are in most cases derived from analysis, although a few are empirical. For example, Meister and Mills (1971) developed a very crude taxonomy based on descriptors catalogued from experimental studies. Abilities taxonomies are often derived from factor analysis of experimental data. In each case, however, these classificatory systems utilize a qualitative (i.e., verbal) approach based on the presence or absence of critical attributes.

Researchers have most frequently attempted to describe tasks in terms of the functions required to perform them. A wide variety of descriptive terms are used as labels for these functions (e.g., problem solving, decision making).

The problem encountered in comparing these taxonomies is the inability to equate one verbal term with another. For example, are decision making and problem solving the same? Some authors use only one term or the other, some use both as two distinct descriptors, others use the terms interchangeably, and still others combine the two terms into the broader category of "cognition." How does one relate experimental data obtained with a task described by one term to those from a task described by a different although related term? Obviously we must answer this question before we can generalize performance data across tasks or predict performance with new tasks.

Taxonomies have also utilized different levels of description ranging from the highly molecular (e.g., move index finger) to the very gross (e.g., bank aircraft). Implicit in all these is the assumption that the more molecular behavior "nests" into (is subsumed under) the more molar.

Consequently a behavioral taxonomy may be appropriate to one level of system performance but not to a higher or lower level. Thus many categories in common use (e.g., "cognitive," "motor," "perceptual") are much "too general" for a particular purpose, whereas others at a molecular level (e.g., throws switch) are "too specific."

In this connection the particular task descriptive phrase one uses may be less important than the identification of the particular system level to which it applies. For example, it may not matter whether one uses the term decision making or problem solving, as long as these terms are reserved for only very gross system functions and are never applied to detailed tasks.

How is one to select among the various taxonomies? Unfortunately there are few experimental studies aimed at the validation of taxonomies, although Fowler et al. (1968) have attempted to validate equipment/task indices and Finley et al. (1970) have done similar work with functional taxonomies.

Other efforts have been made to demonstrate the utility of a particular classification scheme. Of special interest in this regard is the study by Levine et al. (1973) in which 53 studies in the vigilence literature were classified in terms of four abilities—perceptual speed, flexibility of closure, selective attention, and time-sharing. For studies falling within each task category, mean performance computed across studies was plotted as a function of time in vigil. Curves relating time in the vigil to detection accuracy differed as a function of the ability requirements of the tasks. Similarly, when the effects of signal rate, sensory mode, and knowledge of results on vigilance performance were examined, different functional relations were found, depending on the abilities presumably required by the tasks. Apparently, then, task classification on the basis of an ability taxonomy improved generalizations about the effects of independent variables on vigilance performance.

Studies such as these encourage the researcher to pursue taxonomic efforts. However, it is a long way from the investigation of Levine et al. to the use of such a taxonomy for system developmental purposes.

Virtually the only guideline supplied by practical experience is that task description at the very gross function level (use of categories such as monitoring, decision making) affords little help in analyzing the task, explaining why it is performed more or less effectively, or in predicting its performance. Functional descriptors lack sensitivity. It is necessary to utilize behavioral dimensions that cut across the major functions and represent ways in which the functions can vary within themselves.

Agreement on a particular classification system is unlikely at the moment. Nor would such agreement testify to the inherent validity of that system. However, an agreed-on convention would be more useful than the present proliferation of taxonomic systems.

Illustrative Taxonomies

The taxonomies described in this section are illustrative of those developed by HF specialists with varying biases, and they reveal the strong and weak points of such biases. No attempt has been made to list all the taxonomies developed over the years, particularly since it has been a habit for HF specialists working in system development to design their own. Moreover, the reader can note the similarity among many of the classification elements.

General Taxonomies

Berliner et al. (1964)

The well-known taxonomy of Berliner et al. is a three-tier system (Table 3-1). The initial level ("processes") describes the gross function. A more detailed function level is described by the term "activities." The category "behaviors" describes the task. More molecular categories are abstracted from a superordinate category and describe in greater detail the category from which they were abstracted.

Table 3-1 Representative Taxonomies: Berliner et al., (1964)

Classification of Behaviors		
Processes	Activities	Behaviors
Perceptual processes	Searching for and receiving information	Detects Inspects Observes Reads Receives Scans Surveys
	Identifying objects, actions, events	Discriminates Identifies Locates

Table 3-1 *(continued)*

Mediational processes	Information processing	Categorizes Calculates Codes Computes Interpolates Itemizes Tabulates Translates
	Problem solving and Decision making	Analyzes Calculates Chooses Compares Computes Estimates Plans
Communication processes		Advises Answers Communicates Directs Indicates Informs Instructs Requests Transmits
Motor processes	Simple/Discrete	Activates Closes Connects Disconnects Joins Moves Presses Sets
	Complex/Continuous	Adjusts Aligns Regulates Synchronizes Tracks

This classification is reasonably descriptive of behaviors that can be observed in task performance; it is comprehensive and easy to use. However, the terms are not exclusive; "inspection," for example, requires observation, and the analyst may run into difficulties if he feels he must use only one term to describe multifunctional behavior.

Many taxonomies are based on or related to the one by Berliner et al. For example, Alluisi (1967) has isolated seven basic functions found in MMS tasks; these (watchkeeping, sensory-perceptual, memory, communication, intellectual, perceptual-motor, and procedural) correspond to the processes named by Berliner et al.

On the other hand, Chambers (1969) skips the intermediate level and proceeds directly from the gross function to the task. In his taxonomy the sensory function requires one to detect, observe and receive. The perceptual function requires discrimination, identification, interpretation, and so on. It is characteristic of the Berliner/Chambers type of taxonomy that superordinate functions are used only to cluster subordinate task categories, ensuring that only the detailed categories are much used (and are useful). In this type of taxonomy definitions of the categories are not provided; it is assumed that the terms can be differentiated on the basis of common usage. On the other hand, although it may be possible to differentiate these behaviors verbally, in describing a task as physically performed it is difficult to deduce the appropriate category from the observed behavior.

A more detailed, multitier functional taxonomy specifically directed at complex MMS operations has been developed by the author (described in Finley et al., 1970). This taxonomy was unusual because it was supplemented by one based on task dimensions. [See also Blanchard (1973) for a somewhat similar taxonomy directed at human error quantification.]

Miller (1967)

In contrast to the preceding taxonomy, Table 3-2 describes very general functions that are actually complex processes. The categories here can be used to describe behavior only in a very gross sense; they describe abilities required for the performance of tasks.

Miller (1971)

Miller's more recent taxonomy (Table 3-3) is based at least partly on the concept of a human analogy to computer processes (e.g., short-term storage, buffer). There are obvious similarities between taxonomies 3-2 and 3-3. For example, "short-term storage" in 3-3 is related to "short-term memory" in 3-2. Again, we are dealing with relatively gross functions.

Teichner and Fleishman (1971)

The approach of Teichner and Fleishman, based in part on concepts derived from electrical circuitry, includes the following major functional

Table 3-2 Representative Taxonomies: Miller, (1967)

1. *Concept of purpose:* the operator, on some basis, must learn the appropriate stimuli and responses and must apply them accordingly.
2. *Scanning function:* the operator must actively search for task- or self-induced stimuli.
3. *Identification of relevant cues function:* from all the available stimuli, the operator must identify significant patterns of cues and name them.
4. *Interpretation of cues:* the operator must assign "meaning" to task stimuli.
5. *Short-term memory:* the operator must retain all significant information during a given task performance.
6. *Long-term memory:* the operator must retain extended stimulus-response associations over long periods of time.
7. *Decision making and problem solving:* the operator must be capable of applying complex techniques, tradeoffs, and rules by which action alternatives are selected.
8. *Effector responses:* the operator outputs of all forms, both manual and symbolic.

categories: searching (receptor orienting or signal seeking), switching (discrete responses), coding (roughly equivalent to cognitive activities), tracking, and complex tasks, which represent combinations of the preceding categories.

Steiner (1966)

Steiner's taxonomy (Table 3-4) describes various *types* of tasks rather than individual functions or task characteristics. The emphasis is on multiman tasks and the way in which group members combine their individual resources.

Specific Taxonomies

Instead of being all-encompassing, a taxonomy may be developed to deal with a specific function such as communicating or decision making. Such a taxonomy may be developed to fit a particular kind of system situation. For example, Krumm and Farina (1962) developed a set of communication categories based on a content analysis of a four-man aircraft crew: requests information, provides information, volunteers assistance, orders course of action, formally indicates compliance to orders, and irrelevant remarks, and acknowledges receipt of message. Siegel and Federman (1973) developed a much more complex communications taxonomy on the basis of factor analysis, and Sidorsky (1964) has developed one for tactical decision making.

A taxonomy for a single function is advantageous in that it is likely to be more descriptive of the situation from which it was derived than a general

Table 3-3 Representative Task Taxonomies: Miller (1971)

Term	Simplified Description[a]
Message	A collection of symbols sent as a meaningful statement
Input select	Selecting what to pay attention to next.
Filter	Straining out what does not matter.
Queue to channel	Lining up to get through the gate.
Detect	Is something there?
Search	Looking for something.
Identify	What is it and what is its name?
Code	Translating the same thing from one form to another.
Interpret	What does it mean?
Categorize	Defining and naming a group of things.
Transmit	Moving something from one place to another.
Store	Keeping something intact for future use.
Short-term storage (buffer)	Holding something temporarily.
Count	Keeping track of how many.
Compute	Figuring out a logical/mathematical answer to defined problem.
Decide/Select	Choosing a response to fit the situation.
Plan	Matching resources in time to expectations.
Test	Is it what it should be?
Control	Changing an action according to plan.
Edit	Arranging/correcting things according to rules.
Display	Showing something that makes sense.
Adapt/Learn	Remembering new responses to a repeated situation.
Purge	Getting rid of the dead stuff.
Reset	Getting ready for some different action.
Goal image	A picture of a task well done.

[a] Detailed definitions and descriptions of these functions are given in Miller (1974).

Table 3-4 Representative Taxonomies: Steiner (1966)

Tasks	Description
Unitary	Cannot profitably be divided into subtasks and performed in piecemeal fashion by two or more operators.
Divisible	Permit the division of labor among operators.
Maximizing	Make success a function of how much or how rapidly something is accomplished.
Optimizing	Success is a function of how closely the individual or group approximates a predetermined "best" or correct outcome.
Disjunctive	When a task is disjunctive, a group's potential productivity is determined by the resources of its most competent member.
Conjunctive	Task success is determined by the performance of the least productive member.
Additive	Task success depends on the sum of the individual efforts of group members.
Discretionary	Permit group members to combine their individual contributions to performance as they wish.

taxonomy would be. Such a taxonomy may be limited, however; a taxonomy describing communications in system *X* may not be equally descriptive of communications in system *Y*.

In general, taxonomies of greatest use to the system development specialist are those that are most specific and descriptive of tasks as observed. Abstract taxonomies have theoretical but not practical value.

Findings

The topics to be discussed in subsequent sections are as follows:

1. Research task types and methodology.
2. Task difficulty.
3. Task patterning.
4. Task complexity.
5. Task organization.
6. Automation.
7. Equipment-task relationships.

RESEARCH TASK TYPES AND METHODOLOGY

It is necessary to describe the types of tasks used in experiments because conclusions derived from the behavioral literature depend to an unknown but significant extent on those tasks.

If the influence of the task on performance is very great, conclusions from studies performed using a single task must be replicated with other tasks to ensure the generalizability of their conclusions.

An important criterion of the effectiveness of HF principles is their generalizability. The two forms of generalizability—(1) to a particular operational system and task, and (2) across tasks and situations—may appear to be opposed, because the more a conclusion relates to a specific operational situation, the less it may relate to other situations.

Maximum generalization to the operational system can be achieved by using tasks performed in the operational environment to gather the data on which the principle is based. Often, however, the opportunity to do so (in any controlled fashion) is absent, forcing the investigator to retreat to a less realistic condition. It then becomes necessary to develop tasks that simulate (to some degree) the operational task by abstracting its major characteristics. Even here, however, a reasonable degree of task simulation fidelity requires that the real world be the criterion.

For generalization across tasks it intuitively seems that the more abstract the task employed, the easier the application of the conclusions derived from the task. As the task becomes more abstract, however, one can generalize to additional situations but say less and less that is specific to each such situation. In determining the characteristics of the tasks he employs, the investigator is therefore forced into continuing compromises. In practice, most investigators select their tasks on the basis of convenience rather than generalizability.

There are ten main tasks employed in the research described in this book.

1. Tasks derived from "small group" research and constructed largely to reveal the effects of group (social) interactions on performance.
2. Tasks developed to study communication networks.
3. Tasks designed to explore the effects of reinforcement processes on team performance.
4. Vigilance tasks, resembling those found in radar and sonar operations.
5. Perceptual tasks.
6. Psychomotor (e.g., tracking) tasks.
7. Reaction time (RT) tasks.
8. Tasks that are analogues of those in military or civilian systems, such as air defense or air traffic control.
9. Probabilistic inference and decision-making tasks (the former are highly cognitive and often somewhat abstract; the latter are often modeled after military command/control situations).

10. Operational tasks (i.e., tasks studied in the operational environment in which they are ordinarily performed).

These tasks vary on three dimensions: abstract-realistic; simple-complex; individual-team. Operational tasks are highly realistic, small group and probabilistic inference tasks are often abstract. RT tasks involve relatively simple functions; decision making involves complex ones. Any of these tasks may describe either individual or team performance.

We study many of the tasks in a context that involves training subjects unfamiliar with the task to some level of proficiency. The focus of such studies is on how rapidly or accurately personnel learn to perform the tasks. The disadvantage of such a research procedure is that in consequence the available data and principles describe relatively naïve people, not skilled operators.

Illustrative Studies

In this section we describe examples of the major classes of tasks utilized in behavioral research. Our intent is to illustrate only the range of task situations used and to provide a background for subsequent sections of this chapter. Only the outstanding characteristics of these tasks are listed.

Group (Social) Tasks

Group tasks are an outgrowth of what has been called "small group" research (McGrath and Altman, 1966). The variables studied are usually social factors, and three types of tasks have been developed: *(a)* creative tasks, which call for the generation of ideas, such as the listing of all possible methods of solving a human relations difficulty, *(b)* negotiation and discussion tasks, which call for evaluation of a "case" or issue and the achievement of a consensual solution, and *(c)* problem-solving tasks that require the integration of information and the specification of a plan of solution.

Man-machine interaction is of little interest to small group researchers; hence they do not find it necessary to replicate real world tasks. The kind of task they utilize is exemplified by "brain twisters," such as the problem of transporting three married couples across a river, given a single three-person boat, with the requirement that a wife is never left in the company of another man unless her husband is present. Or, as in a series of studies by Shiflett (1973), crossword puzzles are performed in different ways by two-man teams. The popular game of "twenty questions" is another example.

The multiman system has a social context that is only a by-product of system organization but may be important in terms of its potential effect on system outputs. Thus some of the tasks used in small group research may resemble aspects of multiman system situations. Because of their abstractness, however, it is difficult to generalize studies with small group tasks to the MMS.

Since HF specialists model their tasks on real world situations, they have done little to investigate the specific effects on performance of tasks as tasks. Since, in contrast, small group researchers develop their own tasks *ab initio,* they have performed comparative studies in which the type of task is a major variable. It is largely from these studies that we derive the notion that task differences produce differential performance effects (Deutsch, 1951; Carter et al., 1950, 1951; Shaw, 1963). The effects of these differences are illustrated in the following cases.

Case 60. This study examined the effects on group interaction of task difficulty and task type, where the tasks were production (creative), discussion, and problem solving. The 108 quarter-hour tasks used (12 for each combination of task and difficulty) included only those having multiple solutions. Each task required a written group answer from the randomly selected three-man groups.

Differences between the three task types did not affect group activity significantly, but individual characteristics of the various tasks significantly differentiated group performance and affected 60% of the group activities. For example, these characteristics produced major differences in structuring the problem or the answer or proposing a solution. Production and discussion tasks were least alike in these respects. Judged task difficulty had relatively little effect on the group interaction process.

Case 61. The same variables (task type and difficulty) and the same types of tasks and procedures were also studied exhaustively in a follow-on to the previous case. As in Case 60, task type determined up to 50% of the variance of group products. In this study task difficulty moderately affected product characteristics.

Communications Network Tasks

A very extensive literature has been generated to study communications processes and structures in man-man systems. [See Glanzer and Glaser (1961), and Shaw (1964), for a review of that literature.] A number of problem-solving tasks have been used in these studies, but the common element was trading information among team members arranged in a specific structure or network until each man knew the unique information given to the other members of the group or at least one team member identified common items of information. A typical task might be an arithmetic problem requiring 12 items of information for solution.

This type of study is performed infrequently these days, but it should be noted that the variable of interest in such tasks was often the effectiveness of a particular network, created physically by having cubicles interconnected by means of slots in the walls separating them. Network members communicated with one another by writing messages on cards and passing them through the slots. Individual communication nets, for example, the wheel and circle (Figure 3-1), could be created by closing appropriate slots.

Although the network correctly simulated the structural characteristics of real world systems, it was so abstracted from the many factors that affect actual communication that the conclusions produced from network studies have little applicability to communications within the MMS.

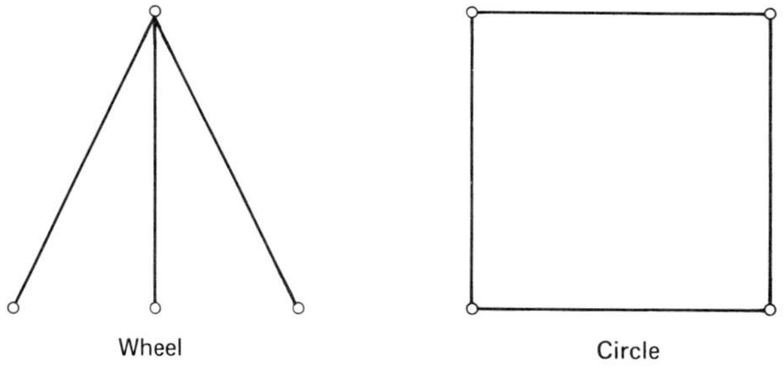

Figure 3-1 Sample communication networks.

Reinforcement Team Tasks

Glaser and his colleagues (Klaus and Glaser, 1970), who examined the role of reinforcement in team operations, created a specific task for their research.

Two team members ("monitors") made choice reaction time responses by depressing a key for either 2 or 4 seconds duration (depending on the visual input signal), while a third member ("operator") observed the durations of these responses as indicated by lights activated by the monitors' key presses. If the operator judged the responses to be of the correct duration, he depressed a key for 4 seconds, but if the response was judged incorrect, he made a 2-second response. In the case of conjunctive parallel teams, correct responses by all three team members resulted in a positive reinforcement of a team point registered on a counter (as well as other lights and a gong), which could be seen by the entire team. In the case of disjunctive teams, a correct response by one of the two monitors followed by a correct operator response resulted in a team point.

The experimental paradigm used in the seven studies performed was to train individually each of the three subjects who were to make up the team until he reached a predetermined proficiency level (.63 or higher). After this the three individuals were combined into a team and assigned to be monitors or operators, and "pattern" training of monitors was begun (two different light patterns associated with the 2-second press, two patterns with the 4-second press). Once monitor proficiency had been achieved, total team training began in which the monitor responses served as inputs to the operator.

This abstract task nevertheless contained the basic elements of real world operations utilizing teams. For example, the timing of responses was critical; simple all or none responses were not utilized as in many abstract laboratory tasks; the monitors did process information, although at a simple level; and the operator was required to make accuracy judgments regarding monitor outputs to be able to generate his own responses, which is a function inherent in supervisory tasks.

Vigilance Tasks

Vigilance tasks are most often performed in the real world by a sonar operator or radarman; as a rule stimuli are only slightly above threshold, the frequency of stimulus occurrence cannot be readily predicted, and the operator's function is to detect the occurrence of a critical (i.e., nonnormal) stimulus and recognize its significance (as in making the distinction between submarine and nonsubmarine).

A well-known experimental analogue of this task is the Clock Test originated by Mackworth (1950), in which the subject has to detect an excessive jump in the continuous movement of a clock hand. Other task variations have involved monitoring the movement of the needle in one or more voltmeters and a series of lights in which a more intense light is the critical signal. Adams et al. (Case 8) made use of a similar monitoring situation based on an air defense analogue; stimuli, however, were well above threshold. Vigilance situations can vary from the highly abstract (e.g., the Clock Test) to a highly realistic air defense simulation (Case 17).

The literature on vigilance is vast because this task has generated great experimental interest. For a review of vigilance data, see Davies and Tune (1969). We do not know exactly why the interest is so great, but what is apparently vigilance behavior is found in many systems, both military and civilian, and the task is superficially easy to set up in a laboratory. Nevertheless, the laboratory vigil is extremely specific, and conclusions reached from it do not necessarily generalize to other types of situations.

Perceptual Tasks

Tasks requiring the analysis of visual stimuli have been used extensively in studies of basic psychological processes, such as short-term memory, or in studies of variables such as exposure time or interstimulus interval. (Auditory stimuli are also sometimes used.) The stimuli (symbols, alphanumerics, figures) are projected for short periods and the subject is required to perform a variety of functions such as counting, updating, or recalling them. These tasks are often analogues of real world information processing situations, and provided they are not too abstract and the variables for which they are studied are not too molecular, the tasks have some applicability to real world systems.

Tracking Tasks

Tracking tasks, whether of the pursuit or the compensatory type, are familiar to almost everyone. Detailed description and review of the tracking literature are outside the scope of this book, but the reader particularly interested in the topic is directed to the book of Poulton (1974). Like vigilance, and for the same reasons apparently, tracking is a favorite experimental situation. The significant thing to remember is that since the tracking task is continuous and emphasizes psychomotor behavior, conclusions secured with this task type *may not* be valid for discrete or cognitive tasks.

Reaction Time Tasks

As with tracking, little need be said about reaction time (RT) tasks, which are defined by the time measure of the operator's reaction to the stimulus presentation. Historically this is one of the oldest psychological tasks, having been invented by Helmholtz in 1850. It can be complicated by requiring a choice between stimuli, but it is perhaps the most molecular task available. Generalization of conclusions reached using RT tasks to more complex activities must be very guarded.

System Analogues

Next to actual operational tasks, system analogues are most applicable (at least superficially) to real systems. Parsons (1972) provides the most comprehensive description of military research performed with such tasks over a 25-year period (approximately 1940–1965). Unfortunately much of the research he cites is not available in the general literature, mainly because of its original security classification. System analogues are abstracted versions of actual systems and their functional processes. The major behavioral characteristics of advanced systems are as follows.

1. They are computer controlled or at least utilize computer processing to aid the operator. Typically the operator enters data and commands to the computer; the computer processes these and displays the processed information to the operator, who then commands the computer further on the basis of the displayed information.

2. These systems often perform in multiple alternative modes, either sequentially or concurrently.

3. System operations are likely to be performed by a *team,* consisting of a number of operators specialized by function and a supervisor.

4. System manning is often flexible, and size of team and distribution of operator specialists vary as a function of operating mode selected.

5. System operations are likely to involve at least two levels of information processing, with information from individual work stations funneling into and displayed at a central command station.

6. The functions performed by individual operators interact; thus actions taken by operator 1 may influence those of operator 2, and so on.

7. A system may make use of a number of data entry methods and display modes, as well as extensive symbology. Communications, too, may be performed through a variety of media.

Simulations of such systems would ideally include the following features:

1. Variable access to information (distributed or integrated).
2. Capability of combining or distributing system functions.
3. Capability of varying the number of system personnel and their arrangement.
4. A variety of data entry and response modes.
5. Flexibility in degree of computer assistance to the operator.
6. Equipment capable of being operated in various interactive modes.

Research applying these characteristics is often found in studies of team operations. Some of the best work on team dynamics has been performed by Lanzetta and Roby (e.g., 1956, 1957), who employed an analogue of an aircraft crew station. The task utilized three-man crews, a total of six input displays (airspeed indicator, manifold pressure, etc.), and six switches. Each man had a procedures card to define the actions required of him by input signals. The tasks called for somewhat complex information processing by the team members, and in some experiments the team relayed information back and forth from one station to another.

The research by Horrocks and his associates (e.g., Horrocks et al., 1960) used a six-man team whose job was to translate inputs from code to English. The input message was submitted to a router who decided which of two decoders would first receive parts of the input for processing. After the two decoders performed most of the translation, their outputs were fed to an evaluator who completed the translation, if necessary; this output in turn was passed on to two integrators who worked together to attempt to assemble the individual parts into a meaningful sentence.

The brilliant series of studies performed mostly by researchers at the Ohio State University (Kidd, 1959; Alexander and Cooperband, 1964; Briggs and Johnston, 1967) focused on air traffic control (ATC). In the Kidd studies, up to four controllers were jointly responsible for directing aircraft entering the simulated terminal ATC area (50 miles) into one or the other GCA gates using minimum flight time and fuel consumption. The Alexander and Cooperband task, utilizing six-man teams, reproduced the San Francisco terminal area. In all cases high fidelity visual (simulated PPI) displays were employed and communications procedures were based on those utilized in the operational context.

Other system analogues have simulated air defense (SAGE) operations. Alexander et al. (1962) employed 13-man crews, and Chapman et al. (1959) utilized teams as large as 30 to 40 people. In some cases the operational system itself was used in a research mode (e.g., Sackman, 1967).

In studies by Briggs and Naylor (1964) and Briggs and Johnston (1965, 1966), two- and three-man teams operated as air intercept controllers in the simulated context of an antiaircraft warfare problem. The teams observed simulated radar returns on PPI displays and, working in real time, attempted to vector interceptor aircraft to the vicinity of target aircraft. Teams issued verbal headings and speed commands to simulated interceptor pilots (assistants who entered the information on signal generator consoles and simulated interceptor returns on the PPI displays).

Among the more complex and realistic simulations worthy of notice are the studies described in Case 42, in which the system was built up of IBM 7090 and 1401 computers, display consoles, closed-circuit TV, telephones, and audio monitoring equipment. In these studies an eight-man team used simulated air reconnaissance reports (descriptions of weapons, vehicles, etc., in enemy territory) to produce threat evaluations of enemy activities.

Studies such as these come closest to the criteria for system studies described in the preface. In particular, they reproduce (or attempt to duplicate) the hierarchical subsystem levels so characteristic of multiman systems.

One of the most significant aspects of tasks that simulate actual systems is the possibility of making use of system output measures (e.g., weapons destroyed, fuel consumed) as descriptors of system performance, as well as the more traditional individual operator measures, such as response time or accuracy. This permits exploration of the crucial question of how the performance of the individual operator or team interacts with or affects total system output. Most abstract artificial tasks do not afford this opportunity.

Probabilistic Inference and Decision-Making Tasks

Probabilistic inference tasks have certain hallmarks. They require subjects to utilize information from several variables in predicting the state or value of a criterion variable; in this respect they resemble concept formation studies. Second, the stimulus, response, and reinforcement cues used are continuous rather than categorical. Finally, they ask subjects to make probabilistic inferences based on frequency (probability) information about environmental events. These characteristics are often found in the real world. Yet since many probabilistic inference tasks are highly abstract, the applicability of their results to real world situations must be qualified.

Probabilistic inference is required, of course, in real world decision-making systems, and such system tasks can be created in analogue form. An excellent example is the work by Sidorsky and his colleagues (1964). In the two-man situation they created, each booth contained an *X–Y* coordinate display in which a moving pip (naval target) was projected during each trial. Located on either side of the console was a digital readout that indicated in 10% increments the momentary probability of success for "own ship" and "enemy ship" as the tactical situation progressed. The two subjects, who competed against each other, were supposed to decide the proper times to fire their weapons.

Operational Tasks

It is also possible to secure data on tasks performed in real world operational contexts. Unfortunately, as Christensen and Mills (1967) pointed out in their survey of such tasks, "very little useable data have been gathered on human activities under operational or field conditions." These authors had "to resort to data from paper and pencil studies, simulation studies, and the results of system tests" to flesh out their conclusions. A further limitation was that "little consistency was found from one study to another."

Observations of operational tasks have most immediate applicability (as one would expect) to operational problems. They are less useful for developing generalizable principles than are system analogue tasks; since the investigator lacks complete control over the conditions during which he makes his observations, acceptable data are more difficult to gather. Although it is possible to manipulate variables in the operational environment (see Fineberg, 1974), some control is lost because of irrelevancies such as test interruptions or changes in schedules.

Moreover, operational experiments are often quite expensive, as illustrated by the Fineberg research involving simulated combat missions flown in a helicopter at nap of the earth (i.e., as close as possible to the ground). The cost of running one subject over a dozen 3-hour flights was approximately $3600, exclusive of salaries for test personnel and pilot subjects; the flights required a minimum of 6 working days.

In summary, to perform research, the investigator must choose at most one or two tasks to be used as a vehicle to test his hypotheses. It is not feasible to test all tasks. This leads to the consideration of the representativeness of the selected task(s)—essentially a philosophical question, since criteria of representativeness depend on the investigator's own "world view" and are not susceptible to empirical verification.

Any task selection must be based on the researcher's model of which task is "typical" or "most important" or "most interesting" or, as is probably most characteristic, which task permits selected variables to be most easily tested.

Since the nature of the task selected may seriously reduce the generalizability of the data secured, the researcher must be aware of the implications of his choice, and a task taxonomy, based on the need to generalize data, might be useful in this area.

TASK DIFFICULTY

Conclusions. The definition of task difficulty presents serious problems because it interacts with other task characteristics and contains both objective and subjective components. The effect of task difficulty is usually reduced performance compared to performance of an "easier" task. The evaluation of task difficulty often varies as a function of the measure employed.

Illustrative Studies

The most complex task characteristic is difficulty, which can be defined—if at all—by interrelationships among and effects of other task characteristics. Difficulty may not exist as an independent dimension; it may be subjective shorthand for any task characteristic that produces negative effects on performance. Even so, difficulty may be fundamental to behavior because as a subjective feeling it may influence performance by causing changes in decision criteria, level of aspiration, and perhaps the amount of effort expended by the operator.

Difficulty is a characteristic that is hard to define because its meaning is apparently self-evident.

Any one or more of the following definitions may be correct. Task difficulty is:

1. The experimenter's *inference* from an operator's differential performance on various tasks. When task output on one task is significantly poorer than on another, the first task is considered more difficult.

2. Some *property* of the task that requires the operator to exert more physical and/or psychological effort than he would otherwise. Note the similarity to one of the definitions of work load (Chapter 2). It is this property the HF specialist seeks to determine by analysis to design the operator's job.

3. The operator's *feeling* (of effort in performing a task) and/or his *evaluation* of the task in terms of experienced success/failure and effort required/expended.

In fact, the notion of difficulty may include all these factors combined.

Several things are immediately apparent from the foregoing definitions.

1. Difficulty has both an objective and subjective component, and it is the only task characteristic possessing this quality. More molecular task characteristics like input rate may not be *experienced* by operators as such; instead their effects may be transformed into a judgment of task difficulty. The same phenomenon applies to work load, with which difficulty is intimately related.

Difficulty may be both the *objective* effect of antecedent task characteristics and the operator's *perception* of the effect of these characteristics. Suppose, for example, that the experimenter varies the input rate from one stimulus every 5 seconds to two stimuli every 5 seconds. The increase in input rate affects the operator's performance; in addition, if he perceives the effect on himself (e.g., in terms of greater effort required), he may change his work effort or strategy. Considered in this light, difficulty becomes a higher-order task characteristic that is created by the effects of lower-order task characteristics like input rate.

2. Since difficulty is a higher-order task characteristic, it must be fractionated to determine the more molecular factors responsible for it.

3. Difficulty can be defined only in comparative terms. Although we can conceptualize difficulty in absolute terms (but not easily), it is impossible in the real world to examine the single task and say that it is easy or

difficult without referring, either explicitly or implicitly, to some standard.

Unfortunately, few experimental studies have been conducted on difficulty per se. Rather, we have studies on numerosity, sequence length, coherence, and other molecular factors in which difficulty is an unacknowledged, confounding variable lurking in the conceptual shrubbery.

One reason for the paucity of studies on difficulty is that such studies may seem pointless, since "everyone knows" that task performance varies inversely with its difficulty. Yet we know almost nothing about the constituent elements of difficulty. The few studies available do little more than verify the common view of task difficulty. Merely to demonstrate that task difficulty degrades performance, however, is trivial.

Difficulty is inherent in every experimental task but is often ignored by the experimenter as being a "given." Difficulty that is not measured or scaled may distort a subject's outputs, and this is also ignored. On the other hand, how can one study so complex a variable? Is it possible that because "difficulty" cannot be defined, no such construct exists? But in that event, why is it impossible for everyone to refrain from using the term?

From a practical standpoint, the determination of the factors that lead to task difficulty are critical for system design and training. Everyone assumes that difficulty "causes" error, even if we can only theorize about error-causing mechanisms (see Singleton, 1973). Therefore, we strive to find clues that will enable us to make tasks "easier," thus reducing error potential.

The need to define difficulty in subjective as well as objective terms leads to a methodological problem in studying the subject. Like workload, difficulty may be experienced by operators in one way when it is manipulated by the experimenter in another way. If we compare two tasks that differ experimentally in terms of objective difficulty indices, and subjects perform both tasks at essentially the same level, do the tasks actually differ in difficulty or is difficulty irrelevant to the performance of that task? If the operator experiences no *feeling* of difficulty, even though his objective performance on one task is significantly worse than on the other, is the first task actually more difficult than the second? Only if both objective and subjective indices of difficulty are collected and coincide, can we be certain that task difficulty exists (which is an argument for never concluding an experimental study without intensively interviewing the subject).

Like work load, task difficulty is also highly dependent on the aptitude, personality, and motivation of the individual operator. For example, a difficult task for the 5-year-old child is usually ludicrously easy for the 15-year-old.

The system designer seeks to reduce difficulty, thereby presumably reducing error likelihood. Yet he seldom is able to eliminate a system task because it is difficult; if it was appropriate to develop the task in the first place, the task was and is required for system operation. Hence the designer must retain the task but change one or more of the properties that make it difficult.

But there is a further problem. When there are alternative ways of measuring performance, objective difficulty may depend on the particular measure employed, as illustrated in Case 62.

Case 62. The task was to track a half-inch vertical hairline controlled by a lateral arm controller. Using a step-function input, the target jumped at predetermined times from one of as many as 15 possible positions to another position in the middle of the CRT screen.

Under these conditions task difficulty could be defined in terms of the number of target steps (5, 10, and 15) employed (sequence length). Subjects were given equal practice time (90 one-minute trials), the sequence being repeated 12, 6, and 4 times per trial (to equalize the time) for 5, 10, and 15 target steps.

Very early in practice integrated error scores were proportional to sequence length; by the second session, however, there was no difference between the 5- and 10-step sequences, although both differed throughout training from the 15-step sequence. Thus difficulty defined in error terms was not equated with sequence length up to a certain value (between 10- and 15-step sequences) after which difficulty appeared to be increased.

If, on the other hand, difficulty was defined as the ratio of (1) number of repetitions of the sequence needed to reach a performance criterion to (2) number of items (targets) in the sequence (i.e., repetitions per item), relative difficulty actually decreased with increases in sequence length.

Performance might also be inversely related to number of possible target positions from which the sequences were drawn. These alternatives were varied (5, 10, and 15 positions) orthogonally with respect to sequence length. Number of alternatives had no reliable effect, nor did it interact significantly with sequence length in affecting error scores.

A second study utilized two training criteria. In one the subjects were run an equal amount of time, in the other the subjects received a fixed number of repetitions of the sequences. Sequence lengths were 8, 12, 16, 24, and 48 targets.

With equal practice time (therefore, unequal repetitions), error scores at the end of 100 training trials were significantly related to sequence length for both the patterned and partially patterned (fixed and repeated elements) conditions.

With equal repetitions of the sequences, however, no significant sequence-length effect or task patterning by sequence-length interaction was found, and there were no differences among task means under patterning conditions.

These results were repeated in a unpublished study (cited in Case 62) using sequences of temporal intervals. Within the limited range of values employed, there was no evidence to suggest that task difficulty increases disproportionately with increases in the length of completely patterned sequences.

This study illustrates one of the problems described in Chapter 2—the criterion employed often determines the kind of conclusions derived.

Case 63. This study considered difficulty in the context of a very different task. Three tasks had been selected to vary primarily on a difficulty dimension quantized by rater consensus, a method often employed in sociopsychological studies. To provide such a task difficulty rating, the task must be fractionated into various attributes. In this case task 1 (which required a group ranking of six cities in the United States in order of population) had a difficulty rating of 2.5; task 2 (a change of work task adapted for use with five-man teams), 4.2; and task 3 (a puzzle problem about dormitory assignments), 6.1.

The tasks themselves are of little interest to us because they have so few real world (in our terms) parallels. It is significant, however, that the task differences attributable to difficulty were statistically reliable ($p = .001$). Mean time to solve each task was very nearly linearly related to the difficulty scale values described previously (task 1, 7.7 minutes; task 2, 18.4 minutes; task 3, 29.2 minutes). The results of this study would be very impressive if the tasks used related to system operations.

Case 64. It is also possible to quantize task difficulty in terms of an information measure H. The task selected for this study involved two communications structures, the wheel and the circle. The other variables were team size (four- or five-man groups) and task difficulty. Task difficulty could be quantized by establishing a set of events whose frequency and probability were known *a priori.* The subject's task, a characteristic one in communications network studies, was to determine in each of 15 problems which of 6 symbols occurred most frequently and that symbol's frequency of occurrence. A problem was considered solved when all group members had found the correct answer. By controlling the frequency/probability of symbol events, two levels of task difficulty were established ($H_1 = 1.6 =$ easy; $H_2 = 2.4 =$ difficult).

Neither network structure or team size proved to be significant. Task difficulty was, however, responsible for significant differences between the networks.

Quantization of task difficulty was possible in this case only because the task was artifically structured. The information content of real-life tasks is not so easily determined.

Case 65. In a follow-on study to Case 64, task difficulty again produced significant effects on operator performance.

Case 66. One of the interesting hypothesized effects of task difficulty is its relationship to criterion measurement. Weitz (1961, 1964, 1966) suggests that when a task is easy, maximal differentiation on a dependent variable will appear relatively early in the task; if the task is difficult, differences will appear late. This relationship should hold regardless of the independent variable being manipulated (but see Case 49).

Tasks may be easy or hard for a number of reasons, including the aptitude or suitability of the individual for the task. Thus if an individual has high aptitude for a particular job, the task for this individual is relatively easy, and vice versa. The effect of an independent variable, therefore, should show up early for high aptitude people, later for low aptitude subjects.

The work of Fleishman and his co-workers (e.g., Fleishman and Hempel, 1954, 1956) suggests some means of classifying tasks on qualitative and quantitative shifts in ability and motivation components at various stages of task performance. Easier tasks might undergo fewer shifts than difficult tasks.

Of course time of measurement, early or late, must be defined relative to the task that is undertaken. For experimental tasks used in a laboratory, "early" may mean the first few trials; for job performance, it may mean the first few months.

(The generalization of laboratory learning principles to the learning of actual tasks assumes that effects from highly accelerated laboratory testing correspond to actual job learning effects, which may appear much more slowly. The assumption seems logical, but we have few hard data to back it up.)

Several investigators have attempted to verify Weitz' hypothesis but only in a sociopsychological context. Hakel (1968) demonstrated some support for the concept, as did the work of Dodd et al. (1970). However these studies, as well as those of Weitz himself, have utilized only very abstract tasks.

The task in Case 66 involved a 7 × 7 matrix in which columns were designated by color names, rows by letters. Cell entries were payoffs ranging from −12 to +15. The game was played by three-man groups, which causes one to ask whether task difficulty for groups (crews) is the same as that for individuals. (It may not be—task difficulty for a crewman may depend on what, when, and how he is required to contribute to the crew output.) The group's task was to select a particular row for each of 15 trials. The object of the game was to maximize payoffs by figuring out the systematic basis on which payoffs were to be made. Task difficulty was determined by alterations in the system of correct column choices for each trial; easy tasks had highly patterned column choices, the hard tasks less regular ones.

As one would expect, easy problems yielded significantly higher payoffs and shorter completion times than did difficult problems. Mean times to completion and mean payoff scores were 31.43 and 291.44, respectively, for the easy task, 44.46 and 178.72 for the difficult task. The results provided some support for the hypothesis, although not without significant discrepancies.

If the hypothesis is valid, experimental conclusions are affected not only by the particular measure the experimenter employs but also by the stage in practice at which he measures. Therefore, failure to take task difficulty (however one defines it) into consideration in experimental design could lead to erroneous conclusions.

TASK PATTERNING

Conclusions. Patterning of stimulus inputs and of sequential tasks (whether defined as input regularity, stimulus redundancy, or task organization) is conducive to reduced error, faster learning, and more effective operational performance.

Illustrative Studies

Task patterning (or "coherence," as it has also been termed in the literature) is defined by the relationship over time among tasks and task elements, that is, frequency of occurrence of task elements over time. In contrast to difficulty, coherence is a unidimensional variable. Patterning is produced by the repetition of stimulus inputs and responses. It may occur in any type of task.

Patterning would tend to counteract difficulty if the critical factor inducing difficulty were the amount of uncertainty associated with stimuli and responses. The effect of patterning should be to make the task (at least subjectively) easier to perform because the operator can anticipate stimulus response requirements more readily. The operator may experience feelings of uncertainty when he cannot anticipate the stimuli he will encounter and the responses he must make, at a high enough level of probability (whatever that level is). Patterned stimulus-response relationships may also be learned more effectively because these are repeated more frequently.

We may say a bit fancifully that task patterning may help to establish "order" or "structure" in the environment. The operator may "prefer" minimal uncertainty, and (if correctly motivated) he may consciously or unconsciously attempt to increase environmental structure by patterning his actions.

Since patterning requires repetition, infrequently performed tasks are probably more difficult to pattern than those occurring more frequently.

It is obviously easier for the system developer to pattern inputs arising from the operations of system equipment than to control inputs arising from outside the system.

The effects of patterning were studied as part of Case 62, which featured a task involving compensatory tracking of a target on an oscilloscope CRT. The target was made to jump at predetermined times from one of as many as 15 possible positions to the middle of the display surface. Patterning was defined in terms of the proportions of fixed and repeated elements in the series of target events. Thus the operator could anticipate in varying degrees the direction of target movements.

The more patterned task produced less error. Error was highly correlated with task uncertainty. Patterning appeared to enhance perceptual anticipation, hence anticipatory responding (thus reducing lag time). Loss in retention was also inversely related to patterning, but intermediate patterning produced results no better than those achieved with a purely random task.

In one of the studies reported in Case 62 a tracking task was developed in which direction, amplitude, and time dimensions could be varied individually or in combination. One, two, or all three dimensions were patterned, while the remaining dimensions were randomly programmed. Neither a fixed time nor a fixed directional pattern alone resulted in improvement better than that achieved on a completely random task. When two dimensions were patterned, however, error scores were intermediate between the completely random and completely patterned task. Thus although redundancy with respect to one dimension of the task did not improve performance, redundancy on two dimensions resulted in some reduction in error scores, although the effect was not proportionate. The extent to which patterned information is used apparently is not a linear function of the proportion of predictable elements in the stimulus sequence, although after considerable practice performance does seem to be closely related to input uncertainty.

Of special interest is the observation that operators utilized the spatiotemporal relationships of the stimulus pattern to organize patterns of responses that improved their performance. In other words, they imposed an organization on their environment. Patterning, therefore, may produce its positive effects partly by permitting the operator to develop more effective reactive strategies.

Subjects developed a matching strategy in anticipating directional choices, and a compromising strategy with amplitude choice situations. These results support the contention that subjects organize their responses in a manner consistent with, though not always proportional to, the kinds and amount of information available in the environmental input.

Task patterning also interacted with other variables (target rate, display mode, and sequence length) to determine task difficulty.

Case 67. In this case the task was modeled after aircraft and aerospace vehicle operations. The operator was confronted with a panel of lights indicating the occurrence of a sequence of nine critical events. If the events occurred in the expected fashion, the operator merely acknowledged the occurrence of each event by pressing an "OK" button. If an emergency occurred, however, the operator had to take over and manually correct the situation by a particular sequence of button presses.

Task patterning was defined as the spatial sequence in which the critical events occurred. For the highly organized task, a simple serial (top to bottom) sequence of light occurrences was employed (i.e., 1 to 9, in that order). For the low task organization condition, the sequence was quite irregular (e.g., 1, 4, 2, 9, 8, 3, 6, 7, 5).

The index of task patterning (organization) was the information metric H calculated on the basis of relationship between changes in light positions rather than on the basis of light position itself. For the high organization task H equaled 0.000 bit, whereas for the low organization task H was 2.808 bits. The high organization condition, therefore, provided more redundancy in terms of the progression of stimulus positions.

The subject also had to perform a three-dimensional compensatory (nulling) tracking task. As one would expect, training on the tracking task was much more effective for the highly organized groups (Figure 3-2). Although differences between groups became smaller with training, these differences persisted over time and indeed increased as the retention interval increased.

Retention was also far superior for the highly organized groups, both for the short (1 week) and long (4 weeks) periods. Additional training tended to compensate the low organization groups for the disadvantage of their condition. Similar results were found for the procedural task (although not to the same degree of significance).

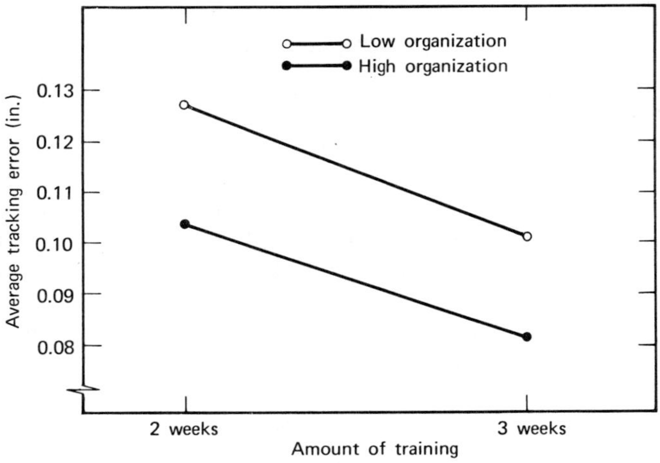

Figure 3-2 Tracking performance as a function of task organization and amount of training. With permission, from Naylor, 1962, Case 67.

Task organization was important only under conditions of lesser degrees of training. Probably patterning has less effect once the task has been well learned, the operator having become able to deal with the uncertainty in his task.

Case 68. The effect of task predictability was also demonstrated in a task involving the processing of displayed instrument readings, the relaying of information to other operators, and the execution of control adjustments based on relayed or directly available readings. Variables manipulated were group organization, rate of input change, and input predictability. Only the latter is of interest to us in this section.

Each of three subjects was seated in a booth constructed to prevent visual and direct auditory communication among occupants. All communication was by means of an intercom between the booths. Pictures of two simulated aircraft instruments were projected on the wall of the booth, and each subject had to adjust two on-off switches to a correct setting based on various combinations of projected instrument readings. Inputs to each subject consisted of slide changes that altered instrument readings; these changes could be either random or predictable (i.e., in order of booths 1, 2, and 3).

Unsurprisingly, error scores were much less under the more predictable rate of input change: 211.75 to 253.50 (stimulus changes every 10 seconds), and 99.75 to 152.75 (15-second changes).

TASK COMPLEXITY

Conclusions. Task complexity increases error probability and reduces performance. In relatively simple, repetitive tasks, the addition of a little complexity tends to stimulate the operator.

Illustrative Studies

Task complexity may be generated in many ways—by increasing the number of required subtasks, by ensuring stimulus-response (S-R) incompatibility, by adding a secondary concurrent task to a primary one, by introducing irrelevant information, and by increasing the number and rate of stimulus changes or the number of procedural rules to be considered in making a judgment—the list does not exhaust the possibilities.

Like difficulty but unlike patterning, therefore, complexity is a heterogeneous variable. In fact it is often considered equivalent to difficulty. Complexity may differ from difficulty by not having a subjective component; one does not experience complexity as such, but rather in terms of difficulty. Conceivably one could define complexity as the objective stimulus component of difficulty; what is ordinarily termed difficulty may be the subjective (felt or observed) component of complexity, which then becomes the independent variable, with difficulty the dependent variable of a phenomenon, tossed together and loosely termed "difficulty." The author does not insist on this; it is only a hypothesis that may be worth following up.

The one factor common to all the variables producing complexity is ability to increase task uncertainty. In each case the number of possible alternatives among which the operator must select is increased. The operator who performs a task containing subtasks must consider (consciously or unconsciously) which of these subtasks is demanded as a response. An increase in the number of subtasks (particularly when they must be performed concurrently) acts to increase error likelihood because of *interference* effects among these potential responses. One would expect, therefore, to find a greater amount of task complexity in indeterminate systems requiring many contingent responses.

Many of the factors generating task complexity are the same as those inducing load; in fact, task complexity can act as a load-inducing condition. Task complexity therefore reproduces many of the load phenomena observed in Chapter 2, including the use of operator strategies to reduce complexity effects. Like load, moreover, task complexity is transformed into the operator's subjective experience of difficulty.

Case 69. In this study complexity was defined in terms of stimulus-response relationships, which were manipulated to create a simple and a complex display-control task. Two-man teams were trained on one or the other. Figure 3-3 plots the effects as differences in learning. Performance differences are very gross because the task was deliberately constructed to maximize the complexity differences. The results shown are precisely what one would expect to find: that complexity reduces performance.

It seems reasonable to suppose that the effect on task performance of an environmental input would vary as a function of the complexity of the task. Broadbent (1964), investigating the relationship between attention required to an RT task and degree of S-R compatibility, found that the incompatible task caused more interference with the secondary task. In an earlier study (1954) he had reported noise-induced decrement on a "difficult" watchkeeping task but not on an "easy" one. Park and Payne (1963), however, observed no interaction between noise conditions and level of "difficulty" of visual problems.

Case 70. This study attempted to test the hypothesized relationship between input effect and task complexity with noise as the input. The subject, after being alerted by a green warning light, responded as quickly as possible to the onset of one of four red stimulus lights by pressing one of four switches located directly below each of the lights. In the simple (S-R compatible) condition, the correct switch corresponded to the light directly above it; in the complex (S-R incompatible) condition, there was no spatial correspondence between the light and switch. The 48 subjects simultaneously performed either a simple or a complex auditory monitoring task. All tasks were performed in quiet and in noise.

Noise had no significant effect on the primary task, although predicta-

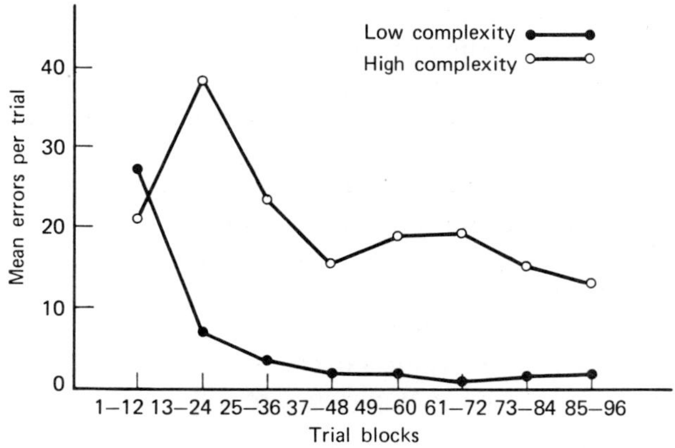

Figure 3-3 Learning as a function of differences in task complexity. With permission, from Zink, 1957, as described in Rosenberg, *Ergonomics,* 1959.

bly RT on the complex task was somewhat longer than on the simple task (808 vs. 763 milliseconds). However on the secondary (auditory) task, a significantly greater number of errors was made on the complex task than on the simple one. That the noise × complexity interaction was significant is shown by the finding that the differences in mean errors between the noisy and quiet conditions were 3.25 for the simple task and 7.29 for the complex one.

The demands of a complex task on the operator are obviously greater than those of a simple task. In the primary task the operator was able to match those demands by utilizing what some researchers have called "reserve capacity"; presumably the secondary task utilized still more of this reserve capacity. The addition of the noise input to the secondary task may have drained more of this capacity, resulting in more errors.

Case 71. In another task that simulated radar control of manned interceptors against target aircraft, task complexity was defined in terms of the number of functions to be performed. In the low complexity task the controller commanded speed and heading adjustments only; the high complexity condition called for adjustments in speed, heading, and altitude. Although task complexity was essentially only a secondary variable (the major variable was task organization, as discussed later), it was the most potent and consistently influential variable affecting performance of the three-man crews. Performance was consistently inferior for the high complexity condition, which is understandable because the third function (altitude) required operators either to remember the item of information or request it from pilots.

Case 72. Task complexity in this information processing situation was defined as the number of rules that had to be considered in making a judgment—a special case of response criteria imposing a load on the operator. Subjects, who were tested individually, worked at the self-paced problem solving task for varying times averaging 30 minutes. Each subject sat alone in a small room facing a screen, both hands resting in front of a panel of six keys. Seventeen practice slides, each containing six symbols, were presented. Subjects were instructed to select a single symbol in accordance with certain rules. These rules were: (1) choose black rather than white, (2) choose large rather than small, (3) choose squares rather than circles, and (4) choose "up" positions rather than "down" positions. Since all rules were equally important, the subject had to select from among the six symbols the one having the greatest number of positive attributes. The time taken to solve each problem and the symbol chosen by pressing a key were recorded automatically.

After subjects had learned the rules, test slides were presented. One set of 20 test slides required application of one rule, the second set of 20, two rules, and so on. Groups received slides in order of increasing or decreasing complexity, or in random presentation. The complexity variable was the amount of information to be processed, and it increased as the number of rules required increased. The "uncertainty" variable consisted of changing the order of presentation of these rules. Thus both the complexity and the "uncertainty" of the task were varied.

Solution times and errors increased significantly as complexity increased—as would be expected. However subjects who experienced high task uncertainty (the random order of presentation) exhibited faster solution times than either of the groups carrying out the task in predetermined sequence. This faster solution time, moreover, was not achieved at the cost of more errors. Subjects perceived the task as being significantly more difficult as task complexity increased, thus tending to verify the proposition that complexity is translated into a difficulty judgment.

The results are all predictable except for the shorter solution times for the random (more uncertain) order of presentation. It is possible that uncertainty tends to reduce task boredom, thus increasing operator alertness and concentration. There is a hint, therefore, that uncertainty may have beneficial as well as negative effects on task performance.

Case 73. The effects on performance of task complexity as a load-inducing variable ought to increase when complexity is combined with another load-inducing variable. In this task complexity was paired with time pressure in a team task. The task was assembly and disassembly of Tinkertoy models (as many as possible during a specified period of time).

Task complexity was defined in two ways: in terms of the number of individual operations required and the proportion of simultaneously coordinated operations. The task was specialized for each member of the three-man team; for example, a member could handle dowels or pegs, but not both. Time pressure was defined by the frequency with which the three-man team was informed of time intervals remaining before the task had to be completed: a signal midway through the task (low pressure); a signal every 3 minutes (medium pressure); a signal every minute, increasing to one every 15 seconds (high pressure).

Productivity scores in terms of the total number of correct operations performed by the group indicated the now familiar curvilinear relationship from low to medium to high time pressures. Productivity was significantly greater for the more complex task, but the time condition was not significant. Nor did task complexity and time stress interact; there were no additive effects. The time signals apparently did not produce the desired degree of stress.

Even though subjects indicated that the more complex task was more stressful, they enjoyed it more. Given highly structured, repetitive assembly tasks, teams apparently are more motivated by more complex tasks. This motivational effect may have helped counteract the negative effect of complexity. This tends to confirm the results of Case 72, where uncertainty acted against the influence of boredom. Obviously the effects of complexity, like the effects of load, vary with the specific task. For an inherently simple task, the addition of a little uncertainty that does not seriously threaten the operator may stimulate him. For an inherently complex task, however, additional uncertainty may be highly stressful.

Case 74. Task complexity can also be defined in terms of the number of channels over which signals are presented (see also Case 37). The operator was required to count either three tones or three lights presented over one, two, or three channels. The four tasks studied were as follows: (1) one-channel auditory—only the middle tone was to be counted and the high/low were to be ignored, (2) two-channel auditory—the middle and low tones were to be counted independently and the high one was to be ignored, (3) three-channel auditory—the three tones were to be counted separately, and (4) three-channel visual, the three lights were to be counted separately.

As one would expect, performance was best on the one-channel auditory (60% correct) and worst (32% correct) on the three-channel auditory task. The three-channel visual task (42% correct) was about equal in difficulty to the two-channel auditory (41% correct).

In summary, the results of studies on task complexity validate the system design principle, "keep it simple."

TASK ORGANIZATION

Conclusions. Task organization is a less significant variable than task complexity or difficulty, becoming important only when the task is otherwise stressful. In multiman systems an independent task organization is superior to one requiring interaction among operators. It is possible to predict total task performance from prior performance on individual components of the task. However performance effects on task components do not combine in a statistically independent manner.

Illustrative Studies

Intertask organization is defined by the manner in which personnel operations are arranged (the degree or manner of interaction among two or more crew members). Intratask organization is defined by the manner in which subtasks or task components relate to each other in their performance by a single operator.

Task organization is a structural characteristic ordinarily built into a system during its development. This differentiates organization from input and response patterning, which may derive from extrasystem operations and may vary during mission performance. Since task organization is objective, it can be more precisely varied during experiments than can, for example, difficulty. Unlike difficulty, complexity or patterning, it contains no inherent implication of performance degradation or enhancement. As with the previous characteristics, the operator is unlikely to be directly aware of task organizational effects—therefore he responds to these in terms of the ease-difficulty dimension.

The major question about task organization is: how does error probability vary as a function of the type of task organization developed?

The amount of research on task organization is pitifully small. This is doubly unfortunate: not only is organizational information needed to understand crew processes, but knowledge of such interrelationships is necessary for quantitative prediction of task/system performance.

For example, it is a reasonable hypothesis that tasks requiring coordination or interaction among operators during training and performance entail greater likelihood of error because the coordination/interaction tends to induce load. Similarly, when tasks must be performed in series (task 2 depending on whether task 1 was accomplished to some degree of correctness), there may be a greater probability of system failure if any task is performed incorrectly or not performed at all. Series tasks may be more likely to build up queues.

Intertask Organization

Case 71 also involved a major variable: task organization. That study featured three independent variables in a three-man task: task organization, task complexity, and the relative amount of training of a replacement radar controller (RC). Task organization occurred at two levels: In the independent condition an RC worked without being able to observe his companion RC and was given no opportunity to communicate with anyone but his supervisor. In the interaction condition the two RCs on the team sat side by side, each with his own radar console but each able to observe the other and to trade targets and interceptors freely.

Task complexity also occurred at two levels. The RCs in the less complex condition were required to control only the airspeed and heading of the interceptor, whereas in the more complex treatment the RCs controlled altitude as well as heading and airspeed. Finally, a man being trained to replace one of the two RCs received either one more (high) or one less (low) session of training on that task than did the man he replaced. The replacement always occurred in the third session on the transfer task.

Performance was measured in terms of transfer of training. An *abstract* version of the experimental task served to train the operators, and testing of any transfer that occurred (at four concurrent transfer sessions) was performed on the experimental task.

There was a significant improvement in performance from session 1 to session 2, during which task complexity exerted a statistically significant effect on performance; however task organization did not influence performance during these sessions. Whereas the less complex task was significantly superior to the more complex task in terms of fuel consumed per intercept, the independent task organization did not differ significantly from the interaction task organization. An analysis of sessions 2 and 3 was performed to determine the immediate effects, if any, of the replacement of an RC during session 3. The analysis showed statistically significant improvement, replacement performance under the less complex condition being again superior to that under the more complex condition. Task organization did not differentially influence the replacement effect.

During transfer session 4, each of the independent variables exerted a significant influence on team performance at some time or other. Task complexity remained the most potent and consistently influential variable affecting performance, whereas task organization did not become important until after the system was stressed by replacing one of the two RCs; thereafter task organization continued to affect team performance to the end of the experiment. The training variable (amount of replacement training) had an extremely short-term but precise and very important effect at the time of replacement (see also Cases 125 and 127).

Case 75. This experiment, a follow-on to Case 71, was designed to test further the importance of task organization, to explore its importance in training, and to examine training task fidelity (to the transfer task). The training and transfer tasks were essentially the same as in Case 71, but complexity was, held constant: the RCs controlled only the heading and speed of the interceptors (the less complex condition of Case 71). Also unlike Case 71, there were no operator replacements.

The three independent variables in Case 7 5 were training task organization, transfer task organization, and fidelity of training to transfer task. Training task organization involved an *independent* condition in which the RCs were trained individually with no intercommunication or direct trading of targets and/or interceptors, and an *interdependent* condition in which the two RCs used the same problem board and interacted directly with each other as a target crossed into the transition zone from one area of RC responsibility to the other. Transfer task organization was manipulated at the two levels in the same way. Half the teams trained under an independent task organization were transferred to an interdependent organization; the other half encountered an independent organization in the transfer task, and likewise for teams trained originally on an interdependent task.

Training task fidelity was manipulated at two levels: high and low. Under the low level of fidelity the RCs carried out their duties in relation to an abstract version of the task that required moving checkers representing interceptors and merely calling out heading and speed changes. Under the high fidelity condition the situation was more similar to what would be encountered in the transfer task—experimenters moved the checkers and repeated back the orders from the RCs over a simulated radio link.

Graphic summaries of performance during transfer appear in Figures 3-4 and 3-5. In the first two sessions team performance was affected significantly by both the training task organization and the transfer task organization. In both cases teams with an independent task organization did better on the average than did teams working in an interactive condi-

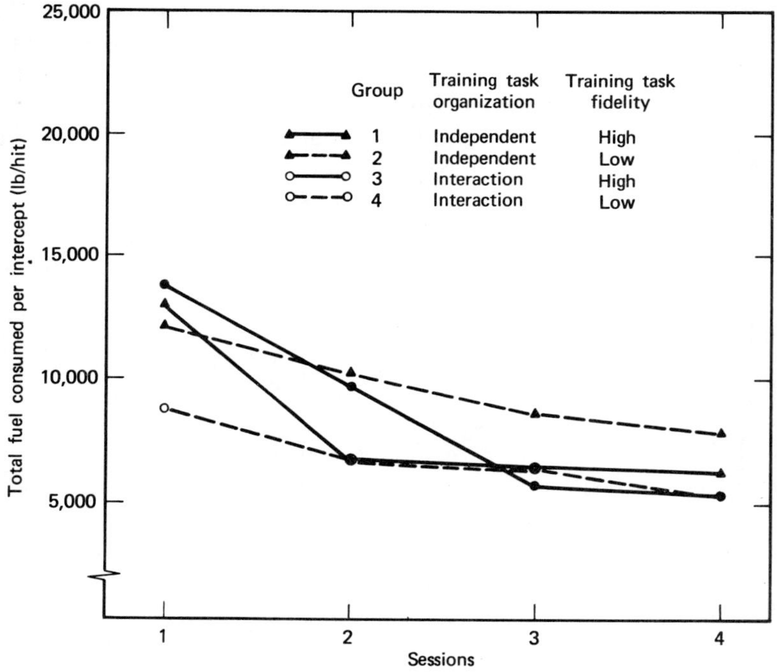

Figure 3-4 Performance as a function of task organization and fidelity (groups 1–4). With permission, from Briggs and Naylor, 1964, Case 75.

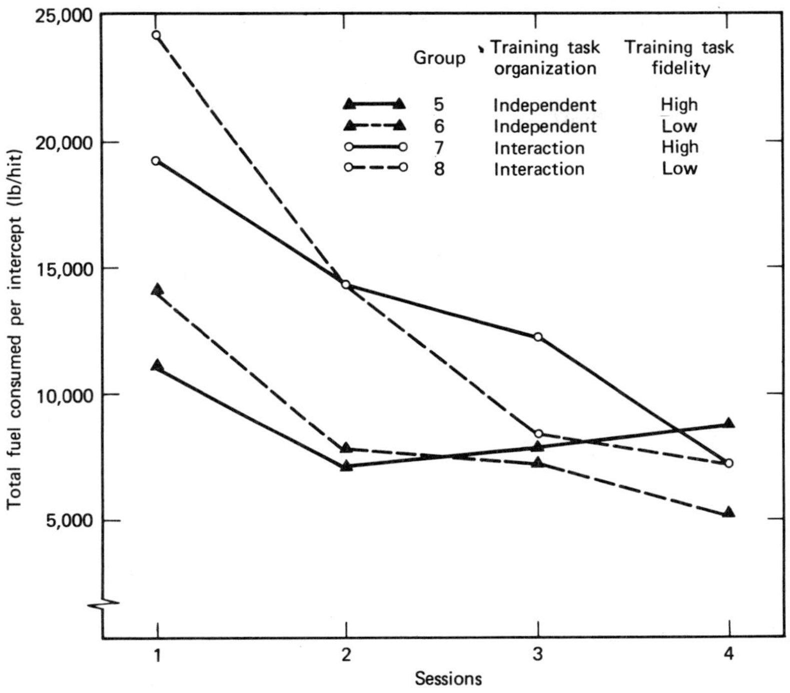

Figure 3-5 Performance as a function of task organization and fidelity (groups 5–8). With permission, from Briggs and Naylor, 1964, Case 75.

tion. We say "on the average" because the two variables interacted statistically; the transfer task organization did not influence team behavior in the same way for each of the two levels of training task organization.

If a subject transferred to an independently organized task, it apparently made little difference which type of training task organization had been experienced previously (19,522 vs. 21,150 lb of fuel consumed per hit). This suggests that an independent task organization is superior to an interactive one. Moreover, if a subject transferred to an interdependent task organization, it was preferable that training have been on an independently organized task rather than one calling for interaction among RCs (20,135 vs. 35,888 lb per hit).

The results seem paradoxical. One would ordinarily train interactively if actual performance required operator interaction. However Case 75 suggests that regardless of the type of task organization in which the operator must work, it is more efficient to train him in an independent mode.

The experimenters admit that these results need further checking, but if substantiated, they would provide a most important training principle. (See also Chapter 5.)

Task fidelity interacted with task organization. Performance under interaction conditions can be improved if prior training is on a high fidelity task emphasizing communication skills.

Cases 71 and 75 both indicate that an independent task organization is superior to an interaction condition. If these results are confirmed, the multiman system should be designed to ensure that minimal interaction is required of operators performing comparable acts. One can hypothesize that since each operator has a finite "channel capacity" for processing information, interaction requirements consume a portion of that capacity, thus leaving less capacity for direct task functions.

From a training standpoint, even if the operator task is organized for interaction, operators should be trained primarily as individuals. (Whether any team training at all should be provided is still an open question.) If a crew member must be replaced, the negative effect of his replacement will be less in an independently organized system.

Despite the effects of task organization revealed in these studies, task organization (at least in terms of independence and interaction) is a weaker variable than the others we have considered. This may be because organization affects the context of task performance, whereas the other variables affect task performance more directly.

Intratask Organization

Understanding the task may be approached through analysis of task components. Information about relationships among parts of the total task and within the total task would obviously be of great value in designing tasks, training personnel, and predicting task performance.

Some of the most significant work in this area has been performed with tracking tasks by Fleishman and his colleagues; the results may or may not apply to discrete and cognitive tasks, but in any event they are highly suggestive.

Chambers (1958a, 1958b) found that proficiency on unpracticed component tasks was increased by prior practice on other component tasks. In addition, there was improved proficiency on the total task when subjects had prior practice on three components simultaneously, rather than on two components simultaneously or on a single component. It did not matter so much which components or combinations of components were practiced, but it did make a difference whether practice was given on only one component or on two or three. (See also Case 62.)

Case 76. Here 204 subjects practiced the components of a complex, multidimensional compensatory pursuit task, both singly and in combination. The device included three dials representing airspeed, bank, and heading indicators. Pointers on each meter varied irregularly, continuously, and independently. The subject's task was to keep all the pointers in the center of his scale by compensatory movements of simulated throttle, stick, and rudder controls. The performance measure was the amount of time all the pointers were lined up simultaneously in a zero position.

A subject first practiced on each of the three components separately, the task being to keep the pointer on one dial centered using only one control. Next he practiced keeping two dials centered by using two controls. Three such double component combinations were practiced, after which the total task was performed. This required the subject to keep all three meters centered using three controls simultaneously.

Results suggest that it is possible within limits to predict total task performance from prior performance on individual components. Predictive correlations for bank, heading, and airspeed were .46, .54, and .61, respectively. Total task performance could be predicted more effectively by combining the three subtasks; the multiple correlation with total task performance was .68 (46% of variance accounted for by prior performance on task components).

The contributions of each previously practiced task to overall performance, however, are not equal. Statistical combination of components may increase the accuracy with which total task performance is predicted, but the amount of increase apparently depends on the individual components combined. In Case 76 it was possible to predict as well from two components as from three.

The principles to be derived from this study are as follows:

1. Prior practice on combinations of subtasks produces better prediction of total task performance than does prior practice on single subtasks.
2. A single score representing prior performance on "double-level" components is at least as predictive of total task performance as a statistical combination of single subtasks.
3. The highest predictions of total task performance result from multiple Rs between prior practice scores on "double-level" components.
4. No one double-level combination is superior to others.
5. Adding a third double-level score to the score achieved with two double-level tasks does not improve prediction.

The primary findings are that the best predictors of total task and multiple control subtasks are other multiple control subtasks; and the particular components involved are less important than the fact that simultaneous practice had occurred. These results argue against any simple stimulus-response view of task components.

Case 77. As one might suspect from the previous study, the order in which subtasks and the total task are practiced also makes a difference in how total task variance can be accounted for. The task in this study required the subject to match light patterns by appropriate adjustments of aircraft-type stick and rudder controls. The apparatus panel displayed two pairs of parallel rows of horizontal lights, separated by a parallel vertical row of lights. In one row of each pair a red light served as the stimulus; in the other row a green light was the response. Using the controls, the subject matched the green response lights with the red stimulus light. A joystick moved left or right to control the top row of green lights, forward and backward to control the middle row of green lights; a rudder bar was moved left and right by the subject to control the bottom row of lights. As in Case 76, the subject was trained on only one, two, or three light rows at a time. The three-row condition was defined as the total task.

The order of part-task practice apparently affected total task proficiency. Part-task scores could be combined in a statistically independent manner to yield prediction scores on a more complex task only when a specific double-level task was combined with a specific single-level task. Observed and predicted total task scores agreed better when double-level tasks were practiced before, rather than after, the total task. Finally, prior practice on double-level tasks led to better total task proficiency than did prior practice on single-level tasks.

The data suggest that the hypothesis that part tasks combine in a statistically independent manner (Bilodeau, 1955) has only limited generality and may apply only for certain types of tasks or only for subtasks rather than double-level components.

Case 78. This study, one of a series preceded by Cases 76 and 77, was performed to investigate subtask–whole task relationships. The apparatus and tasks were the same as described in Case 77. Results indicate that the group that practiced on part tasks performed better on the total task than the group that did not.

All groups improved their performance regardless of the particular part task practiced first. However among single-level tasks, one part task (rudder) was significantly superior to others, suggesting that although practice on any subtask helps, some subtasks are more effective than others in improving criterion performance. (Thus perhaps dependency relationships among tasks must also take into consideration the relative importance or criticality of each task to overall system output. Since task criticality is determined by individual system requirements, it can probably be specified only on a case-by-case basis.) Yet Case 78 revealed no reciprocal transfer; practice on the total task led to poorer performance on the rudder compared to either stick task, and on the stick task compared to either of the two rudder-stick tasks. This suggests that practice on difficult task components may lead to better total task performance.

Conclusions. Various models to explain changes in motor task performance with continuing practice have been proposed, notably the changing task model (Fleishman, 1960, 1966) and the changing subject model (Adams, 1957, 1964).

Illustrative Studies

Performance is a function of operator abilities required by the task and the way in which these abilities contribute to performance. Changes in task performance as a function of practice may therefore be explained in terms of systematic changes in the structure of the task (thus requiring different abilities) or in the way the task is performed by the operator. Fleishman and his colleagues suggest that the structure of the task changes over practice, thus demanding different abilities of the operator.

The changing subject model assumes that the task remains constant (hence the importance of an ability to the task remains constant across trials) but the amount of that ability required may change as a function of practice. The changing subject model conceives of abilities not as stable determinants of behavior but as being learned, forgotten, and so on, like any other performance factor.

The argument between the two models is esoteric and may depend on how mathematical relationships among ability measures are interpreted. The paper by Alvares and Hulin (1972) points out the difficulty of making an empirical comparison between the two models.

AUTOMATION

Conclusions. There is no easy answer to the question, under what conditions should one automate, and how much? Very few empirical studies have been performed to help solve this problem.

Illustrative Studies

The decisions to automate, how much, and at what system level, are particularly interesting to the system developer because they determine how further design will proceed. These decisions also hold significant implications for further behavioral alternatives. As soon as it is decided to automate, the number of alternative ways of performing a function are sharply reduced. If a process is to be performed mechanically or electronically, related operator actions often reduce to discrete, simple, switch behaviors and maintenance actions are increased.

Automation is the substitution of a physical process for one that otherwise would be performed manually. Although most engineers probably have an emotional bias toward automation, the decision to automate is made ostensibly on the basis of engineering feasibility, cost, reliability, and performance criteria. Engineering feasibility and cost are factors outside our consideration. To determine reliability, a comparison is made between the *consistency* of equipment performance over time versus consistency of human performance. Here there is no contest; equipment fails less frequently than humans, makes fewer errors, and shows much lower effects of fatigue.

The performance criterion brings us to the critical point. Can equipment ("off the shelf" or to be developed) perform certain functions as effectively as an operator? If not, the other criteria are likely to be irrelevant. If yes, the decision will probably be to automate, unless cost factors are heavily in favor of the human alternative. The question demands a comparison of operator and equipment capabilities. One cannot, however, make the kind of blanket comparison reflected in the so-called Fitts lists (Fitts, 1951). First, all equipment/operator performance comparisons must be specific at least to a type of equipment or function, if not to a specific alternative system configuration. Second, it is unlikely that all functions and tasks are equally capable of being automated. Third, there are degrees of automation to be considered: *(a)* complete replacement of a manual task by a physical process, *(b)* partial replacement of that task by the physical process, *(c)* physical aiding of a task performed manually.

The overall question, therefore, must be broken down as follows:

1. Which functions and tasks should be automated?
2. How much advantage is gained by automating? Conceivably, the amount of advantage may be so insignificant that the changeover is hardly worth the effort.
3. What is the effect of automating certain functions on those that must remain manual, and on system performance as a whole?

For all the importance of automation, there have been disturbingly few empirical studies addressing these critical questions. There have been a few theoretical works (e.g., Jordan, 1961) and a large number of quasisociological studies, but the latter have emphasized the *effects* of automation on such factors as worker attitudes (Taylor, 1970) and changes in manpower requirements (Crossman et al., 1969) rather than data on which the decision to automate should be based.

We have already encountered in Chapter 2 two of the few studies available.

Case 10 was an air defense problem in which completely manual operations were contrasted with computer assistance. It is apparent from Figure 2-3 that computer assistance was highly effective in increasing number of kills and reducing damage. However the machine, at least in the first session, was far more profligate in expenditure of weapons per number of threats. Depending on one's criterion, therefore, computer assistance may be more or less effective.

In Case 11, also involving air defense, a comparison was made between three levels of "ideal" logical operations: (1) idealized human solution, with two action rates (3 per minute vs. 5 per minute), (2) idealized machine solution, with zero delay in assigning action, and (3) random machine solution, with zero delay and action selections drawn at random. These three solutions were derived for two levels of load (60 and 96 tracks) and compared with empirical (human) solutions.

Idealized machine solutions with zero time delay were superior in terms of average percentage of kills but were not much better than those of the ideal man. Empirical human performance, while substantially below that of the ideal machine or man, was superior to that of the random machine (Table 3-5). But kills by themselves do not adequately describe performance effectiveness; enemy penetration occurred in every idealized machine solution, and damage was assessed in the majority of cases. The successes achieved by the analytic "idealized" solutions came at the expense of nearly total exhaustion of counterweapon resources (similar to Case 10). The experimental commanders' performance, as indicated in Table 3-5, was quite creditable at higher load levels, especially when considered in terms of the real-time characteristics of the problems.

Additional data are available from another early study (Sinaiko, 1958) of a computerized air defense system. Unfortunately only two subjects were employed. For what data from two subjects are worth, the following can be reported. No significant differences in system performance resulted from various degrees of automation (keep in mind, however, that the computer was involved in all modes, however slight that involvement might have been). All modes of automation were more efficient in preventing enemy penetration under light than under heavy loads. As we confirm in Chapter 4, under heavy load conditions (the more difficult task), subjects tended to assume greater control of the system in the mode in which they could overrule the computer. When compared with a completely manual air defense system, the computerized one was significantly more effective, particularly under heavy load.

Computer assistance was also found to be helpful in classifying synthe-

Table 3-5 Comparison of Automated and Manual Problem-Solving Solutions[a]

		Track Load							
		60 Trials				96 Trials			
	Weapon Sites	Kills (%)	Damage (weapons lost)	Cost/Kill Ratio	Weapon Cost	Kills (%)	Damage (weapons lost)	Cost/Kill Ratio	Weapon Cost
Experimental group I	5	65.0	7.0	3.5	138	65.6	13.0	3.0	188
	8	56.6	8.0	3.4	117	59.3	16.0	3.2	182
Experimental group II	8	35.0	12.0	5.4	114	45.8	18.0	3.7	163
Ideal man	5	92.0	2.5	2.8	156	84.6	13.0	3.0	195
Ideal machine	5	92.6	4.0	3.0	163	94.6	3.3	2.7	230
Random machine	5	56.0	4.5	5.0	164	54.0	16.0	4.5	230

[a] From Fox et al., 1961, Case 11.

tic passive sonar signals (Corcoran et al., 1970). During early stages of training, combined operator/computer performance was poorer than that of the computer alone; however at later stages the computer/operator combination exceeded individual computer or operator performance. Use of the computer by the operator increased with task difficulty.

Obviously there is no simple answer to the question, should we automate? This is especially evident when the functions being considered for automation have multiple dimensions.

Other experiments on automation have compared "ideal" (Bayesian logic) solutions of probabilistic inference decision-making problems with manual solutions of these problems. This research is discussed in Chapter 4.

EQUIPMENT-TASK RELATIONSHIPS

Conclusions. Of the many available potential predictors of control-display equipment operation, the most useful appears to be the Design Evaluative Index (DEI, Siegel et al., 1964). A signal figure-of-merit approach to task quantification seems to be unproductive.

Illustrative Studies

Any task description is meaningless unless it specifies the equipment that is operated or acted on; naturally, then, the HF specialist is concerned about equipment-task relationships.

The fundamental principle of human engineering is that different design configurations produce varying operator performances. An extensive literature (e.g. Van Cott and Kinkade, 1972) confirms that principle. From a system design standpoint, however, the general principle is of little use because the designer needs to know *which* equipment variation produces substantial changes in performance. This is a much more difficult question, and so far it has been answered only partially.

The following equipment characteristics may affect operator performance:

1. Total number of controls.
2. Total number of displays.
3. Number of controls of the same type (e.g., knobs).
4. Number of displays of the same type (e.g., meters).
5. Arrangement of controls.
6. Arrangement of displays.
7. Control-display relationships.

8. Number of controls and/or displays that must be activated or read at any one time out of the total number.

9. Number of required operator control *responses;* this may not equal number of controls, because often a particular control must be activated (or a display read) more than once in an operating cycle.

10. Amount of scanning required by tasks (this value during any operating cycle can only be approximated).

11. Number of contingent responses required in operating an equipment.

12. Relative frequency of operation of individual controls and displays.

The list could be extended; it does not, for example, include the physical characteristics of the individual control or display (e.g., switch diameter, length, angle of throw), which might influence the operation of that device. [See Munger et al. (1962) for tables predicting performance as a function of molecular physical control/display characteristics.] However we shall make no attempt to deal with these, because their effect on system output is usually minimal.

The designer must determine which of these indices is most important. Usually in a single equipment he can reasonably modify only one or at most two characteristics; there are too many, and they are too interrelated, to modify them all. Of course each one is itself important when tested individually in an experiment. However this kind of answer, which the HF specialist often gives the designer, is not very helpful. Unfortunately, these characteristics usually do not occur in the design configuration individually but rather are a part of a complex. Which then contributes most to the variance of operator performance?

We know that the designer is interested only in the characteristic that has the greatest influence on performance. Moreover, he cares only about the characteristic *value* that (if included in design) produces a major decrement or increment in performance. For example, assume that the characteristic is *number of displays* on a console and that one measures the operator performance resulting from that characteristic. If there is no discernible change in performance as one progressively adds up to n displays but the $(n + 1)$th display results in a significant drop in performance, the latter is the value the designer is interested in; slight variations of performance for the preceding n displays are of little concern.

To further complicate matters, the significant value changes as a function of the other equipment-task characteristics with which it is combined. For example, display arrangement may be significant for operator performance in a multifunction task situation but insignificant if only a single function is involved.

Despite the importance of the subject, few efforts have been made to develop quantitative predictive equipment-task indices of performance. One of the few major studies of equipment-task relationships is discussed below.

Case 79. The purpose of this three-phase study was to determine the feasibility of quantitatively describing tasks and predicting the effectiveness of Navy training devices. Although the extensive project did not end with a completely acceptable set of indices, it did demonstrate that the goal is a reasonable one and can be accomplished, albeit with much more work.

The general method was to select a wide variety of candidate characteristics (indices) and to refine the list by determining how well they described and differentiated among tasks and trainers. (One might term it a "brute force" technique, in contrast to one based on an underlying model of equipment-operator performance.) The indices selected initially included 13 task characteristics [rating scales developed for a taxonomy project, Fleishman et al. (1970)], the DEI (Seigel et al., 1964), a number of panel layout indices developed by Fowler et al. (1968), and some miscellaneous indices.

To these generic indices that cut across tasks and trainers, 25 descriptors specific to surveillance trainers and tasks were added. In all, 29 generic indices, 15 specific ones, and 10 training technique descriptors were assembled.

These were then applied to task analytic data gathered on three sonar devices. Although the results from applying the 15 specific and 10 training indices were inconclusive, and a number of the generic indices were ineffective, the results of the initial study (Wheaton et al., 1971) were encouraging.

In the second study (Wheaton and Mirabella, 1972), the list of descriptors was refined by deleting ineffective ones and adding new ones. The new list was again applied to task analytic data gathered on devices in the field.

In addition, a synthetic sonar trainer was designed; it could be modified to provide control panels varying along dimensions of the 17 design indices selected for study (e.g., complexity, embedding, and feedback). The goal was to use the indices to predict operator performance on different versions of the device. Subjects then learned to operate each version. If the indices were effective, they should have predicted subject performance on each equipment version.

The results of the sonar validation effort were generally encouraging. Significant multiple correlations were obtained between the quantitative task indices and operator speed and accuracy during learning.

In the third study a further refined set of indices was applied to the experimental learning situation developed in the second study. Again subjects had to learn to operate versions of the experimental device, and predictive indices were correlated with their performance. In addition, regression equations were developed to measure how well various combinations of indices could predict performance during skill acquisition and transfer to another version of the experimental equipment. Definitions of the indices tested are presented in Table 3-6. The indices found to be most effective at the conclusion of the project appear in Table 3-7.

The most important index was the DEI, the most inclusive index (because it describes information flow) and one that has been extensively utilized in analyzing a variety of systems such as surveillance, fire control, and communication. The importance of task embeddedness as reflected by the $E\%$ indices was corroborated. Embeddedness is inversely related to task performance; as embeddedness decreases, performance during acquisition increases.

As Table 3-7 indicates, the pattern of predictors varies across criterion measures and across time blocks within criterion measures. Thus a simple single figure of merit approach to task quantification does not seem to be productive.

Although prior cases appear to have demonstrated the point, it is encouraging to verify that differences in task performance arose as a function of equipment (hence task) complexity. This is illustrated in Figure 3-6, where speed in acquisition is directly related to control panel complexity. (Significant but less striking effects were shown for error.) Performance differences as a consequence of equipment complexity, however, tend to diminish as learning proceeds. It is unfortunate that we do not have comparable data for well-learned tasks.

There is no clear-cut effect of embedding on error scores. Increasing levels of embeddedness resulted in increasing performance time, even though number of task responses was constant across levels of embedding. However embedding did not appear to have any significant effect on errors made during transfer.

Although the results of this study apply only to the relatively simple situation of control panel operation, they support the importance of task complexity as a critical task characteristic. The study also suggests that the relative importance of individual task characteristics will change as a function of training, which points up the necessity of studying task performance *after* the task has been well learned, as well as during the learning process. (See also task performance models, p. 141.)

Table 3-6 Task Characteristics Studied in Case 79[a]

1. MAIN—the number of responses comprising the *main* or dominant procedural sequence in an operations flow chart.[b]
2. CNTG—the number of responses comprising the auxiliary or *contingency* procedural sequences.[b]
3. TA—the total number of responses *(actions)* comprising the procedural sequence in an operations flow chart; it represents the sum of MAIN and CNTG.
4. CONT—the total number of different *controls* manipulated during performance of a subtask.
5. DISP—the total number of different *displays* referenced during performance of a subtask.
6. E—the total number of different equipment *elements* interacted with; this index is given by the sum of CONT and DISP.
7. LV—the *link value* reflecting the relative strength of the sequence of use among the various controls and displays. As used here, it is the sum of the products of the number of times a link is used and the percentage of use of the link.
8. *AA*%—an index reflecting the percentage of *alternative actions* present in an operation. A score of 0% means that the highest number of alternative links are used, each with an equal frequency of use; 100% score means that there is only one link out of and into each control, with the same frequency used for all links.
9. *F*%—another index (Fowler et al., 1968) describing the extent to which all controls and displays are used an equal number of times (0%) or a theoretically defined optimum number of times (100%).
10. DEI—a measure of the effectiveness with which information flows from displays via the operator to corresponding controls. The index yields a dimensionless number representing a figure of merit for the total configuration of displays and controls.

11–13. *D*%, *C*%, and *E*%—respectively, the number of display, control, or combined equipment elements the operator actually employs relative to the total number of such elements available for use.

14–17. CRPS, FBR, INFO, and INST—refer to the frequency with which the operator makes various types of responses during performance of the task. Included are responses involving manipulation of controls (CRPS), securing of feedback (FBR), acquisition of information (INFO), as well as those primarily initiated by the instructor (INST[b]).

---

[a] From Mirabella and Wheaton, 1974, Case 79.

[b] These indices were eliminated prior to analysis of phase III data. Two of them, MAIN and CNTG, correlated almost perfectly with TA and were eliminated for this reason. The third, INST, was invariant and was therefore eliminated.

Table 3-7 Summary of Multiple Regression Analyses for Equipment-task Indices[a]

Criterion	Residual Performance Scores: Multiple Correlation Coefficient	Residual Performance Scores: F Ratio	Residual Performance Scores: Indices in Order of Selection by Regression Program	Unadjusted Scores: Multiple Correlation Coefficient	Unadjusted Scores: F Ratio	Unadjusted Scores: Indices in Order of Selection by Regression Program
Time Measure						
Trials						
1–2	.693	4.92*	*E*%, DEI, CONT	.874	17.30**	DEI, FBR, *E*%
7–8	.673	4.41*	*C*%, *F*%, INFO	.908	25.15**	DEI, *E*%, *C*%
13–15	.619	3.31†	*C*%, DEI, DISP	.920	29.60**	TA, DEI, *C*%
Error Measure						
Trials						
1–2	.474	1.55	*E*%, *F*%, *D*%	.669	4.32†	DEI, LV, *E*%
7–8	.670	4.33*	DEI, FBR, *C*%	.809	10.13**	DEI, CRPS, FBR
13–15	.527	2.05	DEI, DISP, *AA*%	.766	7.56**	CRPS, *AA*%, DEI

[a] From Mirabella and Wheaton, 1974, Case 79.
†$p = .05$.
*$p = .025$.
**$p = .01$.

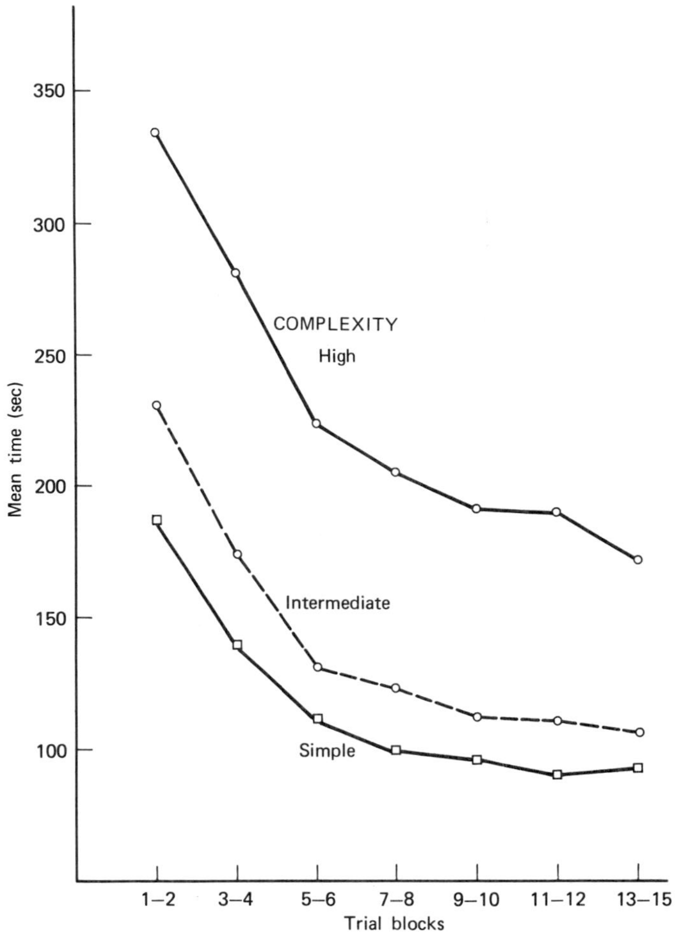

Figure 3-6 Performance as a function of task complexity. With permission, from Mirabella and Wheaton, 1974, Case 79.

SUMMARY OF CONCLUSIONS

1. The amount of experimental work performed to study tasks *as tasks* is comparatively small. In addition, very little information is available about how tasks are performed in the real world context.

2. A wide variety of tasks have been used in behavioral research. Unfortunately particular variables are rarely studied using more than one task type, which makes it difficult to generalize study conclusions beyond the specific task with which the research was performed.

3. As much as 50% of operator performance variance may be determined by task type; however this estimate is based on only one or two studies whose tasks did not involve man-machine relationships.

4. Some task characteristics (i.e., complexity and organization) have multiple dimensions. Only one characteristic (patterning) is described by a single dimension. Task difficulty is a superordinate characteristic that is defined by interrelationships among all other task characteristics. The effects of other task characteristics are often experienced by the operator as a feeling of ease or difficulty.

5. Tasks may vary along the following continua: patterned-random, complex-simple, organized-unorganized, and easy-difficult. Positive task characteristics (patterned, simple, organized, and easy tasks) lead to improved operator performance; negative characteristics (random, complex, unorganized, and difficult tasks) produce poorer performance. The influence of these characteristics diminishes as learning proceeds. There is a suggestion that when a task is easy, maximal differentiation in a dependent variable will appear relatively early in the task, and if the task is difficult, differences will appear later in the task.

6. Operators make use of task patterning to develop strategies to reduce task difficulty. Although the effect of task complexity is to degrade operator performance, the effect is not invariant: when the task is very simple and repetitive, complicating it may improve the operator's performance by minimizing his boredom.

7. It is possible to predict total task performance of complex psychomotor tasks from prior performance on individual components. Prior practice on combinations of components produces a better prediction of total task performance than prior practice on single components. Prediction from two components is as efficient as prediction from three or more. The order in which component tasks are performed influences total task performance.

8. When two tasks are performed simultaneously by the same operator, performance decrement will occur with one or both of the two tasks (see also results in Chapter 2 on multiple task performance).

9. Task organization is a less significant variable than complexity or difficulty. An independent task organization (one in which operators perform individually) is superior to one that requires interaction among operators. Individual rather than team training is to be preferred, even for tasks organized for interaction among operators.

10. The effectiveness of task automation in improving system performance depends on which operator functions are automated.

11. Various quantitative equipment-task relationships show significant correlations with speed and accuracy, suggesting the potential for de-

veloping quantitative predictive indices. However, a single index appears to be incapable of predicting task performance; patterns of indices are required.

DEVELOPMENTAL IMPLICATIONS

A number of recommendations can be made.

1. Tasks should be made as simple as possible by reducing the number of task inputs and required responses. To the extent possible, reduce the number of alternative contingent responses to a minimum. This may be difficult, however, when the nature of the system demands certain operations.
2. In general, a multiman system should be designed so that operators performing similar functions need not interact with each other. The more contact required between operators, the more attention must be divided between task functions and interpersonal interactions.
3. If the system does require interaction between operators, prior training should emphasize individual as opposed to crew operations, yet the training task should be similar in terms of its interaction requirements to the operational task.
4. In arranging the order in which tasks must be performed, care should be taken to ensure that the tasks are patterned as much as possible.

GENERAL REFERENCES

Adams, J. A. The relationship between certain measures of ability and the acquisition of a psychomotor response. *J. Gen. Psychol.,* 1957, **56,** 121–134.

Adams, J. A. Motor skills. *Ann. Rev. Psychol.,* 1964, **15,** 181–202.

Alexander, L. T. et al. The effectiveness of knowledge of results in a military system training program. *J. Appl. Psychol.,* 1962, **46,** 202–211.

Alexander, L. T. and Cooperband, A. S. The effect of rule flexibility on system adaptation. *Hum. Factors,* 1964, **6,** 209–220.

Alluisi, E. A. Methodology in the use of synthetic tasks to assess complex performance. *Hum. Factors,* 1967, **9,** 375–384.

Alvares, K. M. and Hulin, C. L. Two explanations of temporal changes in ability-skill relationships: a literature review and theoretical analysis. *Hum. Factors,* 1972, **14,** 295–308. (See also Dunham, R. B. Ability-skill relationships. *Organ. Behav. Hum. Performance,* 1974 **12,** 377–382.)

Berliner, C. et al. *Behaviors, measures and instruments for performance evaluation in simulated environments.* Paper presented at Symposium and Workshop on the Quantification of Human Performance, Albuquerque, N. M., August 17–19, 1964.

Bilodeau, E. A. Variations in knowledge of component performance and its effects upon part-part and part-whole relations. *J. Exp. Psychol.,* 1955, **50,** 215–224.

Blanchard, R. E. *Requirements, concept and specification for a Navy human performance data store.* Report 102-2, Contract N00140-72C-9351, Behaviormetrics, Santa Monica, Calif., April 1973.

Briggs, G. E. and Johnston, W. A. *Team training research.* Report NAVTRADEVCEN 1327-2, Naval Training Devices Center, Port Washington, N.Y., November 1965. (AD 477–963).

Briggs, G. E. and Johnston, W. A. *Laboratory research on team training.* Report NAV-TRADEVCEN 1327-3, Naval Training Devices Center, Port Washington, N.Y., May 1966. (AD 485–636.)

Briggs, G. E. and Johnston, W. A. *Team training.* Report NAVTRADEVCEN 1327-4, Naval Training Devices Center, Orlando, Fla., June 1967. (AD 660–019.)

Briggs, G. E. and Naylor, J. C. *Experiments on team training in a CIC-type task environment.* Report NAVTRADEVCEN 1327-1, Naval Training Devices Center, Port Washington, N.Y., June 1964. (AD 608–309.)

Broadbent, D. E. Some effects of noise on visual performance. *Quart. J. Exp. Psychol.,* 1954, **6,** 1–5.

Broadbent, D. E. S-R compatability and the processing of information. *Acta Psychol.,* 1964, **23,** 325–327.

Carter, L. et al. The behavior of leaders and other group members. *J. Abnorm. Soc. Psychol.,* 1950, **46,** 589–595.

Carter, L. et al. The relation of categorizations and ratings in the observation of group behavior. *Hum. Relat.,* 1951, **4,** 239–254.

Chambers, A. N. *Development of a taxonomy of human performance: A heuristic model for the development of classification systems.* Technical report 4A, American Institute for Research, Silver Spring, Md., October 1969.

Chambers, R. M. *Effect of knowledge of results on the learning and structure of a complex psychomotor skill.* Paper presented at the Eastern Psychological Association, Philadelphia, 1958a.

Chambers, R. M. *Transfer of training among components within a complex psychomotor task.* Paper presented at the Midwestern Psychological Association, Detroit, 1958b.

Chapman, R. L. et al. The System Research Laboratory's air defense experiments. *Manage. Sci.,* 1959, **5,** 250–269.

Christensen, J. M. and Mills, R. G. What does the operator do in complex systems? *Hum. Factors,* 1967, **9,** 329–340.

Colson, K. R. et al. *Development of an informational taxonomy of visual displays for Army tactical data systems.* Research Memo 74-4, U.S. Army Research Institute for the Behavioral and Social Sciences, Arlington, Va., 1974.

Corcoran, D. W. J. et al. *The cooperation of man and computer in classification.* Royal Naval Personnel Committee, October 1970. (AD 787-024.)

Cotterman, T. E. *Task classification: An approach to partially ordering information on human learning.* Report WADC-TN-58-374, Wright Air Development Center, Wright-Patterson AFB, Ohio, 1959. (AD 210-716.)

Crossman, E. R. F. W. et al. *The impact of technological change on manpower and skill demand: Case study data and policy implications.* Department of Industrial Engineering and Operations Research, University of California, Berkeley, February 1969. (PB 188 694.)

Davies, D. R. and Tune, G. S. *Human Vigilance Performance.* New York: American Elsevier, 1969.

Deutsch, M. Task structure and group process. *Am. Psychol.,* 1951, **6,** 324–325. (Abstract.)

Dodd, W. E. et al. Task difficulty as a moderator of long-range prediction. *J. Appl. Psychol.,* 1970, **54,** 265–270.

Fineberg, M. L. Navigation and flight proficiency under nap of the earth conditions as a function of aviator training and experience. *Proceedings,* Human Factors Society Meeting, October 1974, pp. 249–254.

Finley, D. L. et al. *Human performance prediction in man-machine systems:* Vol. I. NASA Report CR-1614, Ames Research Center, Moffett Field, Calif., August 1970.

Fitts, P. M. et al. (Eds.). *Human engineering for an effective air navigation and traffic control system.* Washington, D.C.: National Research Council, 1951. (AD 133–954.)

Fleishman, E. A. and Hempel, W. E. Factorial analysis of complex psychomotor perfor- *Exp. Psychol.,* 1960, **60,** 162–171.

Fleishman, E. A. Human abilities and the acquisition of skill. In E. A. Bilodeau (Ed.), *Acquisition of Skill.* New York: Academic Press, 1966, pp. 147–167.

Fleishman, E. A. Performance assessment based on an empirically derived task taxonomy. *Hum. Factors,* 1967, **9,** 349–366.

Fleishman, E. A. and Hempel, W. E. Changes in factor structure of a complex psychomotor test as a function of practice. *Psychometrika,* 1954, **18,** 239–252.

Fleishman, E. A. and Hempel, W. E. Factorial analysis of complex psychomotor performance and related skills. *J. Appl. Psychol.,* 1956, **40,** 96–104.

Fleishman, E. A. et al. *Development of a taxonomy of human performance: A review of the second year's progress.* Report AIR-726-1/70-TPR, American Institute for Research, Silver Spring, Md., 1970.

Fowler, R. L. et al. *An investigation of the relationship between operator performance and operator panel layout for continuous tasks.* Report AMRL-TR-68-170, Aerospace Medical Research Laboratories, Wright-Patterson AFB, Ohio, December 1968. (AD 692–126.)

Gagne, R. M. *The Conditions of Learning.* New York: Holt, Rinehart, & Winston, 1965.

Glanzer, M. and Glaser, R. Techniques for the study of group structure and behavior: II. Empirical studies of effects of structure in small groups. *Psychol. Bull.,* 1961, **58,** 1–27.

Hakel, M. D. Task difficulty and personality test validity. *Psychol. Rep.* 1968, **22,** 502.

Horrocks, J. E. et al. *Team Training II: Individual learning and team performance.* Report NAVTRADEVCEN 198-2, Naval Training Devices Center, Port Washington, N.Y., August 1960. (AD 247–147.)

Jordan, N. *The allocation of functions between man and machines in automated systems.* Report P-2310, Rand Corp., Santa Monica, Calif., May 16, 1961. (AD 782-780.)

Kidd, J. S. *A summary of research methods, operator characteristics and system design specifications based on the study of a simulated radar air traffic control system.* Report WADC-TR-59-236, Wright Air Development Center, Wright-Patterson AFB, Ohio, July 1959. (AD 235–126.)

Klaus, D. J. and Glaser, R. Reinforcement determinants of team proficiency. *Organ. Behav. Hum. Performance,* 1970, **5**(1), 33–67.

Krumm, R. L. and Farina, A. J. *Effectiveness of integrated flight simulator training in promoting B-52 crew coordination.* Report AMRL-TR-62-1, Aerospace Medical Research Laboratories, Wright-Patterson AFB, Ohio, February 1962. (AD 273–891.)

Lanzetta, J. T. and Roby, T. B. Group performance as a function of work distribution patterns and task load. *Sociometry,* 1956, **19,** 95–104.

Lanzetta, J. T. and Roby, T. B. Group learning and communication as a function of task and structure "demands." *J. Abnorm. Soc. Psychol.,* 1957, **55,** 121–131.

Levine, J. M. et al. Evaluation of an abilities classification system for integrating and generalizing human performance research findings: An application to vigilance tasks. *J. Appl. Psychol.,* 1973, **58**(2), 149–157.

Mackworth, N. H. *Researches on the measurement of human performance.* Medical Research Council Special Report Series 268, H. M. Stationery Office, London, England, 1950. Reprinted as paper 7 in H. W. Sinaiko (Ed.), *Selected Papers on Human Factors in the Design and Use of Control Systems.* New York: Dover, 1961, pp. 174–331.

McGrath, J. E. and Altman, I. *Small Group Research.* New York: Holt, Rinehart & Winston, 1966.

Meister, D. and Mills, R. G. Development of a human performance reliability data system. *Annals of Reliability and Maintainability,* 1971, American Society of Mechanical Engineers, New York, pp. 425–439.

Melton, A. W. and Briggs, G. E. Engineering psychology, *Ann. Rev. Psychol.,* 1960, **11,** 71–98.

Miller, R. B. *A method for man-machine task analysis,* Report WADC-TR-53-137, Wright Air Development Center, Wright-Patterson AFB, Ohio, June 1953. (AD 15921.)

Miller, R. B. *Task taxonomy: Science or technology?* In W. T. Singleton et al. (Eds.), *The Human Operator in Complex Systems.* London: Taylor & Francis, 1967, pp. 67–76.

Miller, R. B. *Development of a taxonomy of human performance: Design of a systems task vocabulary.* Technical report 11, American Institute for Research, Silver Spring, Md., March 1971.

Miller, R. B. *A method for determining task strategies.* Report AFHRL-TR-74-26, American Institute for Research, Silver Spring, Md., May 1974. (AD 783-847.)

Mills, R. G. and Hatfield, S. A. Sequential task performance: Task module relationships, reliabilities and times. *Hum. Factors,* 1974, **16,** 117–128.

Monty, R. A. Keeping track of sequential events: implications for the design of displays. *Ergonomics,* 1973, **16,** 443–454.

Munger, S. J. et al. *An index of electronic equipment operability: Data store.* Report AIR-C43-1/62/RP(1), American Institute for Research, Pittsburgh, January 1962. (These tables are reprinted in Meister, D. Section 12, Human Factors in Reliability, in Ireson, W. G. (Ed.), Handbook of Reliability. New York: McGraw-Hill, 1966.)

Park, J. F. and Payne, M. C. Effects of noise level and difficulty of task in performing division. *J. Appl. Psychol.,* 1963, **47,** 367–368.

Parsons, H. M. *Man-Machine System Experiments.* Baltimore: Johns Hopkins Press, 1972.

Poulton, E. C. *Tracking Skill and Manual Control.* New York: Academic Press, 1974.

Roby, T. B. and Lanzetta, J. T. Work group structure, communication, and group performance. *Sociometry,* 1956, **19,** 105–113.

Sackman, H. *Computers, System Science, and Evolving Society.* New York: Wiley, 1967.

Shaw, M. E. *Scaling group tasks: A method for dimensional analysis.* Technical Report 1, Contract Nonr-580(11), University of Florida, 1963.

Shaw, M. E. Communications networks. In L. Berkowitz (Ed.), *Advances in Experimental Social Psychology.* New York: Academic Press, 1964, pp. 111–147.

Shiflett, S. C. Performance effectiveness and efficiency under different dyadic work strategies, *J. Appl. Psychol.,* 1973, **57,** 257–263.

Sidorsky, R. C. et al. *Behavioral and operational aspects of tactical decision-making in AAW and ASW,* Report NAVTRADEVCEN 1329-1, Naval Training Devices Center, Port Washington, N.Y., August 1964. (AD 607–888.)

Siegel, A. I. et al. The Display Evaluative Index technique for evaluating equipment systems from an information transfer point of view. *Hum. Factors,* 1964, **6,** 279–286.

Siegel, A. I. and Federman, P. J. Communication content training as an ingredient in effective team performance. *Ergonomics,* 1973, **16,** 403–416.

Sinaiko, H. W. *Artful: An experimental study of an automatic air defense system under varying conditions of human intervention.* Report B-104, Control Systems Laboratory, Univ. of Illinois, June 1958, (AD 162–664.)

Singleton, W. T. Theoretical approaches to human error. *Ergonomics,* 1973, **16** (6), 727–737.

Steiner, I. D. Models for inferring relationships between group size and potential group productivity. *Behav. Sci.,* 1966, **11,** 273–283.

Taylor, J. C. *Advanced technology and work group behavior in a setting of planned social change: A replication study.* Technical report, contract N00014-67-0181-0013, Institute for Social Research, University of Michigan, July 1, 1970. (AD 711-310.)

Teichner, W. H. and Fleishman, E. A. *Development of a taxonomy of human performance: Evaluation of a human performance classification system for generalizing human learning data,* Technical Report, 8, American Institute for Research, Silver Spring, Md., 1971.

Teichner, W. H. and Olson, D. E. A preliminary theory of the effects of task and environmental factors on human performance. *Hum. Factors,* 1971, **13,** 295–344.

Van Cott, H. P. and Kinkade, R. G. (Eds.). *Human Engineering Guide to Equipment Design,* (rev. ed.) Washington, D.C.: Government Printing Office, 1972.

Weitz, J. Criteria for criteria. *Am. Psychol.,* 1961, **16,** 228–231.

Weitz, J. The use of criterional measures. *Psychol. Rep.,* 1964, **14,** 803–817.

Weitz, J. Criteria and transfer of training. *Psychol. Rep.,* 1966, **19,** 195–210.

Wheaton, G. R. et al. *Trainee and instructor task quantification: Development of quantitative indices and a predictive methodology.* Report NAVTRADEVCEN 69-C-0278-1, Naval Training Devices Center, Orlando, Fla., 1971. (AD 722–423.)

Wheaton, G. R. and Mirabella, A. *Effects of task index variations on training effectiveness criteria.* Report NAVTRAEQUIPCEN 71-C-0059-1, Naval Training Equipment Center, Orlando, Fla., 1972.

CASE REFERENCES

Case 60. Morris, C. G. Task effects on group interaction. *J. Pers. Soc. Psychol.,* 1966, **4** (5), 545–554.

Case 61. Hackman, J. R. Effects of task characteristics on group products, *J. Exp. Soc. Psychol.,* 1968, **4,** 162–187.

Case 62. Noble, M. and Trumbo, D. The organization of skilled response. *Organ. Behav. Hum. Performance,* 1967, **2,** 1–25.

Case 63. Shaw, M. E. and Blum, J. M. Group performance as a function of task difficulty and the group's awareness of member satisfaction. *J. Appl. Psychol.,* 1965, **49** (3), 151–154.

Case 64. Morrissette, J. O. et al. Group performance as a function of size, structure and task difficulty. *J. Pers. Soc. Psychol.,* 1965, **2** (3), 451–455.

Case 65. Morrissette, J. O. Group performance as a function of task difficulty and size and structure of group: II. *J. Pers. Soc. Psychol.,* 1966, **3** (3), 357–359.

Case 66. Levine, E. L. and Weitz, J. Relationship between task difficulty and the criterion: Should we measure early or late? *J. Appl. Psychol.,* 1971, **55** (6), 512–520.

Case 67. Naylor, J. C. et al. *The effects of task organization, training time, and retention interval on the retention of skill.* Report AMRL-TDR-62-107, Aerospace Medical Research Laboratories, Wright-Patterson AFB, Ohio, September 1962. (AD 288–890.)

Case 68. Lanzetta, J. T. and Roby, T. B. Effects of work-group structure and certain task variables on group performance. *J. Abnorm. Soc. Psychol.,* 1956, **53,** 307–314.

Case 69. Zink, D. L. The development of role differentiation in dyads as a function of task complexity. *Am. Psychol.,* 1957, **12,** 371. (Abstract.)

Case 70. Boggs, D. H. and Simon, J. R. Differential effects of noise on tasks of varying complexity. *J. Appl. Psychol.,* 1968, **52** (2), 148–153.

Case 71. Naylor, J. C. and Briggs, G. E. Team training effectiveness under various conditions. *J. Appl. Psychol.,* 1965, **49** (4), 223–229.

Case 72. Davies, D. R. et al. The effects of complexity and uncertainty upon performance at a problem-solving task. *Psychonomic Sci.,* 1972, **27** (4), 193–194.

Case 73. Pepinsky, P. N. et al. *Motivational factors in individual and group productivity, III: The effects of task complexity and time pressure upon team productivity.* Ohio State University Research Foundation, Columbus, 1956. (AD 113-197.)

Case 74. Kennedy, R. S. A comparison of performance on visual and auditory monitoring tasks. *Hum. Factors,* 1971, **13** (2), 93–97.

Case 75. Briggs, G. E. and Naylor, J. C. *Experiments on team training in a CIC-type task environment.* Report NAVTRADEVCEN 1327-1, Naval Training Devices Center, Port Washington, N.Y., June 1964. (Exp. II.) (AD 608–309.)

Case 76. Fleishman, E. A. The prediction of total task performance from prior practice on task components. *Hum. Factors,* 1965, **7,** 18–27.

Case 77. Freedle, R. O. et al. Studies of component-total task relations: order of component-total task practice and total task predictability. *Hum. Factors,* 1968, **10,** 283–296.

Case 78. Zavala, A. and Geist, A. M. Component-total task relationships: Simple and sequential practice effects. *Hum. Factors,* 1968, **10,** 333–343.

Case 79. Mirabella, A. and Wheaton, G. R. *Effects of task index variations on transfer of training criteria.* Report NAVTRAEQUIPCEN 72-C-0126-1, Naval Training Equipment Center, Orlando, Fla., January 1974. (AD 773–947/7GA.)

IV
DECISION MAKING

SUMMARY. The most important function an operator can perform is to make decisions. This chapter describes the decision-making literature most relevant to the operator of indeterminate systems. The effects of input variables (e.g., input rate, source reliability) on decision-making performance are summarized, as well as such inherent operator biases as conservatism which degrade performance. Special attention is given to comparisons of human and automated (e.g., Bayesian) decision making in an effort to determine when and how to automate this function.

INTRODUCTION

Although there is a vast literature on decision making *(DM),* the studies in this literature deal mostly with the individual in static, highly abstract situations and stem from the effort to determine whether man is "rational" (i.e., performs in accordance with economic logic).

This chapter attempts to extract the studies describing *DM* in the system context and the ways in which the system affects and is affected by individual *DM.* Hence this is a highly selective, not a general, description of *DM* research; for *DM* in general, readers are referred to reviews by Edwards (1961), Becker and McClintock (1967), and Rapoport and Wallsten (1972), and to the book by Lee (1971).

The material in this chapter is particularly relevant to *DM* systems (e.g., tactical command/control) and subsystems (e.g., system management). As the system becomes more indeterminate, one can expect *DM* to be more heavily involved. Chapter 5 presents additional cases involving *DM* and teams.

Definitions

Like formal definitions of the system (Chapter 1), formal definitions of *DM* are not very satisfactory, if only because *DM* characteristics vary with the task situation. Making a procedural choice between two switches on a control panel and developing an anti-inflation program may both be considered *DM* if a rigid definition is applied; instinctively, however, one feels that the two cases are not quite the same. Nor can one equate *DM* with information processing; the latter is found in many functions (e.g., tracking) that may not involve solving problems or choosing between alternatives. Since almost every observable behavior can be construed to be part of or related to *DM*, a completely inclusive definition becomes synonymous with cognitive behavior in general. It is therefore more helpful to define *DM* in terms of its component behaviors.

In his synthesis of *DM* models, Schrenk (1969) finds three major stages: problem recognition, problem diagnosis, and action selection. These general categories can be broken down into the more detailed behaviors listed in Table 4-1, although no one (or even several) of these describes all *DM*. Each behavior is required at one time or another in different decisions.

Table 4-1 *DM* Behaviors

1. Determining that a problem requiring *DM* exists.
2. Determining the nature of the problem, which involves analysis of inputs to ascertain:
   a. their outstanding characteristics
   b. their relationship to the problem
   c. the cause of the problem
   d. the characteristics of the environment from which problem inputs arise (e.g., that certain environmental events or phenomena will occur).
3. Determining that a number of alternative decision responses to the problem exist or can be developed.
4. Determining the relationship between inputs and these decision alternatives.
5. Selecting one of the alternatives by
   a. estimating its *probability of success* (*Ps*) in solving the problem
   b. aggregating individual response *Ps* where multiple factors exist
   c. estimating the cost and payoff of the decision alternatives
   d. trading off alternatives on the basis of *c*.
6. Making the decision response.
7. Determining whether the selected response was appropriate on the basis of feedback.

These behaviors occur sequentially and logically; each is dependent on its predecessor and in turn initiates the next behavior in sequence. In actual

DM, however, the various stages blend into each other and the decision maker (DM) may not be aware of sequential stages in his behavior.*

Types of *DM* Tasks

The study of human decision behavior has been largely confined to tasks involving either a single choice or a series of independent choices from a number of static alternatives (e.g., Becker, 1962; Pruitt, 1962), each differing with respect to a known P_s, cost, and/or amount of payoff.

Real life decisions usually do not fit within so structured a framework, and conclusions from laboratory research may not apply fully to less structured situations. For example, Meeker et al. (1962) point out that problem elements (at least in a tactical environment) may not be encountered in a compact, temporal unit but may be spaced out haphazardly over time. There is no guarantee that an individual item of information is relevant to a given problem situation. Then, too, it may be difficult to determine where the problem begins or ends. Problems may overlap both temporally and contextually. The problem situation may change markedly over time. Problem solutions typically involve a number of subgoals (protection of targets, minimization of losses, etc., as well as kill ratio); problems may therefore unfold as a series of subproblems, each associated with its corresponding subgoal.

Furthermore, all the information needed to make an optimal decision may be lacking or may be provided only incrementally; P_s, cost, and payoff may be only partially known. All the responses available to DM may not be known to him. Problem characteristics may change over time, and the optimal moment at which to decide may be unclear. In a multiman system, DM may have responsibility for only part of the system decision.

The many variables inherent in *DM* lend themselves to the development of different types of experimental tasks. The most common differentiation is between static (e.g., gambling) tasks and sequential/dynamic tasks. Tasks also vary in terms of whether DM's decisions affect the problem outcome (closed-loop tasks) or do not (open-loop tasks). Tasks also vary in terms of the number of alternatives available and the degree to which these are clearly known. At one extreme "brainstorming" exercises have a theoretically unlimited number of alternative solutions (maximum uncertainty); at the other, tactical (as well as much industrial) *DM* occurs when fairly well-defined, significant events are possible and/or expected and a number of predetermined responses to the situation can be made (Kinkade et al., 1965). Troubleshooting is

*Throughout, *DM* (italics) refers to the process of decision making; DM (roman) designates the decision maker.

intermediate in terms of uncertainty. *DM* may or may not be ancillary to other activities (e.g., selecting a stock while watching a display quotation tape). A satisfactory *DM* taxonomy does not exist, although efforts have been made to develop a taxonomy describing tasks found in tactical *DM* (Sidorsky et al., 1964).

In addition to tasks containing all or most of the behaviors listed in Table 4-1, experimenters have developed tasks based on individual component behaviors: for example, asking the operator to analyze a sequence of numbers to determine their mean and distribution (Irwin and Smith, 1957); estimating the probability that certain events will occur (Fleming, 1970): predicting the P_s of alternative responses (Kinkade and Ranc, 1967). It has become popular to compare human decisions with computerized (i.e., Bayesian) solutions (Howell, 1967). In a procedure termed "optional stopping," DM may be asked to choose after each new input whether to make a decision or to seek additional information (e.g., Fried and Peterson, 1969).

DM Variables

Two types of variables influence *DM*. *Structural* variables (those inherent in the *DM* process) are organized around input characteristics, the desired solution, and response alternatives. *Situational* variables are organized around task characteristics, operator factors, and the problem environment.

In the following section statements made in a positive manner should be considered as hypotheses rather than as assumptions or established fact.

Structural Variables

Input Characterisics

Many input characteristics are analogous or identical to those described in Chapter 2. Their effects on operator performance may differ, however, because of the *DM* context.

Mode of Presentation. The inputs that describe the decision problem may be received "all at once" (statically) or sequentially (dynamically) over time. In the latter (more common) case the operator has at any one time only part of all the information he needs; if he responds before all information is available, his response has a higher probability of being erroneous. With sequential inputs, initial decisions may affect the characteristics of subsequent inputs. In a military situation, for example, the commander may respond to initial enemy probing attacks in sector X by ordering a counterattack in sector Y. This may cause the enemy to switch his attack from X to Z, thus changing the nature of the problem.

All other things being equal, unitary input situations are simpler than sequential ones because the DM does not find it necessary to analyze continuously changing information.

Input Rate of Change. The faster the sequential situation changes, the more difficult the *DM* task may be.

Amount of Information. All other things being equal, the greater the amount of information that must be assimilated to solve the problem, the more difficult the task may be. On the other hand, the more information is available when more information is required, the more likely it is that a correct decision will be made. The role of information in *DM* is, to say the least, obscure.

Proportion of Irrelevant Information. The greater the ratio of irrelevant to relevant pieces of information in the inputs, the more difficult the operator should find the *DM* task, especially when relevancy criteria are imprecise.

Input Value Because of criteria implicit in the *DM* problem or because of prior training, certain inputs may be considered by DM as worthy of more attention and greater weight than others.

Input Variability. The more *DM* inputs vary from each other over time, the more difficult DM should find the task, because of the need to integrate their varying dimensions.

Number, Type, Availability, and Reliability of Input Sources. DM's confidence in the correctness of his input data may be affected by the availability of additional information sources (e.g., photo reconnaissance) to check his data from another source (e.g., ground scout reports). DM should have more difficulty in deciding when several sources provide conflicting information.

Input History. The nature of the inputs prior to DM's decision should affect that decision.

Input Significance. The significance of the input (in terms of its meaning for problem solution) may be obvious or obscure. For example, the meaning of equipment malfunction symptoms in troubleshooting is often unclear; in air surveillance, on the other hand, the identification of enemy attackers as bombers or missiles may immediately direct the kind of weapons released against them. Input significance is one of the factors on which system determinacy depends.

Input Cost. The cost of securing additional information to help in making a decision (effort and delay and sometimes expenditure of men and material) may cause DM to reject the course of gathering additional information.

Nature of the DM Solution

System and Mission. The type of solution required depends on the system and its mission, and the nature of the desired solution may dictate DM's response. The desired resolution of a problem is either to return the system to a state it possessed prior to the problem (e.g., troubleshooting an equipment malfunction), to implement the system mission (e.g., overcome an enemy, develop a system), or sometimes to accomplish both simultaneously. Where system state is highly precise (e.g., circuits must produce specific outputs), there is less uncertainty and task difficulty.

Solution Criteria. In some *DM* problems, such as troubleshooting, only one criterion exists: to restore system function. In many other real world situations, the desired solution may have several dimensions that must be traded off against each other. For example, in tactical operations the destruction of an enemy unit can be traded off against maximum acceptable loss of one's own men and equipment. In allocating functions in system development, degree of automation can be traded off against cost.

If *DM* rules exist, these will in part determine the responses selected. A rule might impose priorities in solution criteria: for example loss of 5% of an attacking force is acceptable; if more than 5% is lost, break off the attack. An increase in unemployment to 8% as a result of a governmental decision to reduce spending may be acceptable; more than 8%, unacceptable.

Time Requirements. In sequential *DM* but less so in static *DM,* the time in which one must decide may affect the solution. If time is short, decision quality may be reduced. If DM takes too much time, the problem may change its parameters. In tactical situations the longer a decision is delayed, the more opportunity for the adversary to vary his behavior. The longer the time period during which one must predict another system's behavior, the more difficult (subjectively) the problem may be.

Response Alternatives

Response Availability. DM's responses in structured (determinate) tasks are largely predetermined by the characteristics of the task; in less structured tasks like brainstorming, DM may have to create his own responses.

Number of Alternatives. The larger the number of possible responses, the more difficulty DM may have in selecting among them, because additional criteria and weights come into play. On the other hand, system efficiency may be increased because the larger number of alternatives permits greater flexibility in responding.

Response Criteria. The criteria that form the basis on which a response is selected include the probability of success (P_s) of the response in securing an acceptable solution, the expected payoff if the response is correct, and the expected loss if the response is incorrect. All other things being equal, and assuming he knows and can integrate payoff/cost values, DM will select a response that maximizes P_s and gain and minimizes loss.

Situational Variables

Task Factors

1. *Task characteristics* such as the abstract/concrete nature of the task or its apparent difficulty probably influence *DM* performance. Context has also been found to be important (Corbin et al., 1975).
2. Whether *feedback* of decision consequences is available, and the degree and specificity of such feedback, may be critical in sequential *DM*.
3. *DM* under severe *load* is probably less efficient because of correlated stress.
4. It is possible that *DM* is influenced by *system organization* (e.g., number of hierarchical levels and distribution of information sources). For example, if DM's responsibility is limited, he may be more ready to decide.

Personal Factors

DM may also be affected by the following conditions.

1. *DM* should be more effective whem DM is *experienced* in dealing with a general class of problems.
2. A more highly *motivated* DM may be more effective (in terms of searching for a solution) than one who is not; if the motivation is anxiety induced, however, DM may concentrate inefficiently on only a subset of alternative responses.
3. A high degree of emotional *stress* may render the *DM* process less effective by permitting internal stimuli and responses to compete with those that are task specific.
4. It is logical to assume that increased *aptitude/skill* in performing a task will improve decisions made relative to that task.

The Environment

The environment outside DM's system includes systems other than DM's own. To the extent that one is attempting to control the environment (including possible adversary systems). the nature of that environment (e.g., its variability, the amount of control one has over it) is important. Unfortunately there has been little research on the environmental impact on *DM,* although see Kinkade et al. (1965).

INPUT-OUTPUT FACTORS

Because research in *DM* has centered on inputs, very little of this section describes output (response) factors. The concept of the "rational" man, which underlies so much *DM* research, assumes that if the environment (the input source) is interpreted properly, the correct decision will almost automatically fall out. This, of course, oversimplifies actual *DM.*

All the input factors described in Chapter 2 probably have some effect on *DM* performance. Those most frequently studied in *DM* have been:

1. Input frequency, variability, patterning, and pacing.
2. Information relevance, cost, and transmission rate.
3. Payoff ratios.

Response factors studied have included the number of and differences between available decision alternatives.

Since the interpretation of information gained from sequential inputs is critical for *DM,* researchers are much concerned with information taking (the rate at which DM asks for further inputs). Stimulus value and payoff ratio are also important research topics because any choice between alternatives implies value and payoff.

In the studies presented below, these variables are often defined in a highly abstract manner. This abstractness, which perhaps derives from the mathematical framework in which many *DM* problems are phrased, may present special difficulties to DM because few operators are mathematically oriented. It is possible that DM's responses would be different if these problems were more like those he encounters in the real world. For example, in one study (Stager and Muter, 1971) of subjects tested with verbal (abstract) and visual (concrete) problems, performance was superior with the latter. Still more definitive evidence stems from a study by Muchinsky and Dudycha (1975), who compared subjects' performance on a multiple-cue probability learning task (such as Case 91); one version of the task was phrased in concrete ("meaningful") terms, and a second version was in the usual abstract format. Subject performance with the meaningful task was significantly superior (strikingly so) to performance with the abstract task. In addition, performance was *qualitatively* different in the concrete version.

Researchers have also tested DM's response to *individual characteristics* of the *DM* input (e.g., input variability). Fractionating variables to examine their influence may cause DM to perform differently than he would if these variables were part of a total problem.

Conclusions

1. Sequential *DM* performance varies with objective task difficulty.
2. DM is sensitive to input variability in very gross, all or nothing fashion.
3. DM is more willing to gamble on difficult decisions and is cautious with easy ones.
4. DM's strategies are not optimal in terms of objective probability and payoff values.
5. Individual differences in *DM* are very great.

Illustrative Studies

Case 80. This case is characteristic of experimental situations devised to study tactical (military) *DM*. DM received a series of updated reports regarding enemy threats to four sites under his command. The threat was initially identical for all sites but increased linearly at different rates for each; differences in threat increased with time. As soon as DM believed the accumulated information permitted him to do so (optional stopping), DM had to estimate which site would be threatened most. He was permitted to continue drawing information until 100 reports were in, or he could terminate the problem at a point 10 reports past his last decision; he could also change his decision as additional reports arrived.

Independent variables were (1) the difference in rate at which alternative threats developed (four levels) and (2) the amount of variability of the information developed (four levels). These variables were assumed to determine task difficulty; as rate differences decrease and variability increases it should become more difficult to distinguish the condition of the locations and to predict their future situation.

Points were awarded for correct decisions and deducted for incorrect ones; scores were also computed on the basis of number of reports drawn. Both procedures are very characteristic of experimental *DM*, however uncharacteristic they are of the real world.

Decisions were permitted at three levels of specificity. At level 1 DM predicted that the most threatened position would be any of three (2 payoff points per report); at level 2 he predicted that it would be either of two (4 points); at level 3 he predicted which one it would actually be (6 points per report).

Information was presented in the form of statistical distributions plotted as deviations above or below a slope line representing the threat buildup. (Note the highly abstract nature of the input. This, too, is characteristic of experimental *DM* analogues.)

Both the independent variables were significant at the .01 level, although their interaction was not statistically significant. As Figure 4-1 reveals, scores increased with rate increases but decreased with increased variability. The number of reports drawn prior to decision (the cost of *DM*) decreased with rate increases and increased with increased variability—a mirror image of Figure 4-1.

As one would anticipate, sequential *DM* varied with objective task difficulty. A highly significant correlation was also obtained between decision quality and information taken and subjects' estimates of error risk and difficulty. This finding, which agrees with reports of others (e.g., Becker, 1958), suggests that DM behaves more cautiously when the decision is an easy one than when it is difficult (i.e., DM is more willing to "gamble" on difficult decisions). The factor involved appears to be payoff. It appears also that the subject is sensitive in a rather gross, all or nothing fashion to variability changes, but he utilizes information regarding differences in growth rate much more efficiently.

DMs apparently weight mean differences more heavily than variability in making probability judgments (Edwards, 1961). Poor utilization of variability information may reflect cultural history: skill in judging variability is probably seldom rewarded.

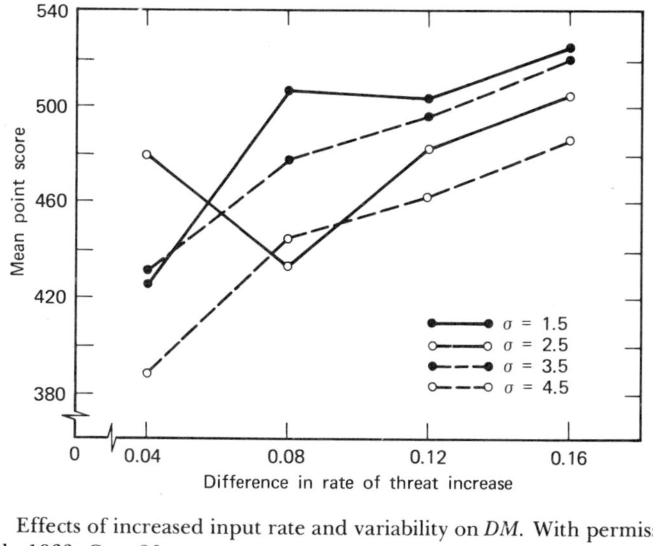

Figure 4-1 Effects of increased input rate and variability on *DM*. With permission, from Howell et al., 1963, Case 80.

In this study, no subject developed an optimal strategy. As in many of the static gambling studies reported by others (Edwards, 1962; Pitz et al., 1969; Pruitt, 1962), and as the reader will see in subsequent cases, DM does not utilize an optimum strategy based on objective probability and payoff values.

A finding common to most studies of sequential *DM* was that individual differences in performance were extremely large.

Conclusions. Patterning of inputs (input variability) can be recognized and used as a basis for decision when patterns are very simple and are presented at either high or low frequencies of occurrence.

Illustrative Studies

Case 81. This study asked four questions. (1) Does the form of the input pattern affect the operator's perception of that pattern? For example, if an enemy pursues repeated attacks after resting two days, is this pattern more or less difficult to recognize than others such as continuous attacks, attack followed by rests, or rests interspersed by attacks? (2) Assuming that operators can recognize patterns presented with high frequency, can they recognize them when their frequency of occurrence is reduced? (3) Is it more difficult to discover such patterns in a continuous flow of events than in discrete units? (4) To what extent does experience (repeated exposure to the patterns) enhance pattern recognition?

The applicability of these questions to military operations is obvious, but they are pertinent also to such diverse situations as scientific analysis and playing the stock market.

Sequences of events representing eight patterns at high and low pattern strength (defined as probability of occurrence), continuous versus discrete occurrence, and experience from 1 to 10 trial periods were presented on a CRT screen to 48 enlisted men. Operators indicated which of two enemy activities (attack or rest) was likely to follow each two-event clue.

All the independent variables were significant. Of greater interest, each of these effects was involved in one or more interactions, particularly as they related to the experience factor. Three of the four patterns were "learned" over 100 trials; this learning was characteristic of high strength patterns but not low strength ones (Figure 4-2). Three of the four patterns were recognized more readily in the continuous than in the discrete sequence.

Apparently a very simple pattern such as repetition of the same event is easily recognized at high and low frequencies of occurrence. More complex patterns required a frequency of occurrence greater than the baseline of 68% before they were perceived. Finally, pattern recognition apparently depends a great deal on continuity in the event sequence.

Subjects improved their predictions of high probability patterns over time (from 60 to 80%), whereas those of low probability did not improve (Figure 4-2). If the event patterns have a high probability of occurrence, complex, automated systems are likely to add little to pattern recognition performance. This implication was also drawn by Howell (1967), who suggested that systems that habitually handle predictive data of high probability have little to gain from automating the decision process.

When the patterns are of low probability, however, computer aids might be quite helpful. Since operators were unable to recognize three of four patterns occurring at 68% strength, they were unable to take advantage of a considerable amount of predictability present in the event

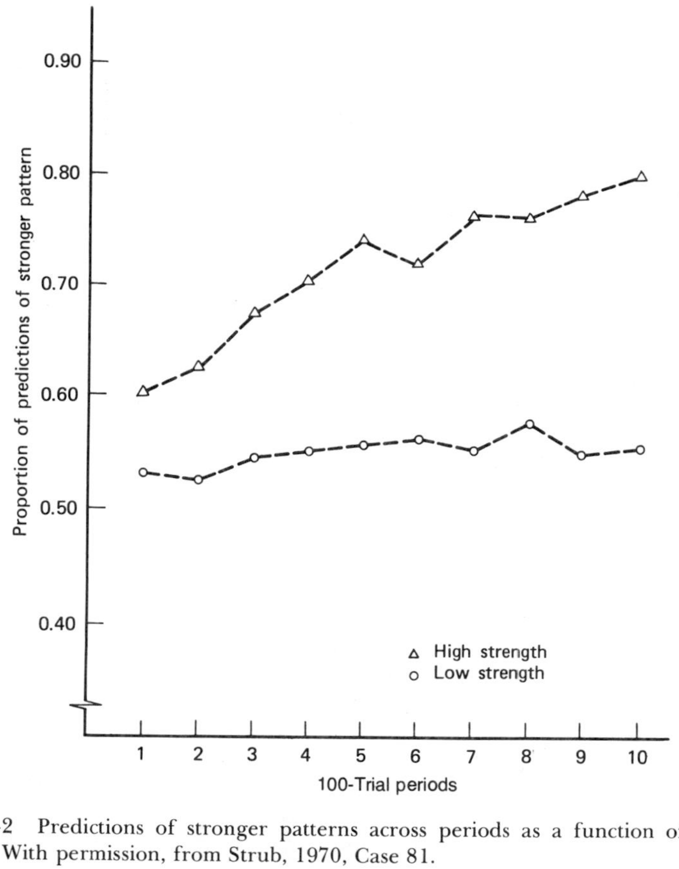

Figure 4-2 Predictions of stronger patterns across periods as a function of pattern strength. With permission, from Strub, 1970, Case 81.

sequence. [Howell (1967) also found this.] Computers could be programmed to detect weak but significant patterns in an apparently random set of events.

Conclusions. Successful *DM* does not require that all data be provided. How much data can be missing, however, before DM's performance is significantly reduced is unknown. On the other hand, *DM* performance is affected negatively by incorrect data.

Illustrative Studies

Case 82. This study was performed to investigate the effects on decision behavior of (1) complete and accurate information, (2) incomplete but accurate information, and (3) incorrect but complete information. Subjects were instructed to identify as early as possible the characteristic features of a new and dangerous satellite, to permit data to be fed into automatic interception devices. Reports were received at 5-second intervals, with one or more of three possible identifying signs displayed by each satellite. Subjects had to detect which of the three signs seemed likely to give the greatest chance of being a useful distinguishing sign and to choose the significant characteristic after each 5-second report.

In this study as in Case 80, low accuracy was reported in difficult situations, higher accuracy in easier ones. There was an apparent recency effect in the subjects' interpretation of data. At several points a substantial (20–30%) proportion of subjects selected one satellite over the other because more recent positive readings in favor of the satellite selected were given undue weight. Subjects also experienced a sudden onset of doubt under conditions of a relatively long sequence of status quo inputs. (This aberration is occasionally experienced under similar circumstances in real life; it apparently reflects idiosyncratic hypotheses about environment variability.)

Successful *DM* was not affected by the problem of missing data—specifically, 10% of the readings missing. [Comparable effects were noted in a study by Connolly et al. (1961) involving an air defense simulation in which the task was to assign weapons against targets. Tracks whose position and identifying descriptive data were only 60% to complete were handled in about the same way as tracks represented by perfect information.] When 10% of the readings were *wrong,* however, *DM* performance was impaired. It may be that when data are missing the operator extrapolates the subjective "mean" of his available (but correct) information to the missing "cells," but when the available data are wrong, the operator may base his decision on incorrect premises.

Conclusions. Decision time increases as a function of the number of input characteristics and number of available alternatives. Increasing the number of input characteristics does not, however, increase decision quality unless the characterisics are relevant to the decision. Training on decision situations that are not favorable to DM should be emphasized.

Illustrative Studies

Case 83. In this simulated tactical situation subjects were required to dispatch one of several aircraft to investigate a reported submarine sighting. Aircraft selection was to be made on the basis of characteristics such as quality of the pilot, aircraft armament and speed, and capability for visually sighting the target.

The number of characteristics varied from two to eight, and each characteristic could assume one of eight values. Amount of information varied: complete and reduced.

No feedback was given, and subjects were not told how to weight the various characteristics. Decision time and decision quality were measured.

Decision time was found to increase as a function of number of characteristics and number of available alternatives. In both cases the operator obviously had more to think about before he made his decision. Reduced information had no marked effect on decision time. Presumably the operator makes do with what he has (see also Case 82). Increasing the number of characteristics did not increase decision quality, which suggests that only input characteristics that are meaningful for decisions have an effect on those decisions (assuming DM can distinguish between meaningful and unmeaningful characteristics).

A second experiment attempted to determine whether decision quality could be improved by training. The task was identical to the previous one, but during training subjects received feedback and they had to continue making choices until they arrived at the correct one. (The author considers this experience rather than guided training.) Training did not improve *DM* over the performance achieved in the previous study in which no training was provided.

A third experiment attempted to determine the effects of limited decision time on decision quality. DM had to respond within 10 seconds, but the time constraint had no effect. As in preceding studies, both information transmitted and decision quality decreased as number of characteristics increased. (The operator apparently adjusts his criteria and strategy to operational constraints, which was suggested also in Chapter 2.)

In a fourth study the task was reduced to choices between two aircraft. The number of information characteristics varied as before between two and eight. The results again indicated a marked increase in decision time as a function of increased number of characteristics. (The additional characteristics required additional examination before a decision could be made.) Decision time decreased, and the probability of choosing the better alternative increased as the rank difference between alternatives increased. (Less competition between potential responses?) When the rank difference per characteristic was held constant, the probability of choosing the better alternative did not increase with increasing information.

These studies suggest that it might be advantageous to provide DM with experience in filtering his information, since attending to large amounts of information does not enhance decision quality but does increase response time.

Case 84. The subject's task was to decide which of nine groups of fighter planes to dispatch against an enemy air attack on the basis of the following characteristics: number of planes in the raid, distance of the airbase from the raid, and type and number of aircraft in each group.

Information was placed in different locations in a display matrix (two conditions), ensuring variability. There were three levels of force adequacy (adequate, inadequate, and mixed). Tradeoffs required for decision were varied by three levels of correlation among the characteristics (positive, negative, and uncorrelated). Decision time was the dependent variable.

Rather predictably, subjects' responses were in closer agreement on positively correlated characteristics than on negatively correlated ones. Decision time was determined by subjects' weighting of variables and display arrangement. Decisions were easier and were made most quickly when the problem variables (raid size, base distance, plane type, and fighter group size) were positively correlated (no conflicting variable characteristics), and the resources were adequate. (See also Cases 87 and 88.)

One possible implication is that training in unfavorable tactical situations (e.g., those with negatively correlated characteristics) should be emphasized. This has been pointed out also by Sidorsky and Houseman (1966).

Conclusions. DM tends to use probability information inefficiently. He is, however, quite logical, taking more information when it costs him less, when the payoff is greater, and when the number of possible responses is greater.

Illustrative Studies

The amount of information a person needs to make a decision depends on the inputs and the task. In a study by Irwin and Smith (1956), subjects had to determine whether the mean of a set of numbers presented sequentially was greater or less than zero. Amount of data taken increased both with the variability of the numbers in the set and with the proximity of the actual mean to zero. The same authors (1957) found that subjects took more data before a decision when the payoff for a correct decision was $1.00 as opposed to 50¢ and when each observation cost 0.5¢ as opposed to 1¢.

Similarly, Lanzetta and Kanareff (1962) found that people took more information when an observation cost nothing and the correct terminal decision was worth 5¢ than when they had to pay 5¢ for each item of information and received a 30¢ payoff.

Case 85. This study was designed to investigate the effect of two additional variables: the number of possible terminal acts (decision alternatives) and the information available to DM before he began his task. Both these variables should be important in determining how much information should be taken prior to the decision. From the Bayesian standpoint, the greater the number of possible terminal acts, the more predecisional information should be needed, and the more prior information is available, the less current information need be taken prior to decision.

Fifty-four subjects sampled from a hypothetical binomial universe in which a proportion *p* of all observations were "top quality." Although the task was formulated as quality control inspection of produce used in canning vegetables, where each item inspected had a cost and inspection accuracy was rewarded with a bonus, inputs were presented in a highly abstract manner. Subjects sampled the universe of products sequentially, stopping after some number of observations had been made to make a terminal decision. There were four experimental treatments: one involved three possible terminal acts, another had five alternatives, and there were two prior frequency distributions—one rectangular, the other, negatively skewed.

The largest and most consistent source of variance resulted from individual differences between subjects in the amount of information taken prior to decision (mean, 9.09; range, 0.25–22.59).

More data were taken when there were five alternatives (mean, 10.16) than when there were three (mean, 8.23). This was to be expected, since the larger number of alternatives created more uncertainty, as a result of which additional information was needed. The shape of the frequency distributions had an insignificant effect that may have reflected the general failure of subjects to use probabilistic information efficiently. Alternatively, the subjects may have felt that the difference in these frequency distributions was not large enough to warrant adjusting their decisions to account for the differences.

Conclusions. The rate at which information is presented is more important than whether it is provided automatically or on request. The availability of opportunities to modify a decision tends to degrade performance on the initial decision.

Illustrative Studies

Case 86. This study evaluated the effects of the average rate at which information arrives and the variability of information arrival times. In addition, self-paced and forced-pace information input conditions were contrasted.

Little information is available on these variables. However Sanders and Ter Linden (1967), using a two-choice task, found that more information was acquired prior to a decision as the rate of information presentation increased.

Another purpose of the study was to evaluate the effect of number of opportunities for decision modification. If DM knows he has a second chance, he may make his initial decisions sooner but with less accuracy than if he were committed to his first decision.

Pitz and Reinhold (1968), experimenting with a binary-decision task, found no difference in the accuracy of terminal decisions when under one condition subjects were permitted only one decision at the end of an information sequence and under another condition decisions could be revised after each information item was presented. Gibson and Nichol (1964) reported that more information was needed to change a decision than to make it originally, and Pruitt (1961) had determined that more information was sampled prior to a final decision when subjects had to make a preinformation decision than when they did not have this requirement.

The task in Case 86 required subjects to determine which one of eight friendly locations was the target of a hypothetical enemy advance (the same situation was used also in Case 80). Information supplied on CRTs showed "pathways" connecting discrete enemy positions, and each report moved the enemy from the bottom of the screen progressively to the target at the top. There were three intelligence sources.

With each enemy movement the subject was also given each source's best estimate of which target was under attack (computer aiding). Subjects had to request information to be able to decide which target was under attack. Each of the eight problems used contained information from sources that *(a)* were partially conflicting, *(b)* were on the average 33.3% reliable, and *(c)* provided updated information either 67 or 86% of the time.

Rates of input arrival were fast (one information item every 5 seconds) and slow (one item every 15 seconds). Arrival variability was defined in terms of the standard deviation of the distribution of arrival times (0, 1.4, and 2.5). In the self-paced condition information was provided only upon subject request. The numbers of decision modification opportunities were 0, 1, and 2. Payoff was an inverse function of the number of information presentations prior to a correct decision or elapsed decision time, number of information updates provided, and number of incorrect decisions prior to a correct decision.

More information was acquired prior to a first decision in the fast (16.00) than in the slow (13.45) condition. The number of information updates requested prior to a first decision decreased as the number of opportunities to modify an initial decision increased, but only for the first test replication. Significantly more information was sought when no decision modifications were permitted.

Significantly more updates were received prior to a first decision when subjects were self-paced (16.35) than when they were forced-paced (14.73). Accuracy was better when input pacing was fast (66.32%) than when it was slow (54.51%). Accuracy was significantly greater when no decision modifications were permitted than when either one or two modifications were presented.

Table 4-2 indicates the percentages of problems correctly decided for various conditions.

It is apparent that modification opportunities are much more important than pacing, which in general was unimportant, like variability of information arrival time.

When there was no opportunity to modify, accuracy of and confidence in the first decision was greater than when modifications were permitted. These results suggest that the availability of subsequent opportunities to

Table 4-2 Percentage of Problems Correctly Decided in Case 86

Modification Opportunities		
2	1	0
93.7%	79.2%	72.4%
Fast-Forced	Self-Paced	Slow-Forced
85.1%	83.3%	78.5%

modify a decision degrades performance on the initial decision. However the probability of obtaining a correct decision in any problem did increase with the number of decision modifications (Table 4-2).

Conclusions

1. Data source reliability, information conflict and update frequency affect information seeking and *DM*. However, DM can apparently integrate information regardless of degree of conflict, relevance or update variability. Less information is sought as conflicting information from sources is increased.

2. Degree of information relevance is an ambiguous variable in *DM*. Some data suggest that increases in irrelevant information degrade performance (Edmonds and Mueller, 1970; Goldstein and Allen, 1971; Hodge and Reid, 1971; Well, 1971). Others have found contradictory results (Landis et al., 1970).

3. As update frequency increases, DM seeks more information and makes his first decision on the basis of a larger number of information updates.

Illustrative Studies

Case 87. The experimental task in this study was essentially the same as that in Case 86. The independent variables were degree of conflict among information sources, degree of information relevance, and information variability. Conflict was defined in terms of intersource agreement about the direction of the enemy advance (0% conflict, all sources agree; 33% conflict, one source disagrees with the other two; 100% conflict, all three sources disagree). At any problem step, all sources could agree or disagree. Information relevance was defined in terms of the proportion of instances in which all three sources provided information (50, 67, 83, and 100%). Information variability was defined in terms of the variance of each pathway about a straight line to the target.

Initial decisions were reached somewhat earlier (except for the 100% condition) as the proportion of relevant information increased (Figure 4-3). There were no significant differences among conditions of conflict or information variability. There was, however, a significant conflict by variability interaction. Under low variability conditions decision times were highest for partially conflicting data, whereas for high variability conditions partially conflicting information resulted in lower decision times. Latencies decreased with increases in information relevance and reduction of information conflict.

Nevertheless subjects integrated information quite capably regardless of these variables. Although they sought more information prior to initial

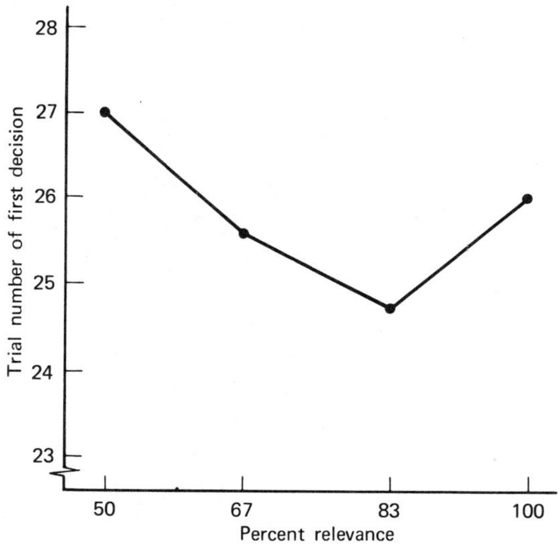

Figure 4-3 Trial number of first decision as a function of degree of information relevance. With permission, from Levine; *J. Appl. Psychol.*, 1973, Case 87.

decisions when relevance was low, they reached correct decisions with the same number of information requests under all relevance conditions. On the other hand, Streufert (1973) found that complex decisions in a tactical game were modified by changes in information relevance, but simple *DM* and information searching were not modified.

Case 88. A follow-on study to Case 87 was designed to investigate information seeking in a more realistic problem situation using *(a)* the same task (eight decision alternatives), *(b)* three different information sources that sometimes conflicted, *(c)* information not always updated with each request, *(d)* information of imperfect and unknown reliability, and *(e)* a payoff that decreased nonlinearly as a function of amount of information received.

Although the previous case (87) showed no significant effect of the conflict variable, the impact of information and DM's strategy had been shown by previous studies (Schum and DuCharme, 1971; Snapper and Fryback, 1971) to be affected by the reliabilities of the information source. Therefore, source conflict, as defined in Case 87, was again studied.

Information update frequency was defined in terms of the proportion of times that updated information was provided (50, 67, 86, or 100%). When information was requested, all three sources either provided new information or repeated the immediately preceding information.

Under low source reliability an average of 16.6% correct target suggestions was presented; under high reliability, an average of 33.5%.

More information was sought prior to the first decision in the low than in the high reliability situation. This is quite understandable. Yet as source conflict increased, less information was sought, which does not make sense unless we assume that the subject had lost confidence in his data sources. Accuracy was greater for high than for low reliability conditions, as one might expect. As conflict increased, accuracy under high reliability conditions increased significantly, whereas no significant change in accuracy resulted from low reliability conditions. Source reliability, information conflict, and frequency of updating information, therefore, have pronounced effects on information seeking and decision performance. The interrelationship among these variables, however, is complex and somewhat obscure.

Conclusions

1. The amount of information taken prior to decision exerts a greater overall effect on sequential *DM* than does the number of problems DM expects to receive, although neither variable is very strong.

2. When DM is presented with very certain or very uncertain situations, he selects a course of action that is less than optimal.

Illustrative Studies

Results of several investigations have shown that DM does not choose the same odds (i.e., the same risk of error) under all conditions (Becker, 1958; Case 80), nor does he consistently choose the decision points that yield the greatest average gain. Subjective probability is not related in any simple fashion to objective measures (e.g., dollars, points). For additional data see studies by Cohen and his co-workers (e.g., Cohen and Chesnick, 1970).

Case 89. This study explored two variables that may affect the selection of odds: *term expectation,* or the number of successive problems which DM expects to encounter, and the *uncertainty level* at which decisions are made (i.e., the amount of information received by DM prior to a decision).

The task featured a 16 × 16 display matrix in which columns defined events (A, B, C, . . .) and rows represented successive time periods (1–16). Elimination of an event was indicated by a cross placed in the appropriate column: one elimination occurred randomly every 5 seconds until all but one of the 16 events remained (the terminal event). DM's task was to predict the terminal event. (Note that this is one of the subfunctions required in *DM* but is not *DM* itself.) The task objective was to earn payoff points through correct prediction; the number of points earned was inversely related to amount of information received before a choice was made. A constant penalty was assigned for a wrong decision.

On any one problem five experienced subjects were permitted to make only one terminal decision. Under some conditions the point of that decision was fixed (prescribed), and in others it was determined by DM himself (ad lib).

Neither of the two variables studied appeared to be particularly powerful, although uncertainty exerted the greater influence. More important, when DM was presented with either a very certain or a very uncertain situation, he was likely to choose a less than ideal course of action. Intermediate levels of uncertainty were dealt with more efficiently. [Again one finds the same curvilinear relationship that runs throughout the behavioral literature, e.g., the relationships between performance and stress (Knowles, 1963) and between information load and performance (Streufert et al., 1972).]

Under ad lib conditions, DM chose an uncertainty level that at the proper risk level would yield very close to the highest possible expected value. However he did not consistently choose the proper risk level for this degree of uncertainty.

When DM was forced to receive either more or less information than he would have accepted in a free choice situation, he preferred higher risks and the chance for higher payoffs. Decisions made under forced levels of information equal those made under freely chosen levels only if the actual information levels are comparable; as either more or less information is forced on DM, his efficiency drops below that for ad lib information.

Conclusions. The physical structure in which inputs to problems are presented influences *DM* effectiveness.

Illustrative Studies

Case 90. To determine the potential usefulness of probabilistic displays in improving *DM* performance, an experimental situation involving a simulated aircraft search-attack mission was devised. Subjects were to attack an enemy target during one of 10 mission stages (time intervals associated with increasing hit probabilities and increasing firing costs). Subjects were told that decreasing range was associated with improvements in target localization information, but each succeeding stage was more dangerous. Monetary payoff and immediate feedback about hits were also provided. Three types of displays were presented: (1) a nonprobabilistic point estimate of object location, (2) a probability density function, and (3) a distribution function–probability contour display.

The differing display types did affect the probability estimates, and the most accurate estimates were obtained with probability contour displays. The nonprobabilistic display group fired most frequently at the fourth stage (equivalent to a hit probability of .20); the probability density and the probability contour groups fired most frequently at the seventh stage (hit probability of .40). Relatively early firing was associated with limited data, not expected to improve in quality over time. When subjects were able to evaluate their environment more effectively, they were willing to wait until their P_s improved, even though this cost them points. A second experiment indicated a tendency for subjects to prefer high probability–low reward to low probability–high reward situations, a tendency toward conservatism frequently found in *DM*.

Conclusions. DM's ability to predict event occurrence is strongly influenced by task characteristics, in particular cue redundancy.

Illustrative Studies

Case 91. Stimuli consisted of 200 triplets of two-digit numbers, the first triplet representing the value for cue 1, the second for cue 2, and the third number the criterion value or correct response.

A trial consisted of displaying the two cues, allowing the subject to make his response, and presenting the criterion while the two cues were still visible. The subjects were told merely that their task was to predict a two-digit number for every pair of two-digit numbers they were shown (again only a *DM* part-task).

Task predictability in this study was defined in terms of cue redundancy. There were three levels of task predictability (R_e=.50, .70, and .90). Subjects achieved better at high levels of total task predictability and at higher levels of cue intercorrelation, as Figure 4-4 reveals. It appears that personnel perform much more efficiently in highly deterministic environments and with cues that are redundant.

Results somewhat contrary to those of Case 91 were obtained by Brehmer (1974). In his study subject performance was initially higher when cues were intercorrelated, but the intercorrelation did not affect the final level of performance. Subjects did not develop optimal strategies, and their inconsistency was an inverse function of task predictability.

The results of Case 91 are attributed by Brehmer largely to statistical factors. It is possible that much of the multiple-cue probabilistic inference study data resist interpretation because the variables they test are defined almost exclusively in mathematical rather than experiential terms.

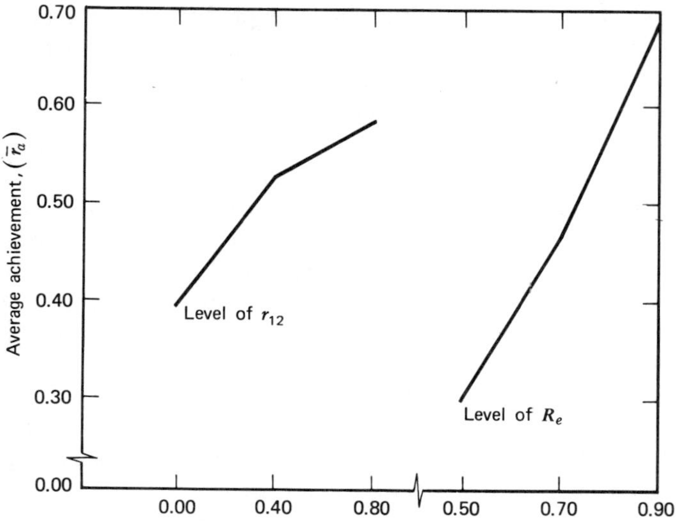

Figure 4-4 Effects of cue intercorrelation (r_{12}) and task predictability (Re) on achievement. With permission, from Naylor and Schenck, *Organ. Behav. Hum. Performance,* 1968, Case 91.

BEHAVIORAL PROCESSES IN DECISION MAKING

To understand how and why DM responds to problem inputs, one must examine the behavioral processes he applies to these inputs.

Formal models of idealized processes for resolving decision problems have long existed in economics and mathematics. [See Edwards (1954) and Lee (1971) for a review.] These models do not, however, describe what DM actually does; they make clearly unrealistic assumptions about the amount and completeness of information available to him and about his innate capacities to utilize this information.

This section asks the following questions.

1. What behavioral processes are actually employed by DM?

2. What biases does he have? If these tend to improve *DM,* perhaps they can be enhanced. If they degrade *DM,* perhaps they can be modified by training or by otherwise changing the *DM* situation.

3. In comparison with optimal solutions achievable by mathematical methods (e.g., the Bayes theorem), how well does DM perform? If human *DM* processes are inferior to computerized solutions, we have a powerful argument for their automation.

Conclusions

DMs utilize relatively simple behavioral processes (see Yntema and Torgerson, 1961). They exhibit a number of biases, primarily conservatism, that reduce their capability to achieve an optimal solution. Subjects have difficulty weighting cues in a compensatory manner and deal poorly with sampling variability. They overestimate the value of results from small samples. Subjective probabilities are biased by considerations of availability. Subjects systematically misperceive the probabilities of compound and highly valued events; their estimates of correlation or contingency between pairs of events are systematically incorrect.

Illustrative Studies

Many factors influence conservatism, which can be defined as DM's slowness in revising subjective probabilities to correspond to changes in objective probability.

Peterson and Miller (1965) found that as *prior odds* and likelihood ratios became higher in either direction (e.g., 1–9 or 9–1), subjects became less conservative. Somewhat different results were reported by Phillips and Edwards (1966), who used a greater range of probabilities. When considering only the range used by Peterson and Miller, however, the results were similar. Generally, as prior probabilities become unequal, subjective estimates become less conservative.

Peterson and Miller (1965) discovered that as the datum has less *impact,* subjects become less conservative. In fact, when a datum was highly undiagnostic (data diagnosticity is the degree to which data can be expected to reduce uncertainty about the hypotheses being considered), subjective estimates were more extreme than Bayesian estimates. This was found also by Phillips and Edwards (1966) and by Schum and Martin (1968).

As the *amount of evidence* increases, subjects become more conservative (i.e., become less capable of extracting all the information in the data). Peterson and Miller (1965) reported this effect in a simple decision-making task and Schum (1966a) found the same in a complex situation. The decrease in efficient use of information as amount of information increased is apparently not due to a decrease in motivation of subjects (Schum et al., 1967).

Using experienced subjects, Schum and Martin (1968) learned that the effect of *conflict* (number of items favoring other than the true hypothesis) was dependent on the items' diagnosticity. As conflict increased, subjects became less conservative with low diagnostic data and more conservative with high diagnostic data.

Phillips and Edwards (1966) examined the effects of *response mode* on conservatism. The various response modes (probabilities on a linear scale, probabilities on a logarithmic scale, odds on a logarithmic scale, and verbal odds) produced varying degrees of conservatism, although subjects were conservative under all conditions.

The same variation in conservatism was found with three *payoff* modes: linear, logarithmic, and quadratic (Phillips and Edwards, 1966; Schum et al., 1967). However payoffs in general decreased the amount of conservatism.

Peterson and DuCharme (1967) observed a clear primacy effect on revision of probability estimates when *data sequences* favored first one of two possible hypotheses, then the other. Subjects' estimates crossed the .50–.50 boundary (indicating a switch in hypothesis) after favoring the first hypothesis much later than Bayes estimates. Subjects also did not favor the second hypothesis as strongly as the first, even though the Bayesian value was the same for the second hypothesis and for the first hypothesis.

Peterson and DuCharme (1967) increased the *complexity* of the *DM* environment and found more conservative estimates, as did Phillips et al. (1966), with similar results. Alker and Hermann (1971) also found that as decision complexity increased, subjects became increasingly conservative as compared with the Bayes model. All these studies seem to agree that the effect of increased decision complexity is a decrease in *DM* efficiency, which completely accords with complexity effects in other behavioral functions.

As *sample size* increased, subjects became more conservative (Peterson et al., 1965).

Experienced subjects were less conservative, determined data sources on the basis of fewer data samples, were more sensitive to prior probability values, and adopted an appropriate strategy more consistently than naïve subjects (Schum et al., 1966). See also Case 106.

As far as information seeking is concerned, subjects do not use an optimal strategy, purchasing either too much or too little information for the situation at hand. Hershman and Levine (1970) found that subjects usually purchased too much information. Pitz and Geller (1970) noted that subjects tended to repeat their previous decision following contradictory information and tended to overemphasize the importance of their most recent data. One reason for the failure of subjects to behave optimally may be that they do not respond correctly to data diagnosticity (Snapper and Peterson, 1971). These findings agree with those of Gettys et al. (1970) that subjects perform poorly when estimating data diagnosticity.

Conclusions

There is evidence that man is a good probability estimator for single-event phenomena but relatively poor at aggregating these estimates (Case 92). Nevertheless, subjective probabilities tend to conform to the combinatorial rules prescribed by probability theory (Case 93). Although DM performance on tactical problems is not optimal from a game-theoretic standpoint, his behavior is entirely rational (Case 94).

In general, agreement between subjective and objective probabilities, when based on event frequencies, is satisfactory (Anderson and Whalen, 1960; Beach et al., 1970). This accuracy is more characteristic of group means than of individual performance, however, suggesting again the importance of individual differences.

Illustrative Studies

Case 92. The purpose of this study was to determine how effectively DMs can combine attack probabilities from three independent sources to arrive at an overall probability of attack (OPA).

The simulated tactical situation involved three factors: altitude, bearing, and type of enemy aircraft. Each factor had a different probability of attacking one of three ships which the subject controlled. The subject's task was to evaluate these probabilities without mathematical assistance and to decide which ship was in greatest danger (i.e., had the highest OPA).

Operators did not know how to aggregate probabilities even though feedback was provided. Most subjects reported using an adding procedure, although correct mathematical strategy was to multiply probabilities. Some subjects followed adding with a grouping strategy (i.e., selecting the ship that had the most consistent probabilities).

There was also strong evidence for response stereotypy and (as is common) conservatism in estimating OPAs.

Case 93. Subjects were asked for estimates of the probabilities of a large number of nonfrequentistic events, both for individual and for interdependent events. Each subject's estimates for the individual events were combined according to probability theory rules and used to predict his estimates for the combination of these events. Median correlations between the resultant predicted and actual probabilities ranged between .86 and .94. These results suggest that despite the results of Case 92 subjective probabilities for combinations of events conform to the combinatorial rules of probability theory and that probability theory is a good first approximation as a model to describe subjective probabilities.

Case 94. The task in this study was a simulated tactical air strike presented on a CRT. Subjects competed in pairs, one as attacker, one as defender. On each trial a blip of light representing the attacker's aircraft was launched from a simulated airfield and traveled at constant speed toward two targets. The subjects had to make a choice of targets by pressing a button before the aircraft blip reached a choice point. After passing the choice point, the blip traveled on a straight path toward the right or left target as designated by the attacker.

There were four kinds of target: aircraft carriers, submarines, destroyers, and single bombers, each having a specified worth (i.e., 100, 50, 20, and 10 units, respectively). The two targets could have the same or different value. Targets were provided with a protective antiaircraft screen of varying degrees of penetrability; information about this penetrability was indicated by digital readouts. In addition, the defender could erect a totally effective defense to one of the targets by pressing a button. Both players made their choice before receiving any information about their opponent's choice. In addition to choosing which target to defend or attack, each subject had to predict his opponent's response, that is, the target he expected his opponent to attack (defend).

Despite the sizable discrepancies between players' responses and the game-theoretic optimum in many cases, the similarity between subject responses and the theoretical standard (Figure 4-5) is quite striking. Obviously, subjects' behavior is highly rational.

The frequent nonoptimal (in a game-theoretic sense) responses do not mean that the subjects' behavior was necessarily wrong. Deviations might reflect their capacity for taking additional factors into account, one of which might be awareness that the opponent was also performing nonoptimally.

Conclusions. DM tends to delay action beyond the optimal time for reacting and often does not take action at all when required.

Illustrative Studies

Case 95. Vaughan et al. (1966) had experienced Army officers monitor a series of messages that indicated the need to change the axis of their simulated advance. With only one exception, the officers did not modify their attack, even in the face of the increasing need to do so. This characteristic may be seen in many descriptions of famous battles.

Conclusions. DM may base his decisions on relatively weak input characteristics. His behavior is often inconsistent. He tends to use simple linear additive models when the diagnostic task is nonlinear.

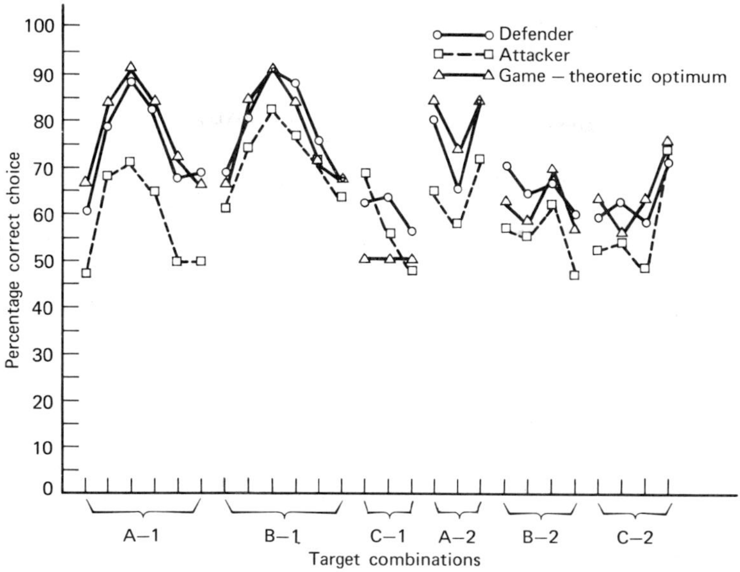

Figure 4-5 Similarity between operator decisions and theoretical optimum. With permission, from Sidorsky, *Hum. Factors,* 1967, Case 94.

Illustrative Studies

Case 96. The task in this study presented teams of Air Force ROTC students or air defense specialists with several decision situations. One situation involved making a choice between areas of high or low altitude radar coverage, because it was impossible to cover both areas simultaneously. The decision ultimately had to be made by predicting whether a low or high track would constitute the greater threat, and by choice of coverage.

Many subjects seized on a weak input characteristic (e.g., slight differences in aircraft position) that had negligible practical consequences before adequately searching other, potentially more significant differences. They frequently selected alternatives involving high gains and losses rather than average ones, apparently with the intention of mitigating the undesirable aspects of the situation. Subjects also exhibited the potentially undesirable characteristic of reversing decisions during the period between initial response selection and its implementation. Such inconsistencies occurred even after the initial choice had been based on careful evaluation of the situation and frequently involved a choice between high-level goals.

Conclusions. Diagnoses, even those of highly experienced DMs, are often of dubious validity.

Illustrative Studies

There is considerable controversy over whether DM uses simple additive, linear, or configural models in making judgments. Goldberg (1968) has presented impressive evidence that diagnosticians typically use simple linear models 90% of the time when the diagnostic task is configural and nonlinear. He reports a large number of studies in which the amount of professional training and experience of the judge does not relate to his judgmental accuracy, nor is the amount of information available to the judge related to the accuracy of the inferences he draws. As a result, diagnoses are unreliable over time and across diagnosticians.

These conclusions have been challenged by Rotter (1967) and by Einhorn (1971) who suggest that under conditions of high information load subjects use complex combinations of models to simplify the situation cognitively; particularly, Mertz and Doherty (1974) point out that there are few judges who function on a purely linear or configural basis—judges utilize a linear additive model on certain tasks but adopt a different strategy when the task changes. The critical question is not which model best describes the human judgment process but "what task conditions elicit what kinds of strategies from what subjects" (Mertz and Doherty, 1974).

Conclusions. DM tends to develop far fewer response alternatives than are required or desirable. He is unable to make use of more than a few decision criteria. He simplifies decision situations by concentrating on only one or two salient characteristics of the problem.

Illustrative Studies

Gagliardi (1965) instructed subjects to work out alternative solutions to a weapons platform positioning problem, to "cover" a set of targets. Subjects generated as many as five solutions in the 30-minute task period, but only with great difficulty, repeating earlier solutions and losing track of prior steps in working out a solution.

Formulation of evaluation criteria presented comparable difficulties. Vaughan et al., 1960, interviewed experienced submarine commanders to identify criteria used to guide choice of action on a variety of control dimensions (depth, speed, weapon mix). On the average, officers offered two criterion considerations per decision and in no case did the number of criteria mentioned exceed four. Schroder (1965) found quality of decision in a weapons assignment task to be degraded by subjects' inability to formulate criterion concepts (such as conservation of resources) and to relate situational data to these dimensions.

Hayes (1962, Case 83) found decreases in *DM* quality and increases in decision time as criteria were increased from two to eight for four and eight alternative problems. In the series of studies cited as Cases 10 and 11, Connolly and his associates assessed the appropriateness of weapon selection decisions by experienced Air Force officers in a simulated air defense problem. Although subjects were instructed to use three criteria in the selection of weapons to targets (minimize damage, destroy maximum targets, and conserve counterweapons), actual selections emphasized the first two alone.

Schumacher et al. (1965) reported that experienced Navy antiaircraft warfare officers used only a single criterion for weapon selection decisions in training exercises. Apparently differential weighting of criteria is difficult; although multiple criteria were used in the selection of aircraft and ships to investigate potential submarine contacts (Vaughan and Gartner, 1965), they were ordered by DMs in priority as a means of narrowing the range of alternatives.

Criterion identification also seems to be correlated with action formulation; DM will think of criteria to support an action alternative on which he has already decided intuitively.

Other evidence indicates that DM unknowingly simplifies the decision situation by concentrating on only one or two salient problem characteristics and by ignoring others. Rigney and DeBow (1967) asked 63 experienced CIC watch officers to judge the similarity in terms of threat value of pairs of air raids presented on slides of CIC air summary charts that varied in range, course, bearing, composition, altitude, and speed. Many subjects were unable to make reliable judgments. Multidimensional scaling analysis indicated that judgments of threat value were based primarily on range and course.

Conclusions. Unorthodox approaches to *DM* notwithstanding, the weights assigned to criteria by experienced personnel are often quite accurate.

Illustrative Studies

Case 97. Vaughan et al. (1964) had 46 experienced submarine officers assign quantitative values to seven criteria affecting the desirability of a running depth for two tactical situations. Criterion weights were highly reliable over time (7–11 months), consistent within subjects (although commanders differed in terms of criteria weights), and differentiated appropriately between tactical situations (the depth selection criterion changed as a function of changes in the situation). Length of experience was not significantly related to ratings of criterion importance.

Conclusions. Specific feedback (e.g., decision consequences in terms of quantitative probabilities) facilitates the quality of decision performance more than less specific (i.e., outcome) feedback. DM modifies his behavior on the basis of feedback when correct instructions are given him; however incorrect instructions cancel out feedback effects. (See also Chapter 6.)

Illustrative Studies

Case 98. The variables studied were the effect of conflicting instructions and feedback specificity on tactical *DM*. Subjects were provided with information about speed and distance, two attributes of a hypothetical air defense environment; each attribute could assume any whole number from 1 to 5. The task was to select one of three decision alternatives (labeled tactics 1, 2, and 3) and to estimate the probability of success (P_s) for the selected alternative on the basis of numerical problem information.

One set of instructions to subjects specified that at low speed and long distance tactic 1 had the highest P_s; at high speed and short distance, tactic 3 had the highest P_s. The other set of instructions stated the exact reverse. Half the subjects received feedback based on an appropriate set of instructions; the other half received feedback based on inappropriate instructions.

The performance of the group receiving correct instructions was superior to that of the group receiving incorrect instructions (mean, 64.2–29.1%). The P_s feedback had no substantially greater effect on performance than outcome feedback under conditions of incorrect instructions, as one would expect; P_s feedback did, however, elicit superior performance with correct instructions (Table 4-3).

Predictably, instructions opposed to the actual tactical situation significantly degraded *DM* quality, regardless of feedback specificity. This is because instructions give the subject more direct guidance than does feedback. In real life one would not expect *DM* instructions to be deliberately incorrect; however in complex tasks instructions may be vague or misunderstood.

Similar conclusions were drawn from a study of feedback and response mode in performing a Bayesian decision task (Martin and Gettys, 1969). Subjects given probabilistic feedback were superior to those given merely nominal or status feedback; those required to respond in a probability mode were more effective than those who merely had to describe status.

Table 4-3 Correct Choices as a Function of Instructions and Type of Feedback[a]

	Instructions (%)	
	Correct	Incorrect
P_s feedback	69.0	27.5
Outcome feedback	59.5	30.7

[a] From Kinkade and Ranc, 1967, Case 98.

Conclusions. Information taking by DM decreases with increasing input history. The amount of information requested by DM increases also with problem difficulty, and as both amount of available information and information resources increase (Levine et al., 1975).

Illustrative Studies

Case 99. The subject in this study played the role of a military commander whose four bases (A, B, C, and D) were under increasing threat of attack. He could obtain successive estimates of the magnitude of threat to each base, but these estimates contained some amount of variable error and consequently varied about the true threat value. The commander's task was to determine as quickly as possible which base had the highest rate of true threat increase. (See also Cases 80, 86, 87, and 88 for illustrations of this procedure.) Points on a graph indicated the sequence of threat estimates for each base. Four such graphs, one for each base, were displayed together for each problem. The subject requested new information displays until he made a final decision.

Several variables were investigated, and one—level of input history—is of particular interest. Input history is the amount of previously obtained information that continues to be displayed after more recent information is obtained and displayed. Input history increases the amount of information available to the subject but also adds to the information load, hence can be considered either positively or negatively. Subjects should require less information before making a decision as more history is provided.

A second hypothesis was that the amount of information taken would increase with problem difficulty. If verified, this would support a finding by Howell et al., 1963 (Case 80) that subjects draw information in rough proportion to task difficulty.

The hypothesis that information taking would decrease with increasing input history was supported statistically (Figure 4-6).

Number of inputs to decision increased significantly as problem

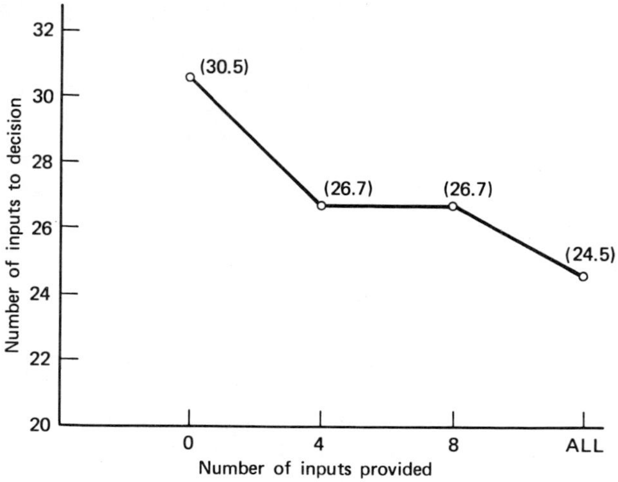

Figure 4-6 Amount of information required as a function of input history. Taken from Schrenk, *Hum. Factors,* 1964, Case 99.

difficulty increased, although the effect was far from proportional. Subjects did not therefore conform to a strictly objective statistical decision model. The number of wrong decisions increased with increasing problem difficulty, as one might expect. An error tended to make the subject more cautious on the following trial.

Conclusions. Decision adequacy tends to vary in accordance with the informational value (utility) of data provided DM.

Illustrative Studies

Case 100. This study tested the usefulness of a method of measuring information utility in tactical problems. The task was to find and sink three out of four enemy submarines operating within range of antisubmarine warfare (ASW) weapons. At the start of the task, the subject was given anywhere from zero to four intelligence reports describing enemy employment, capability, and/or vulnerability. The reports varied in terms of *DM* utility and relevance. The ocean in which the submarines could be located consisted of 128 cells in which the subjects could look for the submarine. The subject had available to him 200 ASW weapons that could be directed at any cell at three depth settings having various kill probabilities. It was found that decision adequacy varied significantly with the information value of the data provided. Information-utility scales predicted performance better than chance. Space does not permit description of the scales themselves.

Conclusions

Personnel do not consult information sources in an optimal manner. As the cost of consulting an information source increases, its use decreases. Operators manifest a reluctance to expend the resources necessary to obtain the "best" information. In general, personnel gather too little information prior to making a diagnostic decision. However, when more information is available, subjects are more efficient in selecting data items (Levine et al., 1975).

On the other hand, personnel tend to consult additional information sources to *verify* information, even when the additional sources are unnecessary.

Illustrative Studies

Case 101. In this study the operator had to decide not only how much information to gather prior to making a two-choice decision but also which of three information sources to consult. The task was phrased in terms of deciding whether a submarine was present, based on information derived from three simulated sensor sources having unequal values. Inputs were presented in the form of cards providing information about the presence of the submarine. Subjects were permitted to stop collecting information at any time—the optional stopping paradigm that has been used in a number of studies (e.g., Schrenk, 1964; Howell, 1966). Payoff and cost were designed to ensure that the penalties for incorrect decisions were severe enough to make it worthwhile to collect a great deal of information.

Performance was clearly influenced by the relative costs of consulting information sources, with some bias toward conservatism. Optimally, sensor source A should have been consulted almost exclusively, even when the cost of following this course was substantially greater than that represented by sources B and C. As the cost of consulting A increased, however, the number of times this source was consulted decreased. Even when the cost of consulting each source was the same, subjects frequently consulted some sources having lower diagnostic value to check information received from another source.

Monte Carlo computer runs of the same task outperformed subjects significantly. Before risking a decision that had a high penalty for error, subjects gathered significantly more evidence. Much individual variability was noted. "Different selection and decision strategies were used, with some subjects performing in accordance with the optimal model, while others were even more cautious than the model prescribed" (p. 385).

Conclusions. Accuracy and confidence in sequential *DM* varies directly with the amount of information provided.

Illustrative Studies

Case 102. Simulated military problems in the form of maps required decisions about which of three friendly units the enemy intended to attack. Varying amounts of information were presented showing four, six, or eight successive enemy moves toward these units.

Accuracy was 46% for final decisions made with four enemy moves, 53% for six moves, and 81% for eight moves. The relations between amount of information and subjects' confidence in their decisions were 52, 60, and 68% for these conditions.

The tendency to judge the information as insufficient for taking action was significantly greater when lesser amounts of information were provided. At each stage, however, some men judged the information insufficient and yet made accurate decisions. [Landis et al. (1969) found only a negligible correlation between decision adequacy and certitude.] Information was judged insufficient for 60% of correct final decisions made with four moves, for 37% of correct decisions made with six moves, and for 21% of those made with eight moves. There were large individual differences, of course.

The relationship between amount of information and confidence in decisions has been supported by other studies (e.g., Hoge and Lanzetta, 1968; Hoge, 1970; Irwin et al., 1958).

The findings in Case 102 suggest that lack of confidence in their ability to make accurate decisions may cause some DMs to delay taking action even when they are able (objectively) to make an accurate decision on the basis of available information. This finding too may be supported by innumerable historical episodes, as well as by the reader's own experiences.

This logical hypothesis is, however, disputed by some sketchy evidence from Ekehammar and Magnusson (1973) whose investigation of the rating of officer candidates revealed that decision time increased monotonically with an increase in confidence (as it did also with amount of information). Confidence in decisions has also been found to decrease with increases in response uncertainty (Sieber and Lanzetta, 1964; Hoge and Lanzetta, 1968).

Conclusions. In a tactical situation DM may change his response criteria as a function of experience. His role as attacker or defender and his advantage or disadvantage relative to his opponent are critical in determining his behavior. When DM is at a pronounced disadvantage, he may allow the enemy to attack first, then wait to maximize his own probability of success.

Illustrative Studies

Case 103. This study used essentially the same experimental procedure described in Case 94. On each trial a pip of light (the attacker) traveled in a straight path from top to bottom or from right to center of a CRT screen. Numerals on readouts indicated to the subject his own and his enemy's probability of success as these probabilities increased over time. The subject was to fire at his opponent at the time he considered best for himself.

The P_s at which firing occurred was computed for each day of the experiment. On each successive day subjects fired earlier (i.e., when P_s was lower) for almost all levels of relative advantage. The reasons for this are not immediately apparent, but they may be related to changes in the operator's response criteria (perhaps increasing self-confidence?).

In Case 94, attackers were better able to predict their opponent's behavior than were defenders. Defenders frequently acted contrary to their own predictions, based on the ratio between the expected values of pairs of targets. They apparently felt that it was better to act contrary to judgment than to accept a course of action having an unacceptably low probability of success.

Although subjects in Case 94 responded consistently when they had the advantage, they vacillated between two extremes when they were disadvantaged. In Case 103 P_s values were normally distributed when the subject's own ship had a decided advantage but were bimodally distributed when the enemy had a similar advantage (Figure 4-7). Subjects indicated that when the odds were heavily against them, they let the enemy fire his weapon in the hope that he would miss, enabling defending subjects to wait until their own P_s reached a high enough value to ensure personal success. Since operators have difficulty dealing with disadvantageous situations, *DM* training should emphasize practice under such conditions.

It should be noted, however, that experimental situations in which P_s figures are given to subjects may not be representative of real world behavior. In a simulated command/control situation (Hanes et al., 1962) in which 10 Navy commanders had to defend against air attack, only one stated that he had kept the probabilities in mind and had acted on them. Most commanders indicated that in actual battle little attention was paid to specific probability values.

Conclusions. DMs often have highly idiosyncratic ways of viewing a tactical situation. Most DMs develop a strategy (intercorrelated response patterns). However the criteria on which these strategies are based are often unrelated to *DM* adequacy.

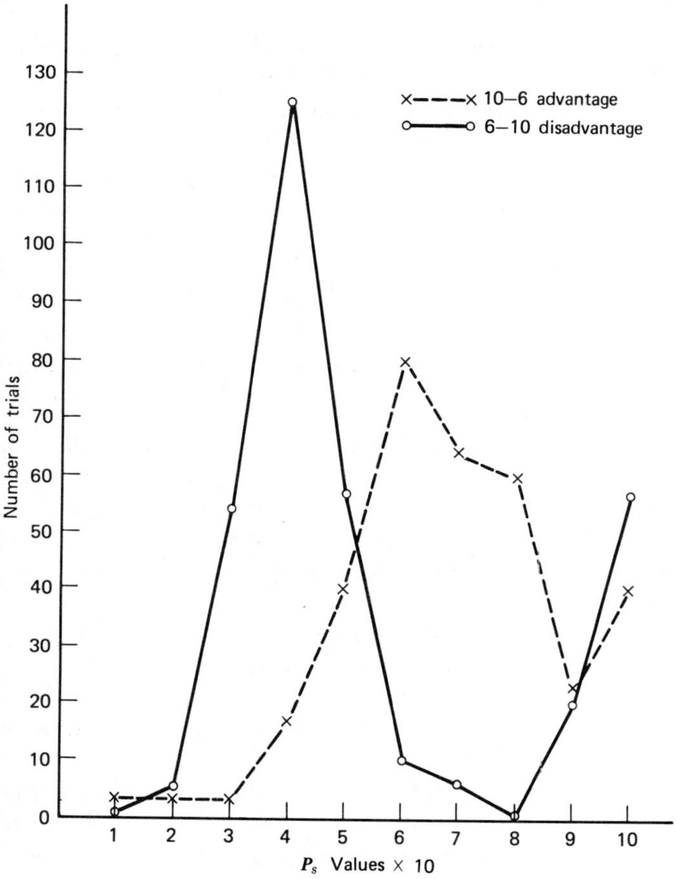

Figure 4-7 Frequency with which subjects fired at *Ps* values when they had a 10–6 advantage and a 6–10 disadvantage. With permission, from Sidorsky and Houseman, 1966, Case 103.

Illustrative Studies

In Case 103 the majority of the subjects (scientists and engineers, half of them with a military background), exhibited significant individual differences in the strategies they employed. For example, one strategy involved firing when the sum of the combined values of the probabilities of success for own ship and enemy ship reached a particular value. Many subjects also modified their behavior on the basis of their success or failure on the preceding four or five trials—a totally inadequate sample.

Similarly, in a command/control situation involving defense against air attack (Hanes and Gebhard, 1964), experienced Navy commanders changed the procedures (strategies) they used as a function of the degree of uncertainty (indeterminacy) in the situation. Under low uncertainty ("easy" attacks), procedures highly specific to the attacks were employed much more than were general procedures. When the attack situation became much more uncertain, the use of highly specific procedures dropped sharply. (See also Cases 31 and 32.)

Target detection capability will also be modified by the set (criterion) the operator adopts for *DM*. Evans (1965) found that operators instructed to adopt a risky set made earlier detections of targets but had a higher "false alarm" rate than when instructed to adopt a more cautious set.

Such biases are not unfamiliar to the *DM* researcher. For example, Ronen (1973) required subjects in two experiments to select one of two mutually exclusive actions, both having the same terminal P_s. Each action consisted of two sequential steps. The P_s of the first step of one action was greater than the first step P_s of the alternative action. Subjects were aware of the individual P_s of each step as well as the terminal (equal) P_s. They systematically and repeatedly preferred those actions in which P_s for the first step was higher. See also the primacy effect noted by Peterson and DuCharme (1967), referred to previously.

These differences in strategy have important implications for *DM* training. All trainees do not necessarily perceive the same tactical situations in the same way. Under such circumstances, since they do not operate on the same conceptual basis as the instructor, they may not be receiving the intended training.

Conclusions

1. More information is needed by DM to change his decision after he has made it than was required for him to make that decision initially (Pruitt, 1961).

2. The longer DM spends analyzing a problem, the more difficulty he has in recognizing changes in problem conditions.

3. The greater the amount of system/environmental change, the more difficulty DM experiences in evaluating accurately the true state of the system/environment.

Illustrative Studies

Case 104. In this case DM had to imagine that he was a field commander and a guerrilla attack could come from either the north or the east. The enemy was known to use various concentrations of troops (e.g., 20% north, 80% east; 40% north, 60% east). The subject was given intelligence reports of guerrilla sightings and told to determine the location of the main body of the troops. Over 30 to 40 trials, he did this for each sighting.

The three independent variablies were (1) initial condition, either 80 or 20% of information samples from the north, (2) time of change, after 10 or 20 trials, and (3) amount of change, 20, 40, or 60% from the initial condition.

Perhaps because the task was quite simple, practice or experience over trials did not significantly affect error scores, nor were the differences between subjects significant.

Results indicated that once DM has decided that a certain environmental state exists, the amount of contraindicative information required to change that decision will exceed the amount of information originally needed to arrive at that decision. This is hardly surprising. When only 2 or 3 items of information have been received, the addition of a single new item will have more impact than will the addition of one item to a backlog of 10 or 20.

The time at which a change occurs also affects DM's error scores. The mean error score of 15.4 after 20 prechange trials was significantly greater than the mean error score of 13.8 after 10 trials. The greater the time before a change was made, the more resistant the subject was to change. This is illustrated in Figure 4-8. Notice also the similarity between DM's performance and those of Bayesian predictions. Although Bayesian predictions were significantly better than those of the DM, there was a close relationship between them, as was found also in Case 94.

Differences due to the amount of information change were significant. The effect attributable to one unit of change was significantly different from the effects caused by both two and three units of change. The greater the amount of change, the greater the error score after change, and the longer it took DM to return to the same level of accuracy obtained before the change.

The interaction between the time of change and the amount of change was also significant. A change in conditions is perceived less readily by an observer who has been developing a conception of the situation over a longer time period than by someone examining the situation over a shorter time period. It is possible that increased experience may reduce DM's flexibility.

So much seems logical. However, Case 105 provides contradictory evidence.

Case 105. The hypothesis was that DMs operating in relatively stable systems would be able to learn and refine rules for operation, thereby enabling them to predict the consequences of different alternatives with greater accuracy than they could have achieved in systems whose state

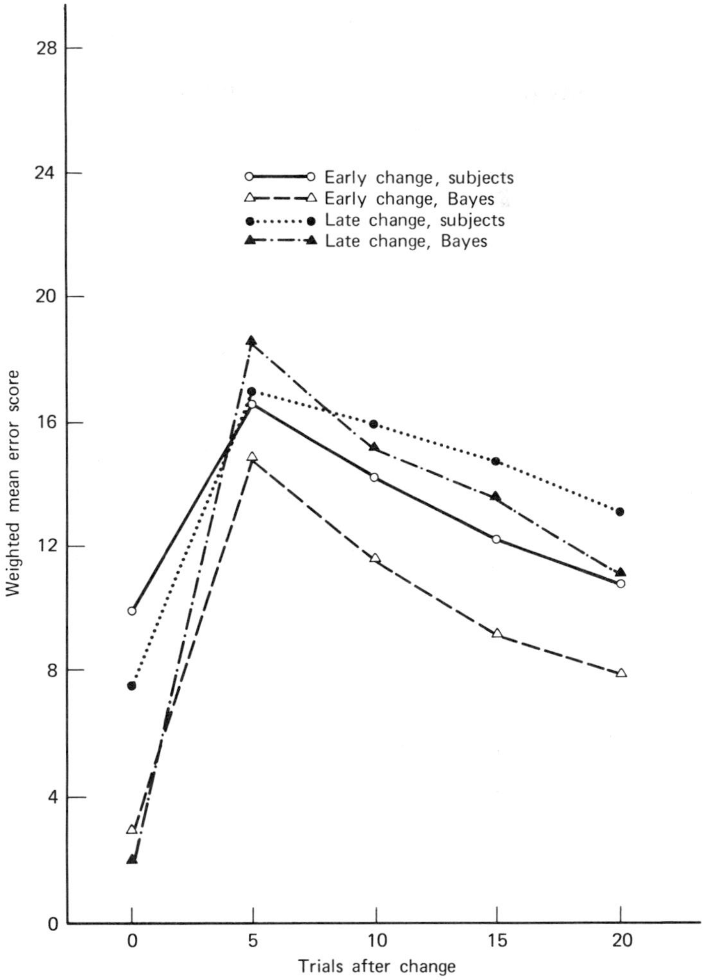

Figure 4-8 Influence of time of change on subjects' performance and on Bayesian predictions. With permission, from Gibson and Nichol, 1964, Case 104.

changed more frequently. The task was the same one employed in Case 98.

Subjects were provided with information about two attributes of a hypothetical air defense situation. Depending on the experimental conditions, speed or distance was differentially weighted by a factor of 3. The task was to select one of three alternatives and to estimate P_s for the selected alternative. Changes in system state were produced by reversing the attribute that was differentially weighted. The result was that differences as a function of this variable were almost nonexistent.

TRAINING AND EXPERIENCE

The training and experience variables should be as important to *DM* as to any other behavior. "Training" refers to a deliberate program of instruction to teach DM ways of coping with his problem, such as learning the relationships between inputs and procedures for solving problems. Training in *DM* rules should logically lead to performance improvement. The question is, how much improvement?

By "experience" we mean the opportunity DM has had to practice problems similar to those he is now encountering. Experience too should improve *DM* performance, but less than specific training, since the former merely provides an opportunity to learn but does not necessarily guide *DM*.

The amount of data testing the foregoing hypothesis is very slight, which accounts for the shortness of this section.

Conclusions. The evidence for learning in improving *DM* is conflicting and probably is affected by the type of task employed. Learning, based on simple exposure to complex diagnostic problems, does not appear to be very effective in improving *DM*, even with feedback.

Illustrative Studies

Goldberg (1968) has reported that clinical psychologists were trained for 17 weeks, during which each subject diagnosed more than 4000 MMPI profiles with immediate feedback. No change was found in ability to diagnose clinical categories accurately.

In contrast, Hammond and Summers (1965) found that training in the use of nonlinear criteria did improve diagnostic performances over those of diagnosticians who were given essentially no training.

Sidorsky and Simoneau (1970) attempted to provide tactical *DM* training by computer aiding. There were poor results in general, perhaps because of the great complexity of the naval problems presented. Subjects who used a sequential process in developing response patterns had difficulty developing the optimal solution. The authors recommend training DM to examine and resolve tactical problems in terms of their ultimate goal rather than by intermediate stages and subgoals.

Conclusions. Experience does improve *DM*, perhaps by increasing DM's sensitivity to diagnostic cues within the problem. Subjective probabilities estimated by experienced subjects are consistently higher (i.e., less conservative) than those of naïve subjects [although see the study by Hanes and Gebhard (1965) described below]. Experienced subjects purchase fewer data than do naïve ones and are much more affected by prior probability values. They also adopt optimal strategies more frequently.

Illustrative Studies

Case 106. Six subjects highly experienced in probabilistic information processing and six naïve subjects performed the familiar ball and urn task in which the subject decides on the basis of the sample of balls drawn which of two urns (red and black) is being sampled.

Experienced subjects produced final higher subjective probabilities than did naïve subjects (Figure 4-9) and purchased somewhat smaller

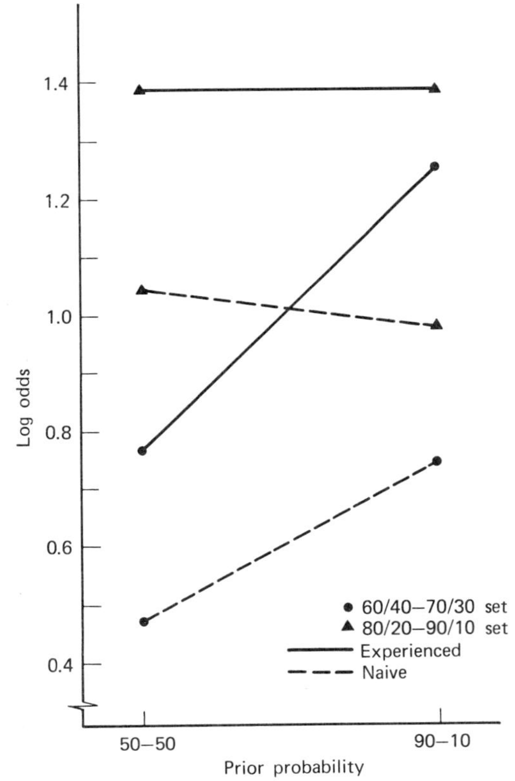

Figure 4-9 Subjective probability values for experienced and naïve subjects. With permission, from Strub, *J. Appl. Psychol.*, 1969, Case 106.

amounts of data. Although this difference was statistically significant, it was not great. However experienced subjects adopted an optimal strategy 78% of the time, naïve subjects, only 35%.

On the other hand, in a somewhat more realistic command/control simulation (Hanes and Gebhard, 1965), experienced Navy commanders selected a more conservative defense against air attack and made fewer shifts in strategy than did inexperienced officers under conditions of high uncertainty.

INDIVIDUAL FACTORS

Large individual differences appear in *DM,* as in behavior generally. Unfortunately, little is known about the meaning of these differences because *DM* research, like behavioral research in general, has tended not to focus on the meaning of individual variability. Only a few, not overly convincing studies, have investigated individual factors.

It would be desirable to know whether there are consistent patterns of response that differentiate groups of DMs from one another. If there are such biases (e.g., particular adaptive strategies, individual risk philosophies), these would have to be taken into account in designing *DM* systems.

Conclusions. *DM* appears somewhat related to personality and cognitive variables, although the variance accounted for by these variables is lower than one might expect.

Illustrative Studies

Edwards (1953) has shown that people have consistent preferences for certain kinds of bets. Block and Peterson (1955) found that fast deciders tended to be passive, suggestible, and conforming, whereas slow ones were ascendant, humorous, and self-assured. Scodel et al. (1959) found that low payoff bettors were more other directed, socially assimilated, and middle class. Long and Zeller (1965) and Pruitt (1961) demonstrated a relationship between the degree of dogmatism and predecisional information search. Presumably, whereas high dogmatic subjects tended to reject or misinterpret additional information, the low dogmatic personality utilizes subsequent information to a greater extent (Rokeach, 1960). An experimental test of this proposition revealed only a very weak relationship (Brightman and Urban, 1974). Statistically significant but low correlations have been found between dogmatism, risk-taking propensity, and intelligence on the one hand, and decision time, accuracy, confidence, and flexibility on the other (Taylor and Dunnette, 1974). A few studies have been performed on something called "decision style," which is defined as characteristic ways DM processes information (e.g., Strub and Levit, 1974; Modrick et al., 1975). Positive attitudinal factors reduce DM's conservatism (Sung, 1975).

Case 107. Other studies have used factor analysis to extract underlying *DM* behavioral dimensions. A factor analysis was performed of a battery of 70 scores, which included a broad sample of *DM* behaviors and cognitive/personality measures. The decisions to be made in these tests were largely "unguided" (e.g., the opportunity to buy or decline to buy stock), and the context of the tasks presented was not particularly mathematical. Subjects were 39 males and females at the high school/undergraduate level. Among the eight factors found were intellectual passivity, constrictive rigidity when required to make too many decisions in a short time, and contemplative appreciation of the structure of events.

Case 108. This study correlated verbal measures of personality and cognition with performance on a verbal *DM* test. Sixteen multiple regression equations were calculated, of which nine were significant at the .01 level and three at the .05 level. Although of course intelligence was related to *DM* efficiency, only 16% of the *DM* variance was accounted for by the equations.

With reasonably intelligent personnel, high correlations between personality variables and *DM* may be unlikely because normal scores on personality tests have a very restricted range. As has been pointed out, normal DMs tend to act rationally, although they may not be as efficient as a computer.

The author has difficulty accepting relationships based on variables so unamenable to objective definition, however logical they may appear. It is possible that a consistent relationship exists between personality and *DM* measures which is small in normal populations but might be larger in more heterogeneous populations.

Conclusions. There is some evidence that operators can be differentiated as being abstract or concrete in terms of the decision strategies they develop. The structure of their perceptual concepts varies curvilinearly with stress (load).

Illustrative Studies

Case 109. Some operators who can be termed abstract thinkers are able to organize information in different ways. Others (concrete thinkers) use very simple and fixed rules.

On the basis of a test of "integrative complexity," 28 concrete and 28 abstract subjects were selected and formed into 14 homogeneous four-man teams. Each group played a complex game in which the task was to conquer a mythical island (see Cases 43, 55, and 56 for further description). The independent variable in the study was information load, represented by messages transmitted to subjects concerning enemy action and answers to requests or instructions from subjects.

Teams composed of concrete subjects produced noticeably fewer integrations under all load conditions than did abstract teams. For all subjects, the greater number of integrations occurred under moderate workload. The inverted-U relationship between information load and *DM* performance is a general finding of Streufert and his colleagues, whose research has been summarized in Schroder et al. (1967).

GROUP ORGANIZATION

There is an increasing tendency in both industry and the military to favor group managerial decisions. The reason is unclear; as we pointed out in Chapter 2, the experimental evidence comparing group and individual *DM* is ambiguous.

In highly indeterminate situations decisions require production of ideas (e.g., how to accomplish an objective), whereas decisions in more structured situations involve choices between already specified alternatives. Because of the high degree of uncertainty in indeterminate problems, DM may find it emotionally supportive to have others associated with him in the *DM* process.

The organization of the group should therefore have an effect on *DM*, even if only a minimal one. The amount of information on this point is, however, sparse.

The number and quality of decisions produced by a group may be greater than the product of an equal number of individuals because the larger size of the group increases the probability of more talented members and may stimulate the individual. Alternatively, the group might exercise a negative effect on *DM* quality because the group decision might overrule a superior individual decision. The division of responsibility (hierarchy) within the group may also have an effect on *DM*, because only those in the higher levels may have a voice. Unfortunately, although a great deal has been written about management structure, most of the material is tenuous, and almost nothing about the structure has been tied to concrete performance outputs.

Conclusions. When *DM* takes place in a highly indeterminate structure (e.g., managerial meeting), *DM* output is likely to be inferior until the group organizes itself for its task.

Illustrative Studies

Case 110. Previous studies had found that when individuals and ad hoc groups were compared, initially individual *DM* was superior, but this superiority was canceled out after 50 minutes. One of the factors influencing decision quality, therefore, may be the time available to arrive at a decision.

In Case 110 officers were designated to work either as individuals or as group members on a problem requiring a plan to improve the morale and efficiency of men in isolated Arctic weather stations (a typical small group research problem). A content analysis of the ideas produced by individuals and groups indicated that increasing the time available from 50 to 100 minutes did not significantly affect individual decisions but did affect group decisions. Individual decisions made within 50 minutes were superior to those of the group, but after 100 minutes, this superiority disappeared.

The relative ineffectiveness of group decisions at 50 minutes may have resulted from inability to "organize" the group for the task before it could proceed. It is possible that after organizing itself, the group can function as effectively as can individuals. However the cost of group *DM* is greater than that of individual *DM*, not only because of the time taken to organize the group but because of the increased number of personnel involved.

Conclusions. Performance in group *DM* tends to vary with division of responsibility among group members and preference for the *DM* role.

Illustrative Studies

Case 111. The task used involved pairs of subjects filtering an ambiguous visual signal to extract its relevant component and selecting an appropriate action. One subject might filter, the other select, or both might perform the joint function. The conditions tested involved the fixed assignment of roles, an alternating role assignment, roles determined by the system on the basis of achieved proficiency, and roles determined in part by subject preferences.

Certain conditions proved to be significantly more efficient than others. Conditions involving adaptive control by the system and preferred subject role were superior to the nonadaptive and nonpreferred conditions. A transient decrement in group performance resulted immediately after a change in role assignment—as one would expect.

Conclusions. *DM* is more effective with greater task structure. The distribution of task responsibilities is less important.

Illustrative Studies

Case 112. This study required prediction of events in highly indeterminate structures. In the model of team performance the study was designed to test, team performance was a function of task, work, and communication structure. Task structure was defined in terms of component complexity, organization, and redundancy. Work structure was defined as the manner in which task components are distributed among team members.

The task used in this study was the same as that of Case 91 and is common in probabilistic inference research. In the two-man teams each subject was shown a series of two-digit numbers, one at a time, and was asked to predict individually which two-digit criterion number was associated with each cue number. Team members then "got together" and made a team prediction of the same criterion. After each of the 200 trials, the team was given feedback.

Task structure in this study reflects the total predictability of the criterion, using the cues both individually and in combination; component complexity was the predictability of the individual cues; component redundancy involved cue intercorrelation; and component organization was the predictability of the cues used in combination with each other.

There were two levels of task structure/predictability (.80 and .40), two of task organization (.00 and .40), and three of work structure. As defined in the study, task structure was determined by task complexity, redundancy, and organization, although from a statistical standpoint task organization was viewed as an independent dimension. The three work structures differed with respect to the number of cues available as a source of information to a team member and/or how many of these cues the subject was instructed to use. It was found that greater achievement occurred with greater task structure (higher predictability); teams performing in an 80% deterministic structure did better than those in a 40% structure. These effects were found also for individuals in Case 91.

Teams having more structured predictive tasks did better relative to the theoretical maximum than did teams assigned less structured tasks. The former were substantially more consistent in their predictions, which is compatible with previous research on individual prediction behavior (e.g., Dudycha and Naylor, 1966).

In contrast, task organization, which was defined in terms of the interrelations between or among task components, apparently detracted from the ability of teams to develop appropriate strategies. The greater degree of component interrelationship that defined organization also induced greater task difficulty, and this result is therefore probably an artifact of the task employed and its abstractness.

The results of this study tend to support the importance of the indeterminacy factor in system operations, at least for *DM* functions.

AUTOMATION

Automation, considered in Chapter 3 as a task characteristic, was defined as the ratio of equipment to manual intervention in performing the task. Studies dealing with automation of *DM* processes were more appropriately postponed to this chapter.

Interest in automating (i.e., computerizing) *DM* may stem from the questions involved. Although the advantages of automation appear clearly evident in the case of psychomotor and motor functions, they are by no means obvious for cognitive functions. Since certain cognitive functions are assumed to be peculiar to the human (see, e.g., the classic comparisons of man vs. machine by Fitts, 1951), their automation might well be a dubious proposition. However evidence now available, indicating that man is far from being an optimal DM, suggests that computer aiding might be of significant help to him.

Note the phrase "computer aiding." No one has seriously suggested the complete replacement of the human cognitive function. There might, however, be special advantages in permitting the computer to *complement* rather than supplant human thought. If so, we must determine how far we should go with *DM* automation, and which of the cognitive functions described in this book are most feasible candidates for this.

Conclusions

1. Human *DM* follows the Bayesian model (i.e., is rational from a mathematical standpoint) but is significantly less effective than the model. This has been a consistent finding (see, for example, Cases 94 and 104).

2. To automate the *estimation* of the probability of occurrence of individual events is not cost effective; humans do this fairly well. To automate the *aggregation* of these probability estimates can improve *DM* accuracy by 10 to 15%.

3. The advantage of automating data aggregation *increases* with the following conditions: *(a)* as data fidelity declines, *(b)* when data are interrelated, and *(c)* as the amount of data to be processed increases.

4. Feedback is useful to both human and automated *DM*. As long as some feedback is available, it need not be provided for all decision out-

comes. Feedback, however, does not enable human *DM* to overcome the superiority of automation.

5. Human proficiency in aggregating probabilities increases with experience and training, but, falls short of automated performance except under optimal conditions.

6. DM tends to resist *DM* automation.

Illustrative Studies

An early air defense study (Sinaiko and Cartwright, 1959), which has largely historic interest, tested the hypothesis that heavy target loads (60 vs. 37) would adversely affect DM but not the performance of an automatic system. There were the usual three levels of automaticity: (1) fully automatic, (2) semiautomatic (the automatic system could be overriden), and (3) manual (the officer made all decisions, the computer controlled the initial vectoring of interceptors). The limiting factor in this study was that a single subject was employed.

Results did not support the hypothesis. The automatic version of the system allowed more target penetrations (18.8) and accomplished fewer kills than did the semiautomatic (10.3). There was no meaningful difference between the manual and semiautomatic modes with regard to raid penetration (the major criterion). The semiautomatic mode, which permitted the officer and the computer to assign weapons, resulted in competition between them. As the experiment proceeded, the man assumed even more of the decision function from the computer. In other comparisons, using other criteria, the automatic mode fared badly compared to the manual mode.

More recent studies provide somewhat different answers.

Case 113. The studies described below (a description based on and owing much to a summary report by Howell, 1967) were performed at the Ohio State University from 1960 to 1966; researchers compared operator performance in *DM* with near-optimal performance as represented by Bayes' theorem.

Bayes' theorem is an optimum procedure for incorporating new information into an estimate of the relative probability of each of a mutually exclusive and exhaustive set of events. It is defined as $P(H/D) =$

$$\frac{P(H)P(D/H)}{P(D)}$$

where P denotes probability, H is hypothesis (an environmental state), D is datum, and / is "given that" or "conditional" (i.e., a conditional probability statement).*

The assumption here was that automation is most desirable where human performance is significantly below the Bayesian standard.

The context in which these studies were performed was a simulated command/control (intelligence threat evaluation) system known as COMCON. Fidelity of simulation to the real world system was excellent, involving CRTs, radio links, and similar equipment. Real-time intelligence data concerning a threat posed by an enemy force were generated from a simulated environment.

The twofold function of COMCON operators was (1) to interpret (process) these data, and (2) to evaluate (decide on) the processed data in terms of their meaning for a present or future enemy threat. The two functions were performed by different individuals.

COMCON was basically probabilistic. Reports of enemy activities and certain environmental states (e.g., pincer or flank attack) occurred with certain frequencies. Operators were required to consider these frequencies in making judgments; DM's typical output was a series of probability estimates for each possible state the environment could assume (e.g., .60 for pincer attack, .30 for flank attack).

In a subsequent series of studies described by Howell and Gettys (1968), COMCON was modified to permit operators to work either with the actual data probabilities or with a transformation of these probabilities, in which the event judged to have the highest probability was assumed to be true; that is, it assumed a value of 1.0 while probability values for all other events dropped to .000. In many real systems operators do not behave as though information is probabilistic, even though it is. They gather information until they feel the environment or the system is in a given state, whereupon they begin to act as though there were no uncertainty at all in the system. The reason for this is that "systems are seldom designed to deal with anything but all or none responses" (Howell and Gettys, 1968). There may well be a bias in the human to transform indeterminate situations to more determinate ones. The effectiveness of probabilistic versus deterministic data aggregation is discussed later.

The studies described start from concepts developed by Edwards (1962), who hypothesized that DM is most effective for estimating the individual probabilities of events and the machine for aggregating these

*The mathematical derivation of Bayes' theorem is of course much more complex than this; it has been simplified for this presentation.

individual probabilities, and deriving their necessary conclusions (e.g., the likelihood of air attack, given that certain events have occurred).

Several basic assumptions underly these concepts. First, operator evaluation of uncertain data is considered essential to diagnostic *DM*. It would be quite unrealistic to try to program into a computer all the experience a commander had gained in making such judgments. Second, the operator is capable of expressing his judgments in terms of numbers (likelihoods, probabilities) as long as they relate to the single, rather than the multiple, event. Finally, computers are capable of producing much better aggregate judgments than is man. Of the available rules of combination (e.g., Bayes' theorem, multiple regression), Bayes' theorem is probably the most straightforward and has certainly received the most attention.

The Bayesian model represents an optimum means for aggregating information in diagnostic decisions. The crux of prediction is the generation from $P(H)$ and D of $P(H/D)$. If data are gathered sequentially, processing merely expands into a succession of $P(H/D)$ estimates, with each serving as the $P(H)$ for the next item of data.

To make $P(H/D)$ estimates, the system must judge the predictive worth of each datum it gathers; otherwise it would be unable to determine how much a prior estimate should be adjusted to accommodate new information.

The diagnostic utility of the datum is assessed by estimating how likely it is that each item of information will be observed when the environment is in each of its possible states ($P[D/H]$). This is done on the basis of observed frequencies (i.e., the number of times in the past a particular datum D has been observed with each environmental state).

Diagnostic *DM*, therefore, is assumed to involve two probability estimates $P(D/H)$ and $P(H/D)$, the latter (which is what we are interested in) being basically a multiplication of $P(D/H)$ by $P(H)$ for each environmental state. The more precisely a system can estimate $P(D/H)$ and $P(H/D)$, the more closely it should approximate optimum diagnosis.

The feasibility of automating the decision process can best be examined by breaking judgments down into estimations of $P(D/H)$ and $P(H/D)$ and rules for relating the two.

Estimating P(D/H) *and* P(H/D)

Logic suggests that the function of estimating $P(D/H)$ should be handled primarily by humans (e.g., a panel of experts), mainly because this judgment often can be made only on logical or intuitive grounds. For example, machines might have difficulty differentiating between "new" and "old" data because of the tenuous nature of the criteria involved. If, however, large quantities of data and/or many hypotheses must be dealt with rapidly, automation of the "bookkeeping" aspect of this estimation might be worth considering.

The weight of the OSU research is that automated aggregation of *P(D/H)* judgments to secure *P(H/D)* is practical. Since a programmed Bayesian process can aggregate large quantities of *P(D/H)* judgments in a very short time, their automation should exploit all the predictive information these data contain. Because the operator, on the other hand, can handle only limited amounts of data, he is more likely to be selective, settling for what he believes to be the *most* predictive data. The overall superiority of automated Bayesian over human aggregation is summarized in Table 4-4. Based on these data, one would anticipate an improvement of around 10 to 15% in decision accuracy as a result of automation.

The critical question, however, is: under what circumstances would automated aggregation be most advantageous? What conditions degrade human *DM* to the extent that automation would seem worthwhile?

Table 4-4 Summary of Comparisons Between Human and Automated Aggregation Effectiveness[a]

Range of Conditions Studied	Decisions Correct % Human	Decisions Correct % Automated
Knowledge of results: 0–100%	34	41
Nonindependent input data: none–high	69	74
Information load: 4–12 data items; prior uncertainty: high–medium	69	80
Payoff functions: linear, log, all-nothing; information load: 4–12 data items; prior uncertainty: high–medium	62	70
Experience: 114–234 hours	47	54
Fidelity (input): complete–medium	56	55
Fidelity: medium–low; time stress: 1–7 minutes/decision	21	31
Automated aggregation: present–absent	25	45
Human control over aggregation: low–high	74	74
Overall averages	45	58

[a] From Howell, 1967, Case 113.

Information Fidelity. DM is often required to deal with contaminated, incomplete, or otherwise deficient information. Since operators concentrate on the "best" information, they may ignore items of low predictiveness that could be used by an automated procedure.

The utility of automated aggregation appears to be greatest when data fidelity is lowest (Figure 4-10); (Schum et al., 1965a, 1965b).

The Role of Feedback in Automation. DM can improve his estimate of *P(D/H)* by observing the outcome of his decisions. However, there is nothing completely veridical about feedback, especially in a situation of uncertainty; it too is probabilistic. In the real world, feedback is often distorted, delayed, ambiguous, or not available (see Chapter 6). What is the effect of this condition on the quality of subsequent decisions?

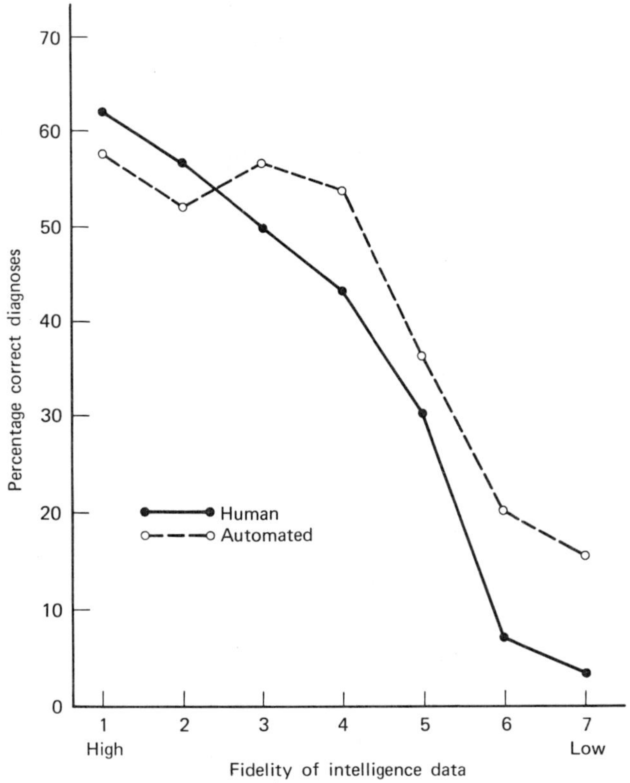

Figure 4-10 Effects of degraded inputs on diagnostic accuracy. With permission, from Howell, 1967, Case 113.

In one study (Goldstein et al., 1967), human and automated diagnoses were compared under conditions in which knowledge of results (KOR) occurred on different proportions of the diagnoses. Figure 4-11 summarizes the most important results of this work. Automated aggregation is superior to human aggregation throughout all levels of KOR above 33%. Below that level there seems to be slight basis for choosing between them. Reduced KOR creates little disruption of diagnostic quality until it drops below 33%. The decision system need not receive feedback on all decision outcomes to maintain diagnostic proficiency at a high level; some feedback, however, seems to be required. (As seen in Chapter 6, this conclusion is true of operators and systems in general.) Thus it may be that when limited data gathering resources are available, more should be allocated to the collection of new predictive data than to verification of past diagnoses.

Stress. The OSU experimenters defined stress when time available for the task is restricted (time stress) and when amount of data at any one time is excessive (information load). Several studies were concerned with this general problem. Some investigators varied the amount of time available for decision (Schum et al., 1965b, 1967); others varied the amount of data to be aggregated (Schum, 1967; Schum et al., 1967). Figure 4-12 illustrates that the automated mode becomes increasingly superior to the manual mode as information load increases.

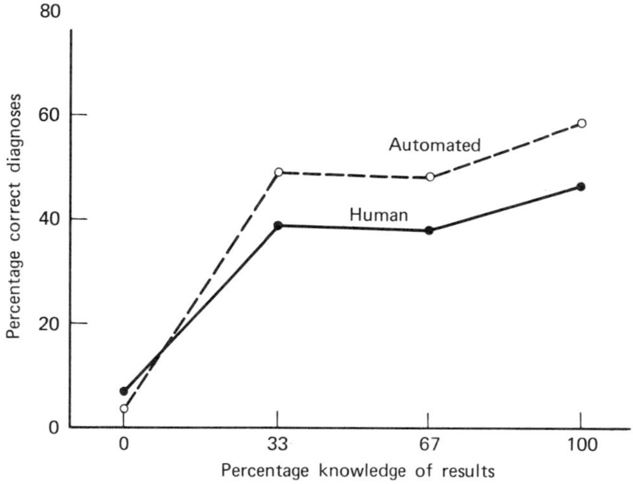

Figure 4-11 Effect of knowledge of results on diagnostic accuracy. With permission, from Howell, 1967, Case 113.

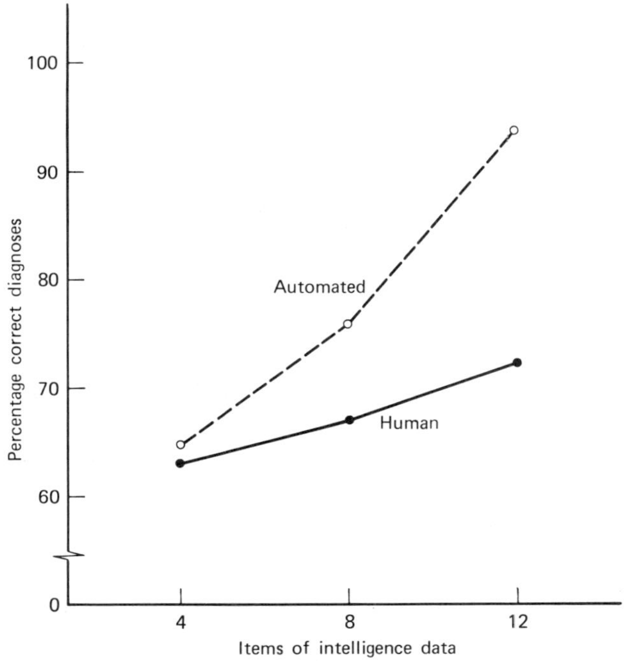

Figure 4-12 Effect of amount of information to be processed on diagnostic accuracy. With permission, from Howell, 1967, Case 113.

Predictive Relationships Among Data Items. Effective *DM* usually requires more than simple aggregation of items of information. Certain items are more likely to occur together when the environment is in one state than when it is in another.

For example, the attack in 1973 by Egypt on Israel was preceded by extensive troop movements on the borders, by the departure of Soviet dependents, and by increased war propaganda. Individually, each of these data items might have low predictive value; together they should have been highly predictive.

Data items, therefore, may contribute to overall diagnosis, both independently and nonindependently (i.e., through their association with other data in a particular environment state).

Studies by Schum (1966b, 1967) compared human and automated capability to handle nonindependence (Figure 4-13). Although humans are capable of predicting from data combinations, they are less proficient than an automated procedure specifically designed to take nonindependence into account. The differences between human and automated

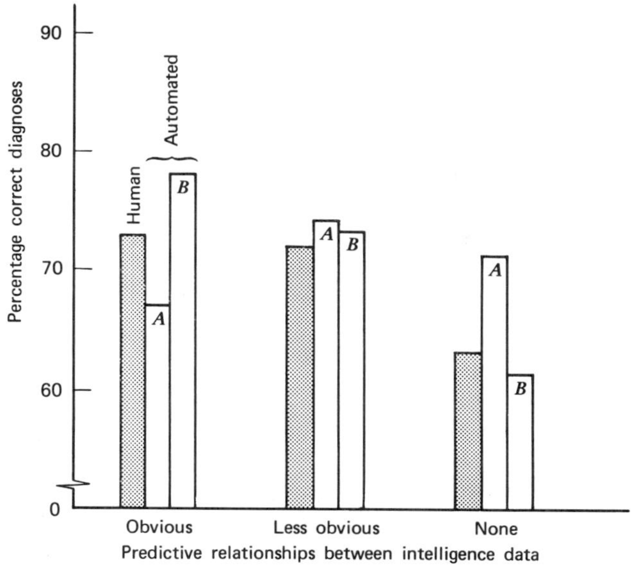

Figure 4-13 Comparison of human and automated diagnosis as a function of data dependency: A = assuming independent data; B = assuming nonindependent data. With permission, from Howell, 1967, Case 113.

proficiency seem rather small, nevertheless. Automated aggregation seems to be most advantageous when data are of relatively low-grade predictiveness and when the human is under some degree of time or load stress.

Experience

Since performance improves generally with experience or training, we would expect the same of diagnostic proficiency. If so, the extent of improvement is critical; if the operator after training can eventually match the proficiency of an automated process, it might be more practical to train DM than to automate.

Studies by Emanuel (1965), Martin (1966), and Strub (1966) have shown that highly trained personnel with experience in making probabilistic judgments are generally more proficient than relatively naïve subjects. Experience here was 9 or more months of half-time employment in probabilistic inference tasks; control subjects received only 4 to 16 hours of instruction. Despite the rather small numerical differences between subjects, experienced subjects showed sensitivity to data subtleties that escaped naïve subjects. The former were also better able to handle low-quality information, and they developed and retained identifiable strategies, whereas naïve subjects were more variable in their responses.

Both automated and human diagnoses improve with experience. Experience plays a somewhat greater role when data are nonindependent. In one study operators were unable to make use of the added predictive power of these data until 24 diagnoses had been made.

Figure 4-14 indicates that the automated process initially improves more rapidly than the human, but later the two functions become virtually identical (Schum et al., 1965a). However the aggregation task in this study was not particularly taxing, complete feedback was provided, and all data were of high fidelity. Under less optimal conditions, differences between human and automated performance might well increase, resulting in greater automation effectiveness.

Where it is not possible to automate, one might still improve human *DM* by requiring the use of quantitative procedures. Operators trained in the Bayesian technique perform more efficiently than those not so trained, although less well than an automated process. This may be partly due to the orderliness of the Bayes process, in part to the requirement that evaluations be expressed in probabilities.

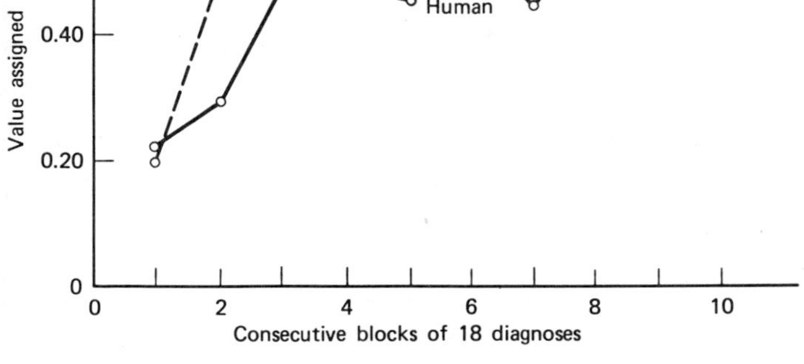

Figure 4-14 Effect of experience on human and automated diagnosis. With permission, from Howell, 1967, Case 113.

Probabilistic Versus Deterministic Information

It will be recalled that the modified COMCON permitted operators to use deterministic as well as probabilistic information. The transformation of probabilistic information into deterministic form (i.e., treating the highest probability data event as if it were true, with a probability of 1.0) degrades the effectiveness of system operations severely, especially when the fidelity of input information (its predictiveness) is about .70 at the point of decision. All the system outputs differed from the optimum by 6 to 10% except where the input information was either extremely good or extremely poor. "The reason, of course, is that when the probability profile is very clear (with good information), there is little difference between *as if* and *true* probability values; likewise, when the profile is completely uninformative (with poor information), neither mode can improve substantially upon chance" (Howell and Gettys, 1968).

Yet when a system's response is not all or none but is dependent on the likelihood of various alternative states of the environment, the importance of probabilistic information increases with the system's overall diagnosticity or predictiveness. "That is, all or none processing assumptions degrade system opinion more seriously, the better the information is that is being processed" (Howell and Gettys, *op. cit.*). Deterministic information is as effective as probabilistic information only when the system objective is to identify which one of the alternatives is best (without regard for how much better it is than the others).

Decision Timeliness

The weight of the evidence suggests that failure to emphasize timeliness of decisions is a consistent shortcoming in human decision performance. Team strategy in COMCON was to stop gathering data and to take action after some intermediate time period, regardless of how good or poor the diagnostic profile was. It may be advisable, therefore, to consider implementing in future systems more rigid decision programming for *when* action should be taken; *what* action to take could remain the operator's responsibility.

Communication in DM Systems

Communication in the modified COMCON had almost zero effect on system performance. The system chose the most appropriate action 81% of the time when communication between DMs was allowed and 79% of the time when it was not. There was, however, a tendency for redundant and irrelevant information to be transmitted when the opportunity for communication existed. (Chapter 5 presents comparable negative results with other types of systems.)

It is also possible to divide the aggregation function between human and machine. If a large number of predictive items must be considered in each decision, it might be more convenient for the operator to deal with them in "batches" rather than piecemeal, which would be tantamount to sharing the aggregation responsibility. Thus human *P(D/H)* judgments could incorporate several predictive items at once, and these values would then be aggregated quickly by the machine to produce *P(H/D)*.

Because of the tendency toward conservatism referred to previously, probabilistic estimates tend to be conservative when diagnostic *DM* is allocated completely to humans. Moreover, that conservatism varies as a function of the payoff value of potential decisions and the ease with which data may be aggregated.

The results described in this section favor automated aggregation. However intermediate conditions—those in which humans do some of the aggregating—may well produce most effective system performance, particularly when one considers "unanticipated" information (i.e., information not incorporated into the original predictive program). Nevertheless there is a limit to the utility of human aggregation because of the operator's inability to extract from large quantities of data all the predictive value of those data. "It appears then that automated systems can profit from manual support in aggregation, even though the result is not as satisfactory as would be total automation of this function (if this were possible)" (Howell and Gettys, 1968).

Recent research (Davis et al., 1975) has emphasized *adaptive* decision-aiding in which a computer model learns DM's utility values as the problem situation unfolds and adapts the computer's responses to changes in those values. The computer then suggests decisions to maximize DM's choices which in turn causes DM to modify his behavior. Experiments with a simulated surveillance task indicate that aided operators are much more consistent and effective than those who are not aided.

Conclusions. In a man-computer system, where the operator monitors the computer solution and may or may not accept its conclusions, the operator often rejects and/or alters optimal solutions.

Illustrative Studies

When automation does not replace but merely aids the operator, the latter often ignores the automated estimates (Southard et al., 1964a, 1964b). Even when provided with control over the automated process, he prefers his own judgments.

Case 114. In this task the type of system illustrated was one in which a computer is programmed to make decisions based on rules, and the operator monitoring can accept, reject, or modify these decisions.

In the first study operators were required to solve problems after the rules and criteria had been stated. In the second study subjects who had developed their own solutions were given those of a computer to monitor and were instructed to modify solutions only if such action would lead to improvement.

Subjects tended to ignore offered solutions and to substitute their own, thus degrading optimal solutions and improving poor ones.

Subject performance was nearly always inferior to the programmed solution. Performance was impaired by increasing the rate of information input and failed to benefit from reduction of uncertainty in the input. KOR helped convergence toward an optimal solution. Uncertainty proved not to be significant, suggesting that subjects failed to exploit all the information available. Subjects tended to make decisions too early.

Other studies (Hanes and Gebhard, 1962; Gebhard and Hanes, 1963) involving a command/control air defense environment indicated that although commanders were unwilling to relinquish complete control to the computer, they were willing to use an automated decision aid when its value was demonstrated in a highly realistic simulation. Although some officers negated the computer by issuing large numbers of anticipatory and substitute orders, no one turned the computer off. Officers accepted computer recommendations 66% of the time for such routine operations as sending an aircraft home when it was running out of fuel, 55% of the time for recommendations of how fast fighters should fly to the target. When problem situations became critical, commanders tended to assume greater control, accepting computer recommendations only 36% of the time. The consensus of subject opinion was that the computer was most valuable as a backup to the commander. The tendency to override the computer suggestion was also noted in further studies involving the same environment (Hanes and Gebhard, 1964, 1965).

SUMMARY OF CONCLUSIONS

1. Humans do not use probability information efficiently; sometimes they ignore it. They estimate the probability of individual data items adequately but perform less acceptably at aggregating (combining) probabilities. Typically DM shows great conservatism and some stereotypy in his probabilistic estimates. The strategies he applies to problems are often nonoptimal compared with objective probability and payoff values.

2. Nevertheless, there is a very close relationship between human and Bayesian (i.e., optimal) predictions, although Bayesian predictions are almost always superior.

3. Increases in task difficulty tend to produce greater variability in the number of inputs DM takes before a decision. The number of incorrect decisions and the number of informational inputs he draws increase with task difficulty. DM is more willing to gamble on difficult (very uncertain) decisions than on easy (very certain) ones. Informational uncertainty exerts a greater influence than other factors on his *DM* behavior.

4. Experience and training improve *DM* performance, training being the more effective variable. Subjective probabilities of experienced DMs are usually higher than those of naïve subjects; they require less information and are more sensitive to diagnostic cues.

5. Feedback is useful in *DM,* and P_s feedback is more useful than outcome feedback.

6. More information is required for DM to change his decision than to make it originally. More data inputs are taken when the number of alternatives is larger.

7. The cost of securing information influences *DM* behavior; DM takes more information when it costs him less.

8. DM is sensitive to the characteristics of the inputs presented (e.g., input variability). However these characteristics must be relatively gross if he is to make use of them. For example, he can utilize input patterning as a basis of *DM,* but only when the patterns are quite simple.

9. The rate at which information is presented is more important than whether it is presented automatically or on request. As information update frequency is increased, DM seeks more information. His sensitivity to information transmission rates is quite limited, however. When the opportunity to modify decisions is available, performance on the initial decision is degraded. DM is sensitive to the reliability (consistency) of his source information; less information is sought as conflicting information is increased. Information that is inconsistently relevant and irrelevant is more deleterious than information which is consistently irrelevant.

10. Information taking decreases with increasing input history. A major change in environmental conditions is perceived less readily when DM has been developing a concept over a long time.

11. Individual differences among DMs are consistently very large.

12. There appears to be a level of information availability at which *DM* performance is maximized.

13. Task characteristics influence *DM* performance more than the structure of the group in which he performs, although division of responsibility has some effect.

14. Humans estimate the probability of individual items fairly well, but automation of the aggregation process (combination of probabilities) can improve performance by 10 to 15%. The advantage of automating data aggregation increases as fidelity of input data declines and stress and data interdependence increase.

15. DM is often unwilling to accept computerized decisions and insists on an active role in the *DM* process.

DEVELOPMENTAL IMPLICATIONS

The data suggest that human *DM* is less than optimal; yet the performance of personnel is rational and approaches, if it does not match, computerized *DM*.

In deciding whether *DM* should be automated, the following principles should be relied on.

1. Wherever possible, individual probability estimates of input events should be made by the operator. However, processing of these estimates, particularly their combination, should be computer aided. Computer aiding will roughly improve *DM* efficiency by 10 to 15%. Whether this increment is worth the cost is a decision the system developer must make relative to the specific system.

2. Computer aiding is of greatest value when the data the system must operate on are contaminated, incomplete, nonindependent or otherwise faulty, and/or when the operator will be under time or load stress.

3. Design efforts to aid *DM* should attempt to fulfill the following functions: *(a)* pattern inputs and reduce input variability, *(b)* ensure that all input data are correct, *(c)* reduce the number of input characteristics that must be considered, *(d)* specify decision alternatives as precisely as possible, *(e)* permit only one decision and indicate precisely when that decision is required, *(f)* reduce any conflicts in information between different information sources, *(g)* quantize wherever possible the data given to the operator, *(h)* indicate quantitatively if possible the relative value of data inputs and the success probability of preferred responses, *(i)* indicate when a decision is required, *(j)* where possible, provide some feedback about the correctness of the decisions made (although in sequential, dynamic *DM*, the frequency of feedback need not be very great), *(k)* in complex situations, display all response alternatives available to DM.

4. Systematic training in the application of decision rules required by the system should be provided. The training should emphasize the ultimate solution to be achieved as well as *DM* stages, the operator's awareness of his biases, and the development of a *DM* strategy. Training for tactical systems should emphasize *DM* in a defensive posture.

5. In the selection of operators, efforts should be made to choose those having greater aptitude for the type of *DM* required.

GENERAL REFERENCES

Alker, H. A. and Hermann, M. G. Are Bayesian decisions artificially intelligent? The effect of task and personality on conservatism in processing information. *J. Pers. Soc. Psychol.,* 1971, **19,** 31–41.

Anderson, N. and Whalen, R. Likelihood judgments and sequential effects in a two-choice probability learning situation. *J. Exp. Psychol.,* 1960, **60,** 111–120.

Beach, L. et al. Probability learning: Response proportions and verbal estimates. *J. Exp. Psychol.,* 1970, **86,** 165–170.

Becker, G. M. Sequential decision making: Wald's model and estimates of parameters. *J. Exp. Psychol.,* 1958, **55,** 628–636.

Becker, G. M. Decision making: Objective measures of subjective probability and utility. *Psychol. Rev.,* 1962, **69,** 136–148.

Becker, G. M. and McClintock, C. G. Value: Behavioral decision theory. *Ann. Rev. Psychol.,* 1967, **18,** 239–286.

Block, J. and Petersen, P. Some personality correlates of confidence, caution and speed in a decision situation. *J. Abnorm. Soc. Psychol.,* 1955, **51,** 34–41.

Brehmer, B. The effect of cue intercorrelation on interpersonal learning of probabilistic inference tasks. *Organ. Behav. Hum. Performance,* 1974, **12,** 397–412.

Brightman, H. J. and Urban, T. F. The influence of the dogmatic personality upon information-processing: A comparison with a Bayesian information processor. *Organ. Behav. Hum. Performance,* 1974, **11,** 266–276.

Cohen, J. and Chesnick, E. J. The doctrine of psychological chances. *Brit. J. Psychol.,* 1970, **61,** 323–334.

Connolly; D. W. et al. *Tactical decision making: II. The effects of threatening weapon performance and uncertainty of information displayed to the decision maker on threat evaluation and action selection.* Report ESD-TR-61-45 and AFCRL 1100, Operational Applications Laboratory, Bedford, Mass., December 1961. (AD 288–878.)

Corbin, R. M. et al. Context effects in optional stopping decisions. *Organ. Behav. Hum. Performance,* 1975, **14,** 207–216.

Davis, K. B. et al., *Adaptive computer aiding in dynamic decision processes: an experimental study of aiding effectiveness.* Report PTR-1016-75-5, Contract No. 00014-73-C-0286, Perceptronics, Woodland Hills, Calif., May 1, 1975.

Dudycha, L. and Naylor, J. C. Characteristics of the human inference process in complex choice behavior situations. *Organ. Behav. Hum. Performance,* 1966, **1,** 110–128.

Edmonds, E. M. and Mueller, M. R. Concept formation and utilization in the presence of irrelevant visual stimulation. *Psychonomic Sci.,* 1970, **18,** 109–110.

Edwards, W. Probability preferences in gambling. *Am. J. Psychol.,* 1953, **66,** 349–364.

Edwards, W. The theory of decision making. *Psychol. Bull.,* 1954, **51,** 380–417.

Edwards, W. Behavioral decision theory. *Ann. Rev. Psychol.,* 1961, **12,** 473–498.

Edwards, W. Dynamic decision theory and probabilistic information processing. *Hum. Factors,* 1962, **4,** 59–73.

Edwards, W. Subjective probabilities inferred from decisions. *Psychol. Rev.,* 1962, **69,** 109–135.

Einhorn, H. J. Use of nonlinear, noncompensatory models as a function of task and amount of information. *Organ. Behav. Hum. Performance,* 1971, **6,** 1–27.

Ekehammar, B. and Magnusson, D. Decision time as a function of subjective confidence and amount of information. *Percept. Mot. Skills,* 1973, **36,** 329–330.

Emanuel, J. T., *An investigation of the effects of four variables influencing performance in a sequential decision task.* Master's thesis, Ohio State University, Columbus, 1965.

Evans, G. W. Risk-taking set and target detection performance. *J. Appl. Psychol.,* 1965, **49,** 243–244.

Fitts, P. M. et al. (Eds.). *Human engineering for an effective air transportation and traffic control system.* Washington, D. C.: National Research Council, 1951. (AD 133-954.)

Fleming, R. A. The processing of conflicting information in a simulated tactical decision-making task. *Hum. Factors,* 1970, **12,** 375–385.

Fried, L. S. and Peterson, C. R. Information seeking: Optional versus fixed stopping. *J. Exp. Psychol.,* 1969, **80,** 525–529.

Gagliardi, V. O. et al. *Man-computer interactions in idealized tactical problem solving.* Dunlap & Associates, Inc., Darien, Conn., May 1965. (AD 618–707.)

Gebhard, J. W. and Hanes, R. M. *Problems of commander interaction with stored logic machines in Navy AAW operations.* TG-528/BSF-123, Applied Physics Laboratories, John Hopkins University, Silver Spring, Md., November 1, 1963. (AD 346-210.)

Gettys, C. F. et al. Human evaluation of the diagnosticity of potential experiments. *J. Exp. Psychol.,* 1970, **83,** 25–31.

Gibson, R. S. and Nicol, E. H. *The modifiability of decisions made in a changing environment.* Report ESD-TDR-64-657. Decision Sciences Laboratory, Hanscom Field, Bedford, Mass., 1964. (AD 610-122.)

Goldberg, L. R. Simple models or simple processes? *Am. Psychol.,* 1968, **23,** 483–496.

Goldstein, I. L. and Allen, J. C. Effects of irrelevant stimuli on the processing of information in complex displays. *J. Appl. Psychol.,* 1971, **55,** 110–113.

Goldstein, I. L. et al., Feedback in a complex multiman-machine system. *J. Appl. Psychol.,* 1967, **51,** 346–356.

Hammond, K. R. and Summers, D. A. Cognitive dependence on linear and non-linear cues. *Psychol. Rev.,* 1965, **72,** 215–224.

Hanes, R. M. and Gebhard, J. W. *Acceptance of standard operating procedures in tactical decisions.* Report TG-455, Applied Physics Laboratory, Johns Hopkins University, November 30, 1962. (AD 343-344.)

Hanes, R. M. and Gebhard, J. W. *Studies of tactical command decision. IV. Commander selection and override of specific and general automated procedures.* Report TG-450-4, Applied Physics Laboratory, Johns Hopkins University, Silver Spring, Md., March 1964. (AD 376-434L.)

Hanes, R. M. and Gebhard, J. W. *Studies of tactical command decision. V. Commander selection and override of automated procedures: effect of program switching under different levels of uncertainty.* Report TG-450-5, Applied Physics Laboratory, Johns Hopkins University, Silver Spring, Md., September 1965. (AD 373-766.) (For a short summary of the Hanes-Gebhard studies, the reader may refer to: Hanes, R. M. and Gebhard, J. W. The computer's role in command decision. *U. S. Naval Institute Proceedings,* 1966, **92,** 61–68.)

Hanes, R. M. et al. *Studies of tactical command decision. I. Effects of attack tactics on the conduct of a high-performance defense in an ECM environment.* Report TG-450-1, Applied Physics Laboratory, Johns Hopkins University, Silver Spring, Md., October 30, 1962.

Hayes, J. R. *Human data processing limits in decision making.* Technical report ESD-TDR-62-48, Operations Applications Laboratory, Hanscom Field, Bedford, Mass., July 1962. (AD 283–384.)

Hershman, R. L. and Levine, J. R. Deviation from optimum information-purchase strategies in human decision making. *Organ. Behav. Hum. Performance,* 1970, **5,** 313–329.

Hodge, M. H. and Reid, L. S. The influence of similarity between relevant and irrelevant information upon a complex identification task. *Percept. & Psychophys.,* 1971, **10,** 193–196.

Hoge, R. D. Confidence in decision as an index of perceived accuracy of information processing. *Psychonomic Sci.,* 1970, **18,** 351–353.

Hoge, R. D. and Lanzetta, J. T. Effects of response uncertainty and degree of knowledge on subjective uncertainty. *Psychol. Rep.,* 1968, **22,** 1081–1090.

Howell, W. C. Task characteristics in sequential decision behavior. *J. Exp. Psychol.,* 1966, **71,** 124–131.

Howell, W. C. *Some principles for the design of decision systems: A review of six years of research on a command-control system simulation.* Report AMRL-TR-67-136, Aerospace Medical Research Laboratories, Wright-Patterson AFB, Ohio, September 1967. (AD 665-469.)

Howell, W. C. et al. *An evaluation of two variables contributing to the difficulty of a sequential decision task.* Report AMRL-TDR-63-58, Aerospace Medical Research Laboratories, Wright-Patterson AFB, Ohio, June 1963. (AD 411–187.)

Howell, W. C. and Gettys, C. F. *Some principles for design of decision systems: A review of the final phase of research on a command-control system simulation.* Report AMRL-TR-68-158, Aerospace Medical Research Laboratories, Wright-Patterson AFB, Ohio, November 1968. (AD 684-548.)

Irwin, F. W. and Smith, W. A. S. Further tests of theories of decision in an "expanded judgment" situation. *J. Exp. Psychol.,* 1956, **52,** 345–348.

Irwin, F. W. and Smith, W. A. S. Value, cost and information as determiners of decision. *J. Exp. Psychol.,* 1957, **54,** 229–232.

Irwin, F. W. et al. Test of two theories of decision in an "expanded judgment" situation. *J. Exp. Psychol.,* 1956, **51,** 261–268.

Kinkade, R. G. and Ranc, M. P. The effect of conflicting instructions and feedback specificity on tactical decision performance. *Hum. Factors,* 1967, **9,** 257–262.

Kinkade, R. G. et al. *A study of tactical decision making behavior.* Report ESD-TR-66-61. Decision Sciences Laboratory, Hanscom Field, Bedford, Mass., November 1965. (AD 478-769.)

Knowles, W. B. Operator loading tasks. *Hum. Factors,* 1963, **5,** 155–161.

Landis, D. et al. Some determinants of certitude judgments in a complex decision-making task. *Psychol. Rep.,* 1969, **24,** 447–460.

Landis, D. et al. The effect of three types of visual irrelevancy on complex decision making. *J. Psychol.,* 1970, **74,** 29–42.

Lanzetta, J. T. and Kanareff, V. T. Information cost, amount of payoff and level of aspiration as determinants of information seeking in decision making. *Behav. Sci.,* 1962, **7,** 459–473.

Lee, W. *Decision Theory and Human Behavior.* New York: Wiley, 1971.

Levine, J. M. et al. Information seeking with limitations on available information and resources. *Hum. Factors,* 1975, **17,** 502–513.

Long, B. H. and Zeller, R. C. Dogmatism and predecisional information search. *J. Appl. Psychol.,* 1965, **49,** 376–378.

Martin, D. W. *Data conflict in a multinomial decision task.* Master's thesis, Ohio State University, Columbus, 1966.

Martin, D. W. and Gettys, C. F. Feedback and response mode in performing a Bayesian decision task. *J. Appl. Psychol.,* 1969, **53,** 413–418.

Meeker, R. J. et al. *A review of the group decision-making research projects.* Report TM-731, System Development Corp., Santa Monica, Calif., June 25, 1962. (AD 281-907.)

Mertz, W. H. and Doherty, M. E. The influence of task characteristics on strategies of cue combination. *Organ. Behav. Hum. Performance,* 1974, **12,** 196–216.

Modrick, J. A., et al. Renew of approaches to cognitive style and implications for human information processing in command and decision situations. *Proceedings,* Human Factors Society 19th Annual Meeting, Dallas, Texas, October 14–16, 1975, pp. 333–335.

Muchinsky; P. M. and Dudycha, A. L. Human inference behavior in abstract and meaningful environments. *Organ. Behav. Hum. Performance,* 1975, **13,** 377–391.

Peterson, C. R. and DuCharme, W. M. A primacy effect in subjective probability revision. *J. Exp. Psychol.,* 1967, **73,** 61–65.

Peterson, C. R. and Miller, A. J. Sensitivity of subjective probability revision. *J. Exp. Psychol.,* 1965, **70,** 117–121.

Peterson, C. R. et al. Sample size and the revision of subjective probabilities. *J. Exp. Psychol.,* 1965, **69,** 522–527.

Phillips, L. D. and Edwards, W. Conservatism in a simple probability inference task. *J. Exp. Psychol.,* 1966, **72,** 346–354.

Phillips, L. D. et al. Conservatism in complex probabilistic inference. *IEEE Trans. Hum. Factors Electron.,* 1966, **HFE-7,** 7–18.

Pitz, G. F. and Geller, E. S. Revision of opinion and decision times in an information-seeking task. *J. Exp. Psychol.,* 1970, **83,** 400–405.

Pitz, G. F. and Reinhold, H. Payoff effects in sequential decision making. *J. Exp. Psychol.,* 1968, **77,** 249–257.

Pitz, G. F. et al. Strategies of information seeking in deferred decision making. *Organ. Behav. Hum. Performance,* 1969, **4,** 1–19.

Pruitt, D. G. Informational requirements in making decisions. *Am. J. Psychol.,* 1961, **74,** 433–439.

Pruitt, D. G. Pattern and level of risk in gambling decisions. *Psychol. Rev.,* 1962, **69,** 187–201.

Rapoport, A. and Wallsten, T. S. Individual decision behavior. *Ann. Rev. Psychol.,* 1972, **23,** 131–176.

Rigney, J. W. and DeBow, C. H. Multidimensional scaling analysis of decision strategies in threat evaluation. *J. Appl. Psychol.,* 1967, **51,** 305–310.

Rokeach, M. *The Open and Closed Mind.* New York: Basic Books, 1960.

Ronen, J. Effects of some probability displays on choices. *Organ. Behav. Hum. Performance,* 1973, **9,** 1–15.

Rotter, J. B. Can the clinician learn from experience? *J. Consult. Psychol.,* 1967, **31,** 12–15.

Sanders, A. F. and Ter Linden, W. Decision making during paced arrival of probabilistic information. *Acta Psychol.,* 1967, **27,** 170–177.

Schrenk, L. P. Objective difficulty and input history as factors in sequential decision making. *Hum. Factors,* 1964, **6,** 49–55.

Schrenk, L. P. Aiding the decision maker—A decision process model. *Ergonomics,* 1969, **12,** 543–558.

Schroder, H. M. *Factors underlying performance in a complex decision making task. Performance in an anti-war warfare context.* Princeton University, Princeton, N.J., 1965. Described in W. S. Vaughan and A. S. Mavor, Behavioral characteristics of men in the performance of some decision-making task components. *Ergonomics,* 1972, **15,** 267–277.

Schroder, H. M. et al. *Human Information Processing.* New York: Holt, Rinehart and Winston, 1967.

Schum, D. A. Prior uncertainty and amount of diagnostic evidence as variables in a probabilistic inference task. *Organ. Behav. Hum. Performance,* 1966, **1,** 31–54. (a)

Schum, D. A. Inferences on the basis of conditionally nonindependent data, *J. Exp. Psychol.,* 1966, **72,** 401–409. (b)

Schum, D. A. *A further evaluation of computer-assisted processing of complex evidence sets in a simulated military threat-diagnosis task.* Aerospace Medical Research Laboratories, Wright-Patterson AFB, Ohio, 1967.

Schum, D. A. and DuCharme, W. M. Comments on the relationship between the impact and the reliability of evidence. *Organ. Behav. Hum. Performance,* 1971, **6,** 111–131.

Schum, D. A., Goldstein, I. L. and Southard, J. F. *The influence of experience and input information fidelity upon posterior probabilities estimation in a simulated threat-diagnosis system,* Report AMRL-TR-65-25, Aerospace Medical Research Laboratories, Wright-Patterson AFB, Ohio, April 1965. (AD 615-758) (a.)

Schum, D. A. Goldstein, I. L. and Southard, J. F. *Further investigation of the effects of reduced input data fidelity upon the determination of posterior probabilities in a simulated threat-diagnosis system,* Report AMRL-TR-65-233, Aerospace Medical Research Laboratories, Wright-Patterson AFB, Ohio, December 1965. (AD 631-781) (b.)

Schum, D. A. and Martin, D. W. Human processing of inconclusive evidence from multinomial probability distributions. *Organ. Behav. Hum. Performance,* 1968, **3,** 353–365.

Schum, D. A. et al. Research on a simulated Bayesian information-processing system. *IEEE Trans. Hum. Factors Electron,* 1966, **HFE-7,** 37–48. (Also see AD 633–904.)

Schum, D. A. et al. Subjective probability revisions under several cost-payoff arrangements. *Organ. Behav. Hum. Performance,* 1967, **2,** 84–104.

Schumacher, A. W. et al. *Operational training research on evaluator decision making in anti-air warfare.* Human Sciences Research, Inc., McLean, Va., 1965 (confidential). Cited in W. S. Vaughan, and A. S. Mavor, Behavioral characteristics of men in the performance of some decision-making task components. *Ergonomics,* 1972, **15,** 267–277.

Scodel, A. et al. Some personality correlates of decision making under conditions of risk. *Behav. Sci.,* 1959, **4,** 19–28.

Sidorsky, R. C. and Houseman, J. F. *Research on generalized skills related to tactical decision making.* Report NAVTRADEVCEN 1329-2, U.S. Naval Training Device Center, Port Washington, N.Y., December 1966. (AD 813–382.)

Sidorsky, R. C. and Simoneau, G. R. *Decision-making study, final report. An Experimental evaluation of TACTRAIN: An approach to computer-aided tactical decision-making training.* Report NAVTRADEVCEN 1329-4, Naval Training Device Center, Orlando, Fla., June 1970. (AD 875–911.)

Sidorsky, R. C. et al. *Behavioral and operational aspects of tactical decision making in AAW and*

ASW. Report NAVTRADEVCEN 1329-1, Naval Training Device Center, Port Washington, N.Y., August 1964. (AD 607–888.)

Sieber, J. E. and Lanzetta, J. T. Conflict and conceptual structure as determinants of decision making behavior. *J. Pers.*, 1964, **32,** 622–641.

Sinaiko, H. W. and Cartwright, G. P. *Careful: A pilot study of the effects of heavy target load on human and automatic decision makers.* Report R-115, Coordinated Science Laboratory, University of Illinois, Urbana, September 1959. (AD 228-109.)

Snapper, K. J. and Fryback, D. G. Inference based on unreliable reports *J. Exp. Psychol.*, 1971, **87,** 401–404.

Snapper, K. J. and Peterson, C. J. Information seeking and data diagnosticity. *J. Exp. Psychol.*, 1971, **87,** 429–433.

Southard, J. F., Schum, D. A., and Briggs, G. E. *An application of Bayes' theorem as a hypothesis-selection aid in a complex information-processing system,* Report AMRL-TDR-64-51, Aerospace Medical Research Laboratories, Wright-Patterson AFB, Ohio, August 1964a. (AD 607-256)

Southard, J. F., Schum, D. A., and Briggs, G. E. *Subject control over a Bayesian hypothesis-selection aid in a complex information-processing system,* Report AMRL-TR-64-95, Aerospace Medical Research Laboratories, Wright-Patterson AFB, Ohio, September 1964b. (AD 608-108)

Stager, P. and Muter, P. Instructions and information processing in a complex task. *J. Exp. Psychol.*, 1971, **87,** 291–294.

Streufert, S. C. Effects of information relevance on decision making in complex environments. *Mem. Cognition,* 1973, **1,** 224–228.

Streufert, S. et al., Information load, group organization and communication frequency. *Psychonomic Sci.*, 1972, **27,** 348–350.

Strub, M. H. *Experience and prior probability in a complex decision task.* Master's thesis, Ohio State University, Columbus, 1966.

Strub, M. and Levit, R. A. Computer compatability with decision style: eddies in a bit stream. *Proceedings,* Human Factors Society Annual Meeting, pp. 46–49, 1974.

Sung, Y. H. Effects of attitude similarity and favorableness of information on Bayesian decision making in a realistic task. *J. Appl. Psychol.*, 1975, **60,** 616–620.

Taylor, R. N. and Dunnette, M. D. Influence of dogmatism, risk-taking propensity, and intelligence on decision-making strategies for a sample of industrial managers. *J. Appl. Psychol.*, 1974, **59,** 420–423.

Vaughan, W. S. and Gartner, W. B. *Human information processing as a source of display requirements in strategic command control systems. Development and application of a method.* Human Sciences Research, Inc., McLean, Va., 1965. Confidential. (Reported in W. S. Vaughan and A. S. Mavor, *Ergonomics,* 1972, **15,** 267–277).

Vaughan, W. S. et al. *Information processing tasks in tactical action selection: Performance of experienced submarine officers in weighting multiple criteria for depth selection.* Human Sciences Research, Inc., McLean, Va., 1964. (AD 433-750.)

Vaughan, W. S. et al. *Study of functional requirements of training equipment for Army command tactical decision making.* Report NAVTRADEVCEN 1341-1, U.S. Naval Training Device Center, Port Washington, N.Y., February 1966. (AD 630–312.)

Vaughan, W. S. et al. *Information requirements for six command decisions: The advanced submarine in an ASW mission.* Human Sciences Research, Inc., Arlington, Va., 1960. (AD 319 838) Secret. (Described in W. S. Vaughan, and A. S. Mavor, *Ergonomics,* 1972, **15,** 267–277.

Well, A. D. The influence of irrelevant information on speeded classification tasks. *Percept. Psychophys.*, 1971, **10,** 79–84.

Yntema, D. B. and Torgerson, W. S. Man-computer cooperation in decisions requiring common sense. *IRE Trans. Hum. Factors Electron.*, 1961, **HFE-2,** No. 1, 20–26.

CASE REFERENCES

Case 80. Howell, W. C. et al. *An evaluation of two variables contributing to the difficulty of a sequential decision task.* Report AMRL-TDR-63-58, Aerospace Medical Research Laboratories, Wright-Patterson AFB, Ohio, June 1963. (AD 411–187.)

Case 81. Strub, M. H. *Perception of military event patterns in a two-alternative prediction task.* Technical Research note 221, Behavior and System Research Laboratory, Arlington, Va., February 1970. (AD 707-379.) (See also M. H. Strub, Pattern type, strength, and continuity as variables affecting the perception of second-order patterns in a binary prediction task. *Proceedings,* American Psychological Association, 1969, pp. 33–34.)

Case 82. Tolcott, M. A. et al. *Human decision making as related to air surveillance systems.* Report 300-3, Dunlap and Associates, Stamford, Conn., December 1960 (reported in Case 103.)

Case 83. Hayes, J. R. *Human data processing limits in decision making.* Technical report ESD-TDR-62-48, Operations Applications Laboratory, Hanscom Field, Bedford, Mass., July 1962. (AD 283–384.)

Case 84. Hayes, J. R. and Smith, E. *Decision making studies. Part I. The tradeoff of variables in decision making.* NRL Report 5418, Naval Research Laboratory, Washington, D.C., January 1960 (reported in Case 103.)

Case 85. Messick, D. M. *Sequential information seeking: Effects of the number of terminal acts and prior information.* Report ESD-TDR-64-606, Decision Sciences Laboratory, Hanscom Field, Bedford, Mass., October 1964. (AD 614-708.)

Case 86. Levine, J. M. et al. Information seeking with input pacing and multiple decision opportunities. *Hum. Factors,* 1974, **16,** 384–394.

Case 87. Levine, J. M. Information seeking with conflicting and irrelevant inputs. *J. Appl. Psychol.,* 1973, **57,** 74–80.

Case 88. Levine, J. M. and Samet, M. G. Information seeking with multiple sources of conflicting and unreliable information. *Hum. Factors,* 1973, **15,** 407–419.

Case 89. Tate, J. D. and Howell, W. C. *Term expectation and uncertainty in human decision behavior.* Report AMRL-TDR-63-118, Beh. Sciences Laboratory, Aerospace Medical Research Laboratories, Wright-Patterson AFB, Ohio, November 1963. (AD 431-634.)

Case 90. Herman, L. M. et al. Operator decision performance using probabilistic displays of object location *IEEE Trans Hum. Factors Electron.,* 1964, **HFE-5,** (1). 13–19.

Case 91. Naylor, J. C. and Schenck, E. A. The influence of cue redundancy upon the human inference process for tasks of varying degrees of predictability. *Organ. Behav. Hum. Performance,* 1968, **3,** 47–61.

Case 92. Fleming, R. A. The processing of conflicting information in a simulated tactical decision-making task. *Hum. Factors,* 1970, **12,** 375–385.

Case 93. Beach, L. R. Studies in the psychology of decisions. Final report, ONR contract N00014-67-A-0103-0011, University of Washington, Seattle, December 1972.

Case 94. Sidorsky, R. C. Predicting the decision behavior of a knowledgeable opponent. *Hum. Factors,* 1967, **9,** 541–554.

Case 95. Vaughan, W. S. et al. *Study of functional requirements of training equipment for Army command tactical decision making.* Report NAVTRADEVCEN 1341-1, U.S. Naval Training Center, Port Washington, N.Y., February 1966. (AD 630–312.)

Case 96. Shure, G. H. et al. *Group decision making under conditions of realistic complexity.* Technical report SP-344/000/01, System Development Corp., Santa Monica, Calif., September 1961 (reported in case 103).

Case 97. Vaughan, W. S. et al. *Information processing tasks in tactical action selection: Performance of experienced submarine officers in weighting multiple criteria for depth selection.* Human Sciences Research, Inc., McLean, Va., 1964. (AD 433-750).

Case 98. Kinkade, R. G. and Ranc, M. P. The effect of conflicting instructions and feedback specificity on tactical decision performance. *Hum. Factors,* 1967, **9,** 257–262.

Case 99. Schrenk, L. P. Objective difficulty and input history as factors in sequential decision making. *Hum. Factors,* 1964, **6,** 49–55.

Case 100. McKendry, J. M. *Utility of information as a predictor of decision adequacy in ambiguous choice situations.* Report 567-R-3, HRB-Singer, Inc., Science Park, State College, Pa., July 1965. (AD 466-844.)

Case 101. Kanarick, A. F. et al. Multi-source information acquisition with optional stopping. *Hum. Factors,* 1969, **11,** 379–385.

Case 102. Hammer, C. H. *Timeliness and accuracy in a sequential decision making task.* Technical research note 162, U.S. Army Personnel Research Office, Washington, D.C. October 1965. (AD 625–223.)

Case 103. Sidorsky, R. C. and Houseman, J. F. *Research on generalized skills related to tactical decision making.* Report NAVTRADEVCEN 1329-2, U.S. Naval Training Device Center, Port Washington, N.Y., December 1966. (AD 813–382.)

Case 104. Gibson, R. S. and Nichol, E. H. *The modifiability of decisions made in a changing environment.* Report ESD-TR-64-657, Decision Sciences Laboratory, Hanscom Field, Bedford, Mass., December 1964. (AD 610-122.)

Case 105. Kinkade, R. G. The effect of situational factors on tactical decision making behavior. In R. G. Kinkade et al., *A study of tactical decision making behavior.* Report ESD-TR-66-61, Decision Sciences Laboratory, Hanscom Field, Bedford, Mass., November 1965. (AD 478-769.)

Case 106. Strub, M. H. Experience and prior probability in a complex decision task. *J. Appl. Psychol.,* 1969, **53,** 112–117.

Case 107. Singer, E. and Roby, T. B. Dimensions of decision-making behavior. *Percept. Mot. Skills,* 1967, **24,** 571–595. (Monog. Suppl. 3-V20.)

Case 108. Townsend, J. C. and Smith, W. J. *Predicting decision making behavior from personality and cognitive variables.* Report ESD-TDR-64-619. Decision Sciences Laboratory, Bedford, Mass, November 1964. (AD 610-521.)

Case 109. Streufert, S. et al. *Performance and perceptual complexity in a tactical decision making task.* Princeton, N.J., Princeton University, Report 1, Contract Nonr 1858-(42) June 1964. (AD 603–177.)

Case 110. Fox, D. J. and Lorge, I. The relative quality of decisions written by individuals and by groups as the available time for problem solving is increased. *J. Soc. Psychol.,* 1962, **57,** 227–242.

Case 111. Pask, G. et al. *Group decision making and communication patterns under conditions of stress and overload, when the participants are permitted to function as a self-organizing system.* Contract DA-91-591-EUC-3216, System Research Ltd., Surrey, England, February 1965. (AD 665-469.)

Case 112. Naylor, J. C. and Dickinson, T. L. Task structure, work structure and team performance. *J. Appl. Psychol.*, 1969, **53**, 167–177.

Case 113. Howell, W. C. *Some principles for the design of decision systems: A review of six years of research on a command control system simulation.* Report AMRL-TR-67-136, Aerospace Medical Research Laboratories, Wright-Patterson AFB, Ohio, September 1967. (AD 665–469.)

Case 114. Shaffer, L. H. Problem solving on a stochastic process. *Ergonomics,* 1965, **8,** 181–192.

V
TEAM FUNCTIONS

SUMMARY. Since the team is the most obvious feature of the multiman system, one cannot understand that system without considering the team. It follows then that one must know how individual operators are integrated into the team. These processes determine what team training will consist of and how it will be conducted. Concepts about team processes and team training have been fuzzy, focusing on crew communications. A good deal of research effort has been expended on determining whether team training results in performance improvement, and the results have been disappointing. The literature describing structural team variables (e.g., crew size, organization, and composition) is also reviewed.

INTRODUCTION

Any MMS containing two or more operators working in interaction involves a crew or *team* (for simplicity, only the latter term is used). The team is the essential element of the multiman system. In many respects, however, the team performs like an operator: it works to a mission requirement, performs tasks, receives feedback, holds goals in common, adjusts its behavior to changing demands, and so on. What then is distinctive about team behavior? How does team behavior differ from that of individual operators? These are questions that have preoccupied many researchers.

The interest researchers have in team processes is based on the assumption that those processes are characterized by something that is (1) essential to the accomplishment of the system mission and (2) not to be found in individual operator performance. The system developer's interest is in designing the team (its composition and tasks) and in training the team to work efficiently.

What the distinctive team element is remains quite elusive. Superficially the added element appears to be the interaction among team members, commonly referred to as "teamwork." Researchers who have attempted to isolate this element usually fall back on "communication," because any mechanism by which two or more operators work together must involve some form of communication. We have seen, however (Chapter 3), that unless the system *requires* communication, this activity is usually detrimental to system performance. Moreover, one operator may convey information to another (e.g., through displays) without that communication involving teamwork.

Definitions

The team must be differentiated from the group, which has been extensively studied in social psychological research. The distinction is a matter of degree, of course, but in general the critical difference is that team behavior is directed by external goals (e.g., procedures, instructions, mission requirements) imposed by a system developer, whereas group behavior is much more self-directed (although completely unstructured group behavior is rarely found).

The distinctions drawn by Klaus and Glaser (1968, pp. 1–2) are pertinent:

"A *team* is usually well organized, highly structured, and has relatively formal operating procedures . . . Teams generally

"1. are relatively rigid in structure, organization and communications networks;

"2. have well defined positions or member assignments . . .

"3. depend on the cooperative or coordinated participation of several specialized individuals whose activities contain little overlap . . .

"4. are often involved with equipment or tasks requiring perceptual-motor activities;

"5. can be given specific guidance on job performance . . .

"A *small group* . . . rarely is so formal or has well defined, specialized tasks. . . . Small groups generally

"1. have an indefinite or loose structure, organization and communication networks;

"2. have assumed rather than designated positions . . . each individual's contribution to the accomplishment of the task is largely dependent on his own personal characteristics;

"3. depend mainly on the quality of independent, individual contributions;

"4. are often involved with complex decision-making activities;

"5. cannot be given much specific guidance beforehand, since the quality and quantity of participation by individual members is not known . . ."

Many of the variables influencing the group also influence the team, although one suspects that their effects may be somewhat different because of the differences cited previously. Lorge et al. (1958) also draw a sharp distinction between ad hoc groups and teams, cautioning readers about the generalization of principles from one to the other. There is sufficient similarity between the team and the group, however, to make it worthwhile to include in this chapter relevant examples of group research.

Defining the team is much like relating the conceptual definition of a task to the task as observed (Chapter 3). For example, a team is not necessarily a stable unit. Its composition and the distribution of its personnel may vary during different tasks or at different times. "The team working in the CIC during surface plot is a very different organization from the same team during general quarters. Their behavior and interaction is very different despite the considerable overlap of personnel" (Glanzer and Glaser, 1955).

One must also differentiate between the operator's *immediate* team (which may comprise a relatively small number of members with whom he interacts) and the *extended* team (e.g., a ship's company) in which the immediate team is embedded. Since experimental research is difficult to perform with the extended team, almost all our data relates to the immediate team. The factors impinging on the operator may differ, however, depending on which context is involved.

How does one determine in large multiman systems whether a given operator is a member of a team? By his presence during a team activity? By the degree to which he interacts (e.g., communicates) with other team members? It is possible to have teams in which there is little interaction. If one uses an overly rigid criterion, personnel (e.g., team supervisors) who do not display much overt activity might be eliminated. On the other hand, if one broadens the definition of what constitutes interaction (e.g., an operator from another subsystem who inputs a message to the team), the team so defined may become unacceptably large.

We also encounter the question of what constitutes a measurable *unit* of team activity. "Team activity" might be defined as everything the team does during a specified time, in which case we might be mixing critical with merely incidental functions. Or we might define the term as the performance of discrete functions or tasks only, in which case interactive dimensions might be overlooked.

The point is that the team as an observable entity may be very different from the team as a construct.

Team Variables

Structural variables are inherent in the team itself; *situational* variables describe the environment in which team activity is embedded. Both of course influence the way in which the team performs. As with other system aspects, some of these variables are the same as those influencing individual behavior.

The structural variables to be described in this section involve

1. Team size.
2. Team organization.
3. Team interactions (e.g., communication among team members).
4. Team composition.
5. Team processes.

Team Size

Team size limits the manner in which the team can be organized and how it can interact. "Small group" researchers have asked whether the size of a group has an effect on the group's output, particularly in problem solving. This question implies the capability of adjusting group size to achieve optimal performance. From our standpoint, since team size in MMS is largely fixed by the nature of the tasks to be performed, this question is somewhat irrelevant except for very large systems having provision for redundancy (as in Navy ships) or very indeterminate systems (e.g., management and command staffs) whose tasks are relatively flexible and autonomous and whose size therefore can be adjusted by redefining division of responsibility. The system developer is interested in determining the *minimum* team size required to perform tasks efficiently.

Team Organization

Team organization is largely a matter of dependency relationships between tasks and operators. As indicated previously, tasks may be performed in series, in parallel, or independently. Team organization may be modified by combining or distributing system functions among the same or a different number of operators or by adding redundant personnel to perform a compensatory or fail-safe function.

Organizational effects therefore vary as a function of dependency relationships. When dependency is low, the organization is relatively unimportant; inadequate performance at one position or subsystem will have little effect on other operator positions or subsystems. When dependency is high, inadequate performance in one operator position will seriously affect another (dependent) position and this effect may be amplified by the nature of the organization. For example, the more centralized the system, the more seriously will performance failures in the central position affect other system elements.

Team organization is also related to indeterminacy; the more indeterminate the system, the more flexible its organization must be to accommodate uncertainty and contingent responses.

Because of the intimate relationship between task requirements and team organization, the system developer does not have complete freedom to select the organization he would like to have. The nature of the system/task enforces certain relationships among operators.

Team organization reflects not only the interaction *within* the team, it is also necessary to consider the interrelationship *between* teams and between subsystems. These last, however, are properties of the extended team or system as a whole rather than of the immediate team.

Team organization is important in predicting the performance of the team quantitatively. Given error data in probabilistic terms, it is entirely possible to quantize team performance (see Meister, 1973), but only if the mathematics applied to the performance prediction takes into account dependency relationships among team personnel.

Team Interactions

It is commonly assumed that what makes a team out of a collection of individual operators is their interaction. With the exception of communications, mostly verbal, sometimes transmitted by displays, the interactions researchers talk about (see, e.g., Hackman and Morris, 1974) are extremely nebulous. The HF specialist needs the answers to the following questions:

1. If these interactions involve more than communications, what precisely are they? (If they cannot be objectively specified, nothing can be done with them.)
2. Are these interactions trainable?
3. Assuming that communications are an essential element in team interaction, how do they reflect team performance, and what is the effect of communications on system output?

Some communications are obviously essential, others irrelevant to mission accomplishment. The former can be specified as part of the system operating procedure, the latter is "noise," however important it may be from an individual (e.g., tension-relieving) standpoint. In practice, however, it is difficult to make such a clear-cut distinction except in highly proceduralized systems.

A communications network consists of (1) the individual operator positions and (2) the links (direct verbal or equipment) between operators and subsystems. A communications network can presumably be plotted by simply determining who talks to whom. However, this approach conceals considerable complexity.

The network may be formally established as part of design or it may be developed in part by trial and error and usage (the latter is more characteristic of indeterminate systems). The very considerable amount of research on communications networks (see Shaw, 1964) has demonstrated that network structure does influence performance (at least in laboratory situations), but we still do not know how this occurs and to what extent.

Team Composition

Team composition can vary along several dimensions: (1) attributes such as sex, age, race, aptitude, training, experience, and personality; (2) the distribution within the team of these attributes; and (3) the stability of team composition over time.

The ideal team, being composed of operators each having the highest possible aptitude and training demanded by the job, would be much more homogeneous than it is in real life. Since team members almost always vary (sometimes markedly), we must ask what is the effect of heterogeneity on team performance? Rosenberg et al. (1955) coined a phrase "the assembly effect" to describe the variation in performance attributable to the *patterning* of the just listed dimensions.

The system manager also faces the problem of "turnover," since , however it is composed, the team must be expected to replace one or more members over time. The new operators have not performed with the more experienced ones and moreover may differ from them in relevant characteristics. The problem is particularly acute in military systems, where personnel are available only for a specific enlistment period, especially if certain skills must be learned while the operator is working as a team member.

Also of interest to researchers is the extent to which team output is related to the performances of the individual operators in the team. The more individual performance determines team output, the less important become the special dimensions of "team processes," and the less necessary it becomes to provide special team training.

It also would be interesting to ascertain the relationship between team composition and the task required of the team, as for example Steiner's (1966) conjunctive and disjunctive tasks. Conceivably the importance of team composition could vary depending on the task performed.

Team Processes

The question of how to describe team operations is an important one, because if proper description is impossible, it is difficult to train an effective team. As we pointed out at the start of this chapter, team operation can be described by many of the same general processes that describe the individual (e.g., responding to inputs, performing functions, and producing an output). Such a formulation does not suggest, however, the special characteristics of a team. Glaser et al. (1955) believe that the following process variables are important in describing how the team performs.

1. Sequence or flow (type and consistency) of team behaviors over time.
2. Communication flow between team members.
3. Task duration.
4. Team-specific behaviors. Glaser et al. (1955) describe 14 measurement variables, which the author has condensed as follows:

(a) Communications complexity, defined by the number of communication links between team members and the frequency of communication among them;

(b) Communications centralization, defined by the extent to which certain team members act as central points for receiving and transmitting messages;

(c) Amount of concurrent activity, described by the extent to which various team members perform at about the same time—presumably, the more such concurrent activity, the more control (supervision) is required;

(d) Process differentiation, the extent to which a team operation can be differentiated into different classes of activity (e.g., functions);

(e) Team autonomy, the extent to which a team generates its own inputs;

(f) Intrateam coordination, defined by the amount of cooperation required of team members in performing their tasks.

No one can be very happy with this listing; it lacks something—but what? A satisfactory taxonomy of team dimensions has not yet been developed. Perhaps team members *respond* to inputs primarily as individual operators but *act* upon them as a team.

Attempts to study team processes experimentally, as Glanzer and Glaser (1961) indicated, have taken two avenues. Studies such as those by Kidd and his colleagues (cases 1 and 2), Johnson and Briggs (see Cases 133, 139, 142, 143, and 148) and Lanzetta and Roby (Case 144) have utilized a type of military or civilian work situation (e.g., air traffic control, aerial interception) as a model. On the other hand, studies such as those by Glaser and his colleagues (Cases 130, 131, 138, and 140), Rosenberg and Hall (1958), and in particular research on communications networks (Shaw, 1964), have attempted to abstract the critical elements of the real world situation, model them in analogue form, and apply a theoretical formulation (e.g., reinforcement theory, network theory) to the experiment.

Both approaches have advantages and disadvantages. The actual task situation may contain unknown elements the experimenter may find difficult to control. The analogue task presumably cuts the situation down to its fundamentals, but as a result subjects may perceive it to be somewhat remote from real life.

Situational Variables

The situational variables are the same variables that influence individual operator behavior (e.g., the mission, performance criteria, amount and type of training, feedback). The system developer asks whether these variables affect the team and the operator in the same way. If their effect is different for the team, team training must take the special effects of these variables into account. For example, feedback is related by the individual operator to his own performance. Therefore feedback describing team outputs will not describe the operator's performance adequately. However it seems logical to suppose that if the team is to operate effectively as a team, it must receive team feedback. Training in the use of team feedback may therefore be required.

TEAM SIZE

It seems logical to assume that size will have a significant effect on team performance. At the very least, size is usually correlated with increasing behavior complexity (e.g., additional operator interactions). Researchers have been interested in two main questions:

1. How much more (or less) effective than the individual (operator) is the group (team)?
2. Does the effectiveness of the group (team) increase (or decrease) with increasing size?

The first question is of little interest to the HF researcher, since the system and the task determine whether the system developer will have a team or a single operator for a given job. Rather, the problem for the system designer is to determine the minimum feasible team size. Consequently, in this section much of the extensive (and confusing) literature comparing individual with group performance is ignored.

The second question has greater significance for the HF researcher. It has been pointed out that in large multiman systems, like Navy ships, and in systems having indeterminate characteristics, it is possible to expand or contract the team by combining or distributing functions to maximize output and reduce cost. The industrial or military management of a MMS, in particular, has the capacity to change its size more readily; management staffs correspond in this respect to the experimental ad hoc teams developed by small group researchers to investigate the size variable.

With increasing team size, a social environment may be created for the individual operator by other team members. This may produce potentially undesirable social factors that interfere with the system output. For example, interpersonal relationships not required by the MMS, such as attitudes toward other team members, are developed. Team interactions produce verbal communication, some of it irrelevant to the mission. These social factors may be healthy or not, but because they are irrelevant to the system goal, they may hinder accomplishment of that goal.

Group Versus Individual Comparisons

A considerable amount of research, going back many years, has been performed on this subject (e.g., Faust, 1959; Husband, 1940; Lorge et al., 1958; Marquardt, 1955; Taylor and Faust, 1952). In critical review of these studies, Thomas and Fink (1963) point out that most of the work contains methodological flaws.

Conclusions

The evidence for group versus individual superiority in problem solving (where most of the work has been done) is quite confusing (see reviews by Anderson, 1970; McGrath and Altman, 1966). Much theory has been written [Davis (1973) gives a representative sample], but factual data are scarce.

1. One the one hand, groups solve problems more rapidly and correctly than do individuals. Yet when the input per individual per unit time is considered, the advantage of groups is less evident; as Davis and Restle (1963) comment: "The result of any experiment which compares groups and individuals depends upon the problems to be solved and the social organization of the groups." The superiority of the team versus the individual depends greatly on the type of task and the criteria employed. For example, productivity is fostered by larger groups, speed by smaller ones.

2. The superiority of group performance is manifested primarily in very small groups. Larger teams are not significantly superior to the two-man team. Gibb (1951) found that number of problem solutions increased negatively as group size increased.

Illustrative Studies

Kinkade and Kidd (1959) contrasted individuals and two-man groups in a complex decision-making game derived from radar air traffic control. They found that team performance was superior to that of single individuals but not proportionately so. The latter result was hypothetically attributed to the diversion of time and productive capacity away from the primary task and into integrative and coordinative behaviors. Productivity in the individual subject was significantly superior to that of the individual in the team. Teams encouraged to communicate within the team were slightly but not significantly superior to teams in which intrateam communication was not permitted.

Steiner (1966) argues that the effects of group size are contingent on type of task and more specifically on the way in which member abilities are integrated into the task solution. Frank and Anderson (1971) reported some support for Steiner's hypothesis; increases in group size increased the number of disjunctive problems solved and reduced the number of conjunctive problems completed. However, qualitative measures did not reflect this finding.

Much of the confusion arises because different criteria produce different results. This is illustrated graphically in Case 115.

Case 115. The task situation was the classic "Twenty Questions" game in which the subject must ask a series of questions to identify an object. Individuals were compared with groups of two and four. The learning curves for groups and individuals were the same; however in each case the groups were significantly more effective (Figure 5-1), although there was no appreciable difference between groups of two and four except in number of failures. Group performance was superior to individual performance in terms of elapsed time to solution. But if the analysis is made in terms of number of man-minutes for solution (a cost-effectiveness measure that takes into account the number of team members involved), the results change sharply: with this criterion the mean scores were 5.06 for individuals, 7.40 for two-man groups, and 12.60 for four-man groups.

The superiority of the group over the individual might be assigned to the performance of the best member of the group; hence with a large group, the possibility of selecting a more competent member would be increased. This has been the basis of attempts to develop equations predicting the effect of group size on performance. For example, Lorge and Solomon (1955) suggest that the probability that there will be at least one person in the group arriving at the right answer to a problem is $1\text{-}(1\text{-p})^n$, where p is P_s of the individual and n is the number of individuals in the group. Davis and Restle (1963) have criticized the Lorge-Solomon formulations and have suggested more complex models. See also Davis, 1969.

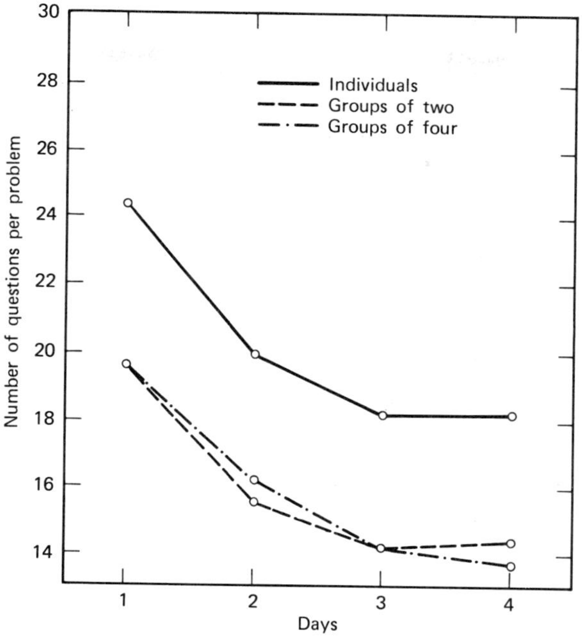

Figure 5-1 Number of questions per problem as functions of days of practice and size of group. With permission, from Taylor and Faust, *J. Exp. Psychol.*, 1952, Case 115.

Subsequently Taylor et al. (1958) developed a method in which they compared the performance of individuals with "nominal" groups formed statistically by randomly combining individual performances. This methodology has tended to refute the superiority of groups over individuals. Osborn (1957) found that the nominal groups (i.e., individual performances) were clearly superior to an equal number of real interacting groups. Similar effects were observed by Kidd (1958), Dunnette et al. (1963), Meadow et al. (1959), and Campbell (1968). Rotter and Portugal (1969) also found that the individual production of ideas was superior to either group production or a combination of group and individual production. Indeed, under certain conditions a group may inhibit the best efforts of individual members. On the other hand, studies by Hall et al. (1963) and Schoner et al. (1974) tend to favor group solutions.

Case 116. This study is characteristic of the results in this field. Five-, seven-, and nine-man brainstorming groups were compared with nominal groups composed of individuals who brainstormed alone. The real groups were superior to the individuals, but the nominal groups were more effective than the real groups. No major increase in performance took place in the real groups as their size increased from five to nine. The reason for this may have been ineffective management of time in the group situation [as suggested by Fox and Lorge (1962) in Case 110].

In Case 2 the effect of adding one or two individuals to an already existent crew performing an air traffic control task did not significantly improve performance.

Conclusions. Team size effects occur not only in cognitive (problem-solving) tasks but also in perceptual ones. In monitoring situations, the addition of an observer improves detection performance if the two observers monitor independently of each other.

Illustrative Studies

A special case of group versus individual comparison arises in the very practical question of whether the addition of personnel will improve monitoring performance. Two observers in a watchkeeping task should jointly miss fewer signals than a single observer. This conclusion follows logically if each observer monitors independently.

Case 117. In this study a small spot of light swept horizontally across a 6-inch screen about once every 3 seconds. The critical signal was a momentary displacement of the blip to a position approximately 1.5 inches above the central axis of the screen.

Eight subjects worked singly over a 30-minute period, and 16 worked as pairs. The proportion of signals detected in the paired condition (.906) was significantly greater than in the single condition (.735), but the estimated probability of an *individual* observer's detecting a signal was essentially identical in each of the two conditions (.693 and .735 for the two- and one-man systems). This suggests that there is no effect (good or bad) on the *individual* performances of the two observers. Bergum and Lehr (1962) reported essentially the same result.

The mean numbers of missed signals in Case 117 were 3.175 (about 27%) in the single subject condition and 1.125 (about 9%) in the paired condition.

Schafer (1949) also found that the combined performance of two observers working together produced from 11 to 20% more detections than either individual working alone, and the combined performance of three watchkeepers produced 6 to 15% more detections than two.

More recently, Waag and Halcomb (1972) discovered in their vigilance situation that increases in team size facilitated detection. Detection increased when team size was increased, regardless of the original size of the team. Thus there was an almost linear increase in detection performance whether one began with a team size of one, two, three, or four. Team organization was a crucial factor: "Maximum detection performance resulted whenever only one member of the . . . team was required to respond correctly. As more members were required to respond (the series rule) detection performance deteriorated. . . ."

Case 118. The system designer must decide not only how many monitors to place in the system but whether they should monitor separately, in a parallel circuit, or together as a team. In this case, there were four groups: one-, two-, and three-man teams, and another three-man team in which the members monitored in isolation but had their responses combined as if they were switches in a parallel circuit. The monitoring task was the detection of an abnormally large deflection (the signal) of a voltmeter needle that made 50 deflections per minute over a period of 48 minutes; 32 signals were presented.

Figure 5-2 indicates that as team size increased from one to two men there was a significant increase in probability of detection, but the increase from two to three men was not statistically significant. The combined

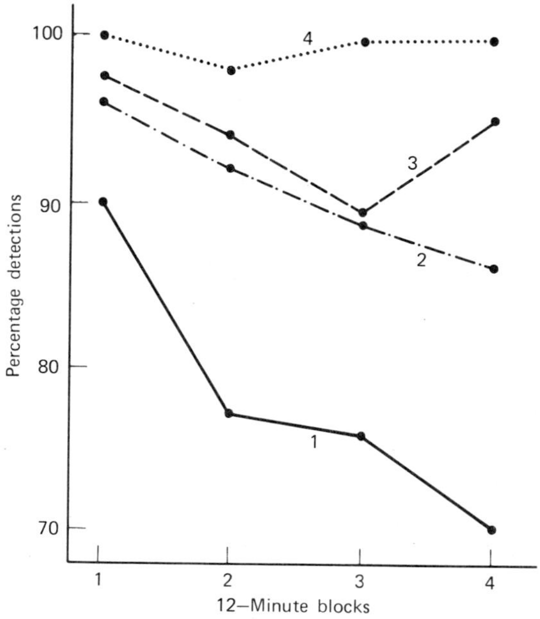

Figure 5-2 Percentage of signals detected as a function of team size. With permission, from Wiener, *Hum. Factors,* 1964, Case 118.

performance of the three monitors in isolation (group 4) was superior to that of the three monitors working together (group 3), indicating that independent monitoring is superior to team efforts. Here, as in other multiman situations to be discussed, team interaction tended to depress individual operator performance.

Conclusions. As with teams in general, monitoring teams adopt various strategies for combining their performance, but these strategies are often nonoptimal. Models of team performance often predict better performance than that actually achieved.

Illustrative Studies

When studies have used real groups to investigate conformity to models of group detection behavior, variations in strategies have been found as a function of task variables. Hornseth and Davis (1967) used visual target-detection tasks with individuals and two-man teams, and in two tasks group performance was no better than individual performance for the better member of the team, supporting a model of independent search by the members of the team. In a more difficult task, however, the same investigators noted a shift toward an apportioning strategy. In this task performance was better than with the independent search model, but not as good as that predicted for completely effective use of an apportioning strategy.

Others have used models that combine individual performance in some arbitrary manner to obtain predicted (pseudogroup) performance and have compared these predictions with actual group performance in the given task. In Case 118 group performance (as measured either by probability of detection or by a detectability index) was better than individual performance, but not as good as an independent events model of monitoring would predict. Clement (1973) verified these results. However, neither an information-integration nor a decision-threshold model adequately represented group performance. Real groups showed characteristics of both independence and dependence. Similar findings have been derived from auditory vigilance tasks (Schafer, 1949).

Case 119. Each stimulus consisted of a square field containing 16 black outline circles in a 4 × 4 array. Fifteen of the circles contained a black outline "upright" equilateral triangle (base down), the sixteenth circle (the target) contained the same triangle (base up). Group performance was better than individual performance for all target locations, displaying some characteristics both of independence and dependence and some apparent use of logical decision rules in pooling individual information. No model of group performance proved adequate. The particular strategies employed influenced overall and differential sensitivity of groups to target locations.

Case 120. A team situation in which individuals simultaneously search for the target permits alternative search strategies. An efficient strategy for a two-man team to adopt would be to assign half the search area to each individual. Alternately, each team member could independently scan the entire search area.

Three tasks were used in this study. For task I the stimulus material consisted of aerial photographs and their corresponding target cue cards, consisting of a cutout of a target to be found in the aerial photograph. The stimulus materials for tasks II and III consisted of a 15 × 15 matrix of Is and Ts. The target for task II was a single I; for task III the target was a 2 × 2 matrix of Is.

The influence of task characteristics was again the most important single factor in performance. Team performance was not significantly faster than individual performance for task I, but it was for tasks II and III. As in other studies, the teams failed to adopt the optimal search strategy.

The Effects of Group Size on Member Behavior

The groups referred to are those found in large establishments like hospitals and factories ("blue collar workers") rather than the smaller teams typical of military systems. Consequently the following materials describe what we have termed the extended team rather than the operator's immediate team. Data from the factory situation can be merely suggestive of effects originating in the immediate team environment, but with reference to team size and output, such reports constitute the only material available. Because of the size of the organizations involved, the data are derived mostly from correlations of size with output, ratings of member satisfaction, and so on.

Conclusions. Organizational size is negatively or curvilinearly related to output, to "working morale," and to quality of work; it is positively related to absence rate, but data are inconclusive with regard to turnover.

Illustrative Studies

Ingham (1970), in a study of eight British firms of varying size, found organization size to be highly related to absenteeism but hardly at all related to turnover. This conclusion is supported by a number of earlier studies.

Of four investigations into the relationship between unit size and turnover (among blue collar workers), three studies determined that turnover was greater in larger units than in smaller ones (Indik and Seashore, 1961; Kerr et al., 1951; Mandell, 1956). The fourth (Argyle et al., 1958) found no such relationship.

Ten of 12 studies dealing with the impact of unit size on absenteeism revealed a positive linear relationship between increased absenteeism and increases in size (Acton Society Trust, 1953; Baumgartel and Sobol, 1959; Hewitt and Parfitt, 1953; Indik and Seashore, 1961; Kerr et al., 1951; Metzner and Mann, 1953; Revans, 1958). These results were found only among blue collar workers, however; in the only study dealing with both blue collars and white collars (Metzner and Mann, 1953) no relationship appeared between unit size and absenteeism among white collars. Argyle et al., (1958) reported a curvilinear relationship between unit size and absenteeism.

Illustrative of the kind of data that emerge is the study by Acton Society Trust (1953), giving a correlation of .447 between factory size and absence rates. Revans (1958) noted that in five randomly chosen gas works ranging in size from 67 to 3,430 employees, the average duration of absences due to accidents and the log size of the organization correlated +.91. He also found that size of plant was correlated .62 with a composite measure of absence due to sickness, injury, and other factors. British data (quoted by Revans, 1958) show that size of general hospitals is related to accident rates within the hospital.

Two other British studies (cited by Indik, 1963) show the logarithmic relationship between size of factory and absence. Absence figures represent a sort of personnel cost. In terms of cost, therefore, organizations become increasingly expensive as their size increases.

Revans (1958) indicates a curvilinear relationship between organizational size and output. He showed that the output of coal per man-year reached a maximum in the size range of 1500–2000 men and thereafter tended to decline. He also reported that in retail sales establishments the average sales rate per person increases to reach a maximum in medium size working units and declines thereafter in larger establishments.

In much the same vein, Porter and Lawler (1965) reviewed the available literature on organization size and productivity and found either negative or curvilinear relationships between these variables. However they point out that "the impact of structural variables appears to be clearer on attitudes than on behavioral variables"; in general, research performed by organizational theorists fails to tie their variables closely to system outputs.

Case 121. Revans' output trends are graphically illustrated in a British study of two automobile factories. The workers, men on the assembly line, were paid on the basis of the total earnings of the group in which they worked, the amount being prorated to the base rate of each worker and the number of hours he put in. These earnings were directly proportional to output and can, therefore, be considered as an output measure.

Table 5-1 indicates that except for the largest groups (50 or more) output decreases as the size of the group increases. Correlations between output and group size over five work periods were low but uniformly negative (range, −.203 to −.491).

In factory B in Table 5-1, a number of men worked and were paid individually. Their earnings were higher (41.0) than that of the smallest groups.

Managers and foremen were asked their opinion on the sizes of groups. The main reasons given for favoring small groups (between 20 and 40) were the workers' ability to see one another at work, more concentrated and efficient work, ease of control by supervisors, and better relations within the groups.

The nature of the task performed may be a crucial determinant of the significance of team size. In a study using the Klaus-Glaser operant reinforcement task, Short et al. (1968) learned that the number of team operators to which an individual was added significantly influenced the individual's proficiency. Even if all team members had the same proficiency as the subject, the latter's reinforcement schedule was changed by the increased number of team members with whom he interacted; the larger the team, the greater the loss of individual proficiency. However these results seem to apply primarily to tasks involving series relationships.

TEAM COMPOSITION

To demonstrate that team performance depends on the capability of the individual team members would merely prove the obvious. There are, however, more meaningful questions.

1. How homogeneous does the team have to be? Does team member variability determine the nature and variability of system output?
2. What is the effect of turnover, and does this effect vary with skills required and type of task?
3. Are the contributions of individual team members to system output equal or differentially weighted, and how can we explain this?
4. How well can we predict team and system output from the combined performance of individual team members?

Each of these questions must be examined in terms of the task being performed. The effects of turnover, for example, could be minimal in a highly structured system and crucial in an indeterminate one.

Table 5-1 Output as a Function of Group Size[a]

Factory A				Factory B		
Size of Group				Size of Group		
Number of Workers	Average	Number of Groups	Output (15 months average)	Number of Workers	Average	Average Earnings
Under 10	6.5	20	133.8	Under 10	4.0	36.4
10–19	14.7	41	132.2	10–19	15.8	35.4
20–29	24.3	35	129.5	20–29	23.8	33.5
30–39	33.6	24	128.8	30–39	34.6	31.2
40–49	43.5	13	126.1	40–49	45.6	31.0
50 and over	88.9	20	127.1	50 and over	69.2	31.7
All groups	30.9	153	130.0	All groups	13.0	35.5

[a] From Marriott, 1949, Case 121.

It is desirable to distinguish between team composition and team organization, because the two are easily confused. Composition factors are properties of the individual that remain invariant when the individual transfers from one team to another. Organizational factors are related to the way individuals are utilized in a specific team.

The impact of team composition on system output, however, may be a function of the type of team organization employed. When tasks involve a high degree of dependency, composition variables may be significantly more important than when there are relatively few task interrelationships.

This section deals with the following topics: (1) the effect of team member characteristics, (2) changes in team composition, and (3) the predictability of team from member performance.

The Effect of Team Member Characteristics

To determine whether variability in team composition affects team output, let us study teams composed of operators having the following skills:

	Teams		
	1	2	3
Composition	$A_1B_2C_3$	$A_1B_3C_2$	$A_2B_3C_1$

where A, B and C are task-required characteristics and 1, 2, and 3 represent mean level of capability of these characteristics within team members. Will these teams produce the same output?

Conclusions. Although the results of the research on this question are by no means unequivocal, and the data emphasize personality variables and subjective measures, it appears that group composition in part determines group member "behavior" (which, however, is not the same as system output). Much depends, in addition, on the type of task being performed and the team dimensions being measured.

Illustrative Studies

Haythorn's review (1968) of the relevant literature indicates that crew composition significantly determines crew member behavior. He reports the following. Stafford et al. (1955) studied the effects on achievement when students in laboratory teams were allowed to work in mutually acceptable pairs. These students achieved significantly more than students paired to avoid individual partner choices. Van Zelst (1952) found that rearrangements of work partners in construction groups (carpenters and bricklayers) on the basis of personal preference resulted in reduction of material and labor cost and in labor turnover. The job satisfaction of employees also increased significantly following this rearrangement. Havron et al. (1952) discovered that infantry rifle squads tended to perform better if the squad members were similar with respect to levels of aspiration, confidence in physical prowess, time in the Army, and amount of combat time. Adams (1953) found that the effectiveness of bomber crews had a curvilinear relationship with degree of similarity among crew members on various status measures. The more effective crews had an intermediate level of status congruency.

In contrast, studies of homogeneously and heterogeneously composed problem-solving groups (Hoffman, 1955; Hoffman, 1959; Hoffman and Smith, 1960; Hoffman and Maier, 1961; Hoffman and Maier, 1966) have consistently revealed that heterogeneous groups produced a higher proportion of high-quality solutions than homogeneous groups.

Similar results (i.e., superior performance by groups composed heterogeneously with regard to attitudes) have also been reported by Triandis et al. (1965), but this was most true for groups homogeneous with regard to abilities. Whether heterogeneity is important seems to depend on the operational definitions used, the factors that are the same or different, the nature of the task, and the dependent variables being measured.

Task factors may be important here. Heterogeneity in problem-solving groups may be desirable because the indeterminacy of the task may call on disparate abilities (see Case 115). In non-cognitive tasks requiring coordination, homogeneity may enhance coordination, provided overall capability is reasonably high.

For example, Tuckman (1967) created three-man groups composed of members exhibiting different degrees of abstractness/concreteness and varying in terms of homogeneity/heterogeneity. There were two types of tasks—one a concrete, structured task, the other, abstract and unstructured. Group performance tended to differ as a function of the interaction between group composition and task.

What combinations of individuals having different skills and proficiency levels produce the most effective groups, and under what circumstances with regard to what criteria? Relatively little research in this area has been reported in the literature. This is in part understandable, since answers to the question may be largely situation specific.

In studies of the relative effectiveness of individuals versus groups in problem solving, it has frequently been noted that the usual superiority of groups is due to the more frequent checking of errors by different members of the group (e.g., Shaw, 1932; Murphy and Newcomb, 1937). This suggests that to a degree, members of groups can complement one another's weaknesses.

A similar hypothesis is presented by Hoffman and Maier (1961), who obtained problem solutions of high quality and high acceptance from groups in which the members had substantially different perspectives on the problem and in which these differences were expressed and used by the group in arriving at the final decision.

Pelz (1956) found that frequent contact with other scientists was positively related to an individual's performance only if the contacts were with scientists unlike him in motivation and experience. Analogous phenomena were reported under more controlled conditions by Hoffman (1959) and by Hoffman and Maier (1961).

Case 122. A study was performed of groups varying in cognitive characteristics, college class composition, and number of team members. The effect of team composition was measured in terms of team profit and loss in a financial game. Neither team size nor college class (freshman, sophomore, etc.) appeared to be a significant factor in profit and loss measures. However a test of cognitive complexity differentiated significantly between teams varying in number of "abstract" players. There was some suggestion that the effect of the composition variable increased over time spent playing the game.

Case 123. Rosenberg et al. (1955) introduced the concept of "assembly effect" in small group task performance. Case 123 attempted to extend these findings, which were based on a motor task, to a cognitive problem in which subjects had to determine which controls affected which displays and to note the extent and direction of movement of signal lights when a button was pressed.

Subjects were categorized on the basis of initial tests into one of three proficiency levels: level 1 (best performers—fewest errors), level 2 ("average" performers), and level 3 (poorest performers). Subjects were then run in three different groups in three sessions; each group contained an individual from levels 1, 2, and 3. Most of the variance was accounted for by the particular individuals comprising the groups. The assembly effect was not secured. When setting up the groups, however, efforts were made to equate them in terms of ability, which may have canceled out the effect.

The effect of sex composition on group performance has also been studied. Men have been consistently shown to be superior to women in problem solving, even when aptitude was controlled (Berry, 1958; Sweeney, 1953), perhaps because of sex-linked attitudes to problem solving.

Case 124. In a study that made use of a "brain twister" problem, 32 four-person groups of the same sex and 33 groups of mixed composition were established: 53.1% of males solved the problem, 25.8% of females.

More males solved correctly after discussion of the problem than did females under the same conditions; more males than females changed from the incorrect to the correct solution. In mixed sex groups, however, the proportion of males and females who solved the problem after discussion did not differ significantly. In other words, whereas males profited as much from group discussion in mixed sex groups as in all-male groups, females profited much more in mixed sex groups than in unisex groups. The hypothesis is that women in mixed sex groups were more motivated to work on the problem than were the women in all-female groups.

In a study involving a different task, Clement and Schiereck (1973) investigated the effect of sex differences in a group (four subjects) signal detection task. There were no essential differences between all-male or all-female groups, but mixed sex groups performed more poorly. This contrasts with the previous findings on heterogeneous groups in problem solving and reinforces the importance of the task variable in explaining differential performance.

Changes in Team Composition

That changes in team composition occur is obvious, but how important are these changes? Real world experience suggests that turnover can be a critical problem, but a number of variables can determine the significance of the phenomenon.

Comparatively few experimental data on behavioral factors in turnover exist (Hinrichs, 1970), although Lefkowitz (1970) has related turnover to inability to do the job—a situation that might be remedied, he feels, by more "vestibule" training. Porter and Steers (1973) in their review of the literature found a (predictably) strong relationship between employee dissatisfaction (e.g., pay and promotion, inadequate supervisory style, lack of recognition) and turnover. As was pointed out previously, organization size is only weakly related to turnover.

The effects of turnover are likely to vary depending on whether the man to be replaced is a member of a team (and on the role he plays in that team). If he is not a team member, these effects are more likely to be economic (e.g., cost of training a replacement) than performance oriented. Industry's concern with this problem is probably largely economic.

Conclusions. When turnover is moderate and does not involve critical personnel, the effect on team output is not significant. For example, Kidd and Kinkade (1959) found in the OSU air traffic situation that a significant decrement occurred at controller changeover; it was short-lived, however, and prechangeover participation by the replacement reduced the amount of performance loss at changeover.

Illustrative Studies

Case 125. Twenty-eight four-member groups performed a team typewriting task using pushbuttons. Only one pair of team members could communicate at any one time, only one member could communicate with the other three members during any one trial, and only one member knew at the beginning what was to be typed. Teams experienced varied amounts of membership replacement.

Team performance suffered significantly if (1) number of members replaced was large relative to the past history of the team, (2) succession (replacement in the control or more important position) was involved, and (3) the successor's intelligence was lower than that of his predecessors. Performance did not suffer if (1) number of members replaced was large absolutely but not relatively, (2) replacements occurred in subordinate positions, or (3) performance was unrelated to intelligence except following replacement.

A relatively large drop in leader intelligence due to succession was accompanied by large drops in performance. Average intelligence in nonreplaced team members showed no relation to performance. Performance suffered more if replacement was heavy and if succession was involved ($p=.05$, Table 5-2).

Case 126. The effects of changes in the composition of bomber crews were measured on such bases as ratings of proficiency, aborted missions, and time error in arriving at control points. Crews with a medium number of membership changes had better combat performance scores than crews with high or low changes. For teams that have operated together for long periods of time and have perhaps become bored with one another, a moderate amount of turnover may stimulate the remaining team mem-

Table 5-2 Team Performance Following Different Kinds of Member Replacement[a]

	Turnover		Succession	
Effect	Heavy	Light	Heavy	Light
Performance better than predicted	9	3	2	9
Performance worse than predicted	13	5	7	5

[a] From Trow, 1964, Case 125.

bers. Heavy turnover, however, may disrupt smoothly functioning team procedures.

Case 71 (Chapter 3) also provides information on the effect of replacement on crew performance in a simulated radar-controlled aerial intercept task. Three-man teams were composed of two radar controllers (RC) and a supervisor-coordinator. Replacement of an RC occurred in a terminal transfer of training session. The replacement had received either one day more or one day less experience on the transfer task than the man he replaced. Teams receiving a more highly trained replacement improved (48%) in performance immediately following; teams assigned a less skilled replacement deteriorated insignificantly (2%) but recovered subsequently.

This relatively weak replacement effect was intermediate between that of Glaser et al. (1962) and Egerman et al. (1962) who found a marked effect, and that of Horrocks et al. (1960, 1961) who observed no effect in their decoding tasks (see below). The difference may have resulted from the different tasks employed in the studies.

Case 127. Horrocks et al. (1961) utilized three-man teams in a version of their decoding task (Chapter 3 gives a description of this task). Training occurred over three daily sessions for a team of two decoders and one integrator, the latter serving to synthesize the work of the decoders. On the fourth day either no, one, or two replacements of personnel occurred, and work continued for two more daily sessions. In terms of time to complete the decodings of messages, there appeared to be a slight deterioration in performance of teams that received one new member, a somewhat greater deterioration in performance of teams having two members. However none of these performance differences was statistically significant.

The Predictability of Team Outputs from Individual Performance

If, as Case 123 suggests, team performance is largely dependent on the proficiency of individual operators, team performance should be relatively predictable from individual performance, unless team performance contains a unique dimension not to be found in individual performance. In the latter case, it would be difficult to predict team performance from individual performance with great accuracy.

Conclusions. Team performance variance is accounted for in some degree by individual (particularly leader) performance, but at least 50% of that variance cannot be attributed to individual member contributions. There is some slight evidence that intelligence and personality characteristics influence group task performance.

Illustrative Studies

Hackman and Morris (1974), citing a number of studies, make the point that when the team task levies specific demands on member skills and knowledges but does not involve substantial interaction among members, it is possible to predict team performance from individual performance quite well. For example, in the task employed by Klaus and Glaser (1968) (Case 138), the learning curves of the individual operator and of the team were almost identical. Yet when considerable member interaction is required, the ability to predict from measures of individual capability is reduced (O'Brien and Owens, 1969; Jones, 1974).

Case 128. Sixty-five pairs of subjects were tested on the Purdue Pegboard Test of psychomotor ability. After testing, two-man groups were formed by combining less with more proficient members (see also Case 123). These groups then worked on the same assemblies they had worked on individually. Group performance was related to the performance of the less capable group member .53, to the more able member, .50. The combination of the two yielded a multiple correlation of .66. This, however, accounts for only 44% of group performance variance, suggesting that the remaining variance was determined by some unique team dimension. Rohde (1958) obtained similar results with a maze task.

Case 129. Similar effects appeared in a study of B-52 bomber crew output measured by simulated bombing exercises. That output could be predicted most reliably by the aircraft commander's experience level (total flying hours or service) plus his educational level. The crew's bombing accuracy improved as that experience level increased to approximately 4500 hours or 11 years service, then declined at an increasing rate until at 6700 hours and 16 years it equaled that of a new crew. Why? No adequate explanation can be advanced. Although 17% of the crews had a substitute for one of the five crew members, there was no statistically significant difference in scores between crews with and without substitute members, thus tending to verify previous conclusions relative to turnover.

In a related study in which performance was rated by judges, leaders of effective Army squads were found to have a significantly higher mean intelligence score than leaders of ineffective squads (Greer et al., 1954). This relation was supported by Havron and McGrath (1961), who reported that the correlation between unit effectiveness and leader intelligence ranged between .35 and .50 in their studies. Intelligence plus job knowledge best predicted group effectiveness.

On the other hand, Spector and Suttell (1957) found no significant relationship between average intelligence of members of three-man groups and crew performance.

Conclusions. It is the level of attainment the team member has achieved prior to entering the team situation, not his individual learning ability, that is predictive of team learning. For some tasks, at any rate, it is possible to predict team proficiency on the basis of the proficiency of individual members.

Illustrative Studies

Case 130. This is another in the series by Klaus and Glaser (1965), involving an operant conditioning task with a series team of three members. Twenty-eight such three-man teams varied in the proficiency individual members had manifested before being formed into a team.

Four variables were investigated: the *proficiency* (low, medium, high) of individual members on entering team training, the *learning ability* of members (whether they were fast or slow learners), the effect of a *delay* in the initiation of team training, and the *homogeneity* of the team (i.e., how widely their individual proficiencies differed; that is, the team composition variable).

Only the entering proficiency of team members had any significant effect on team performance, the low, middle, and high teams varying significantly in terms of achieving the performance criterion. A team whose members were fast learners—those who had achieved the criterion during individual training in fewer trials—did not display rapid learning as a team.

It was possible at the time the team was formed to calculate a probable team proficiency based on the proficiency of the individual members measured at the end of their individual training. The predictive value of this probability measure was determined by Spearman rank-order correlations with the means of each group of three teams having identical member characteristics. Predicted team proficiency correlated $-.70$ and $-.77$ ($p = .05$), respectively, with number of trials required to achieve 100 correct responses and number of trials to achieve the acquisition criterion.

TEAM ORGANIZATION

Team organization is highly related to task organization, which was discussed in Chapters 2 and 3. The nature of tasks to be performed and their interdependency in large part determine how the team will be organized. Even so, there is some flexibility in the way in which teams can be organized, and these variations can influence output. This has been demonstrated more in industrial production lines (involving large number of workers, many of whom function in parallel) than in other, smaller teams that are highly interdependent.

A study by Van Beek (1964) demonstrates what can be done on assembly lines. He divided an assembly line for 104 workers into five smaller groups, assigned inspectors to shorter lines, and supplied racks for stockpiling finished products awaiting inspection. This resulted in increased output, shortened waiting time, and higher wages and morale. The question researchers need to answer, at least for smaller, interdependent teams, is: how much does reorganization affect team output, and which organizations are most effective?

Conclusions

1. The effect of series team organization is to reduce the probability that the team will accomplish its task.
2. Team organization in part determines team output. The strength of the organization effect, however, depends somewhat on the output criterion used.
3. Previous experience in self-organizing teams influences organizing activities in new teams composed of the previous team members.

Illustrative Studies

Case 131. In this study (another in the Klaus-Glaser series involving operant conditioning), three two-man team arrangements were compared: a series team (both members had to perform correctly), a parallel team (correct team performance was a function of correct performance of either member), and an "individual" team, in which only one preselected member's performance determined team output.

In a series arrangement the probability of correct performance is a multiplicative function, $p = f(x)(y)$. If one team member has a performance probability of .50 and the other a probability of .75, the probability of team performance is .38: (.50) (.75). In a parallel arrangement the probability of correct team performance is described by the addition theorem: $x + y - (x)(y)$. The parallel performance probability of the team would then be .87: $.50 + .75 - (.50 \times .75)$. In the individual team performance probability would be either .75 or .50 depending on the operator selected.

Subjects were given individual training and assigned on the basis of equivalent proficiency to 18 teams (six in each of the three arrangements). Initial team performance was predicted by inserting each subject's proficiency level into the appropriate probability equation for the type of team to which he was assigned. These predictions correlated .73 with actual proficiency over all 18 teams. Parallel team performance was best, followed by individual and serial teams in that order. This remarkable example of performance prediction is probably achievable only for comparatively simple tasks.

Similar results were found by Zajonc and Taylor (1963), who demonstrated that in a series-type reaction time task, as more series members (two to seven) were incorporated into the crew, the less likely the crew was to accomplish its mission.

Case 132. The task here was image (photo) interpretation performed by two-man teams. Four team methods involved one operator checking the other's work. The system output included: (1) all responses made or approved by the checker without discussion, (2) only responses agreed on prior to discussion, (3) all responses approved by the checker after discussion, and ()) only responses agreed on by both team members after discussion.

Although statistically significant differences appeared, these were not very large, at least to the untutored eye, and were not consistent for various measures of system performance. In other words, one method was most effective for accuracy, another for completeness, a third for efficiency. The implication (to the extent that one can generalize these results) is that team organization must be tailored to mission criteria.

A secondary purpose of the study was to determine how performance varied as a function of the proficiency and aptitude of individual team members. Somewhat surprisingly (or perhaps not, in view of Cases 128 and 129), no significant differences were found as a result of team composition.

Case 41 (Chapter 2) also indicates differences in performance as a function of team organization, and Cohen et al. (1962) found that in self-organizing communications networks previous experience in one type of network influenced later organizational activity in a subsequent network.

Conclusions. Team training effectiveness is in part a function of the type of organization for which the team is being trained.

Illustrative Studies

Case 133. This task, radar-controlled aerial intercept, has already been described in Cases 71 and 75. The variables studied were organization of the task in training and transfer, and fidelity of training to the transfer task; the last has been described in more detail elsewhere. Training task organization occurred at two levels: an *independent* condition in which radar controllers were trained individually with no intercommunication or direct trading of targets and/or interceptors, and an *interaction* condition in which the two radar controllers used the same problem board and interacted directly as targets passed from one area of controller responsibility to another. Transfer task organization was manipulated in the same way. Half the teams who were trained under an independent task organization transferred to an interaction organization; half the teams who were trained in interaction transferred to an independent organization.

Team performance, measured in terms of system measures (fuel consumed per raid and number of successful interceptions), was affected significantly by both training and transfer organization. Teams that had an independent task organization did better on the average than teams working with an interaction organization.

The study indicated that if the team transfers to an independently organized task, it makes little difference which type of training organization has been experienced previously. If the team transfers to a task requiring interaction, however, it is highly desirable for training to have been on an independently organized task rather than one that required interaction. It is possible that the verbal communication necessary in the interaction training resulted in a lack of discipline that became disruptive if the team transferred to another interactive condition.

Conclusions. In relatively unstructured situations, such as managerial groups developed to solve problems, interaction among group members during the solution phase of the process may be dysfunctional, although interaction during the evaluation of solution responses may be helpful.

Illustrative Studies

The quality of problem solutions produced by groups often exceeds that produced by individuals (Lorge et al., 1955, 1960; Davis and Restle, 1963; Taylor and Faust, 1952). This superiority occurs partly because in these groups more than one individual was working as individuals on the same problem, rather than having members interacting with each other. Taylor et al. (1958), and Dunnette (1963), found that nominal (noninteractive) groups generated more and superior solutions than real (interactive) groups. It is therefore possible that interaction among problem solvers is dysfunctional.

Case 134. Twelve groups of four subjects each had to generate solutions to human relations (e.g., morale) problems presented in written form. Problem solving took place under the following conditions: (1)

group member interaction during problem solution but not during solution evaluation, (2) interaction only during evaluation, (3) interaction during both generation and evaluation of solutions, and (4) no interaction during either solution or evaluation. Highly significant differences in number of solutions were found in favor of groups not allowed to interact during solution. Given a task that can be performed independently, interaction apparently interferes with task performance. These results support the previous findings of Dunnette et al. (1963) and Taylor et al. (1958).

Conclusions. The effectiveness of a particular team organization varies as a function of task characteristics.

Illustrative Studies

Case 135. Thirty two-man teams either shared or divided their work on two crossword puzzles (easy and difficult). Results were ambiguous: shared labor was more effective on the easier task than on the difficult one. One might hypothesize that interactions involved in shared labor are more disruptive if the task is difficult. Divided labor resulted in greater efficiency (amount of work per man-hour), although shared work produced greater group effectiveness (total performance, disregarding time). Similar cost benefit results were reported in the section on group versus individual comparisons.

Conclusions. Project team organization tends to be superior to one that is specialized.

Illustrative Studies

Case 136. Twenty-four groups composed of three subjects each worked on a word-construction task. The variables tested were mode of organization (specialized vs. project) and individual versus group feedback. Three rules governed performance of the task. In specialized teams each subject could apply only one rule; in project teams anyone could apply all three rules. Performance on each task trial was measured by three indices: score on final word constructed, solution time, and rated creativity of the word. Project teams were superior on solution times and also rated higher on a measure of task satisfaction. Individual feedback was slightly superior to group feedback.

Conclusions. The limiting factor in the efficiency of crews performing information processing tasks is the difficulty they have in gathering information from various sources and in communicating information changes.

Illustrative Studies

In Case 68 (Chapter 3) each of three subjects in a group performed in a separate booth isolated from his fellow crew members. Communication was achieved through interphone. Subjects were asked to adjust switch settings on the basis of slide presentations of simulated aircraft instruments. Correct settings for each control could be determined only by reference to joint readings on two instruments. The experimental variables were two organizational conditions: low (LA) and high (HA) autonomy, defined by the amount of information that had to be secured from external sources. The LA groups had to secure four information units from two different external sources; HA groups were supposed to secure only one input and that from a single source. Errors were lower for the HA structure (Figure 5-3). Thus the most difficult structure for the operator is one in which a larger proportion of information has to be relayed from different sources. Case 52 in Chapter 2 also suggested that information accessibility was a major factor in team proficiency. Similar conclusions can be derived from studies of communications networks.

Conclusions. When task elements are shared (division of responsibility), system performance is generally superior to that obtained when the operator is involved in every detail. Division of responsibility is effective

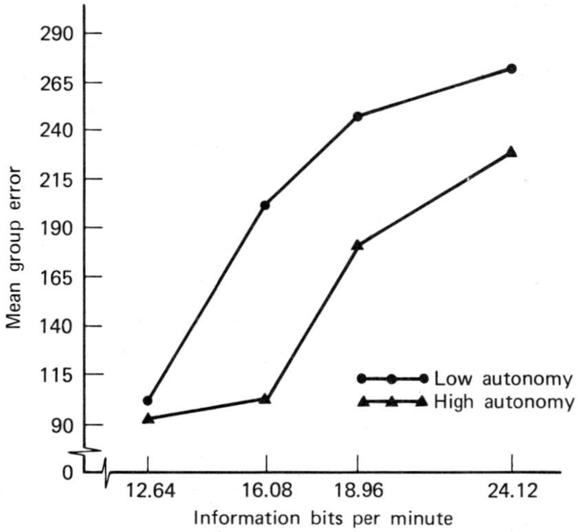

Figure 5-3 Relationships between mean group error scores and information rate for two experimental structures. With permission, from Lanzetta and Roby, *J. Abnorm. Soc. Psychol.*, 1956, Case 68.

when the reapportionment of effort is performed under functionally meaningful lines and when it leads to net reduction of input load to the operator.

Illustrative Studies

Case 137. In this study of air traffic control the responsibility for establishing the aircraft approach to the airport was systematically varied between pilots and ground controllers.

"The pilot was able to set his own course . . . and . . . make his own decisions regarding letdown and speed adjustments. The ground controller . . . still retained responsibility for conflict avoidance since only the ground controller could observe the relative position of all aircraft in the system. . . . Thus, the controller assumed, essentially, an overall monitoring function, while the details of approach path configuration were dealt with by the individual pilots. In both 'normal' and emergency situations . . . system performance was superior when the task elements were shared . . ." (Kidd, 1959, p. 20).

Conclusions. The effects of procedural flexibility interact with stress. In normal operations the operator performs most effectively if he can manipulate system elements on a momentary basis, freed of arbitrary restraints. When stressed, on the other hand, he may be hampered if he has too many options.

Illustrative Studies

In the air traffic control situation of Case 137 it was found that controllers tended to follow an iterative mode of *DM*; rather than making a small number of large heading, altitude, and speed corrections, they made many small corrections, postponing commitment to one course or the other until the last moment. Given this iterative tendency, procedural rigidity (limits on the number of available options) reduced efficiency. In emergencies, however, a reduction in the number of options tended to reduce confusion because the controller did not have to search through a large number of contingencies before selecting a response.

Kidd and Hooper (1959) used two-man teams in the same simulated air traffic control task. In their study controllers could operate under no restraint, partial restraint (a handoff between controllers could occur only within 20 miles of the GCA gate), and complete restraint (no handoff permitted). Although differences between these procedures were not statistically significant, results favored the no-restraint condition.

Case 40 used six-man air traffic control teams under strict flight plans and wider discretionary procedures in which controllers could organize their traffic flow. Results favored the flexible procedure.

Howell et al. (1959) hypothesized, however, that procedural flexibility would be less desirable under emergency conditions. They used two-man teams as in Kidd's air traffic control situation. The emergency resulted from a simulated radar failure. Controllers operated in the emergency under a low flexibility condition (only 3 approach paths) or a high flexibility condition (17 paths available). In terms of system efficiency during the emergency period (mean control time and mean fuel consumption), team performance was superior under the more flexible operating procedures; however in terms of safety (number of missed approaches to the GCA gate and number of separation errors), performance was superior under the low flexibility condition.

Conclusions. A hierarchical team structure is preferable, provided it does not employ an exclusively serial structure. As indicated previously, team performance in a series structure is determined by the least effective team member, whose overloading can seriously damage system output. Given some degree of centralization, it is critical that the most effective team member occupy the central position.

Illustrative Studies

Leavitt (1951) reported that accuracy and speed of performance were superior in a network of five men in which one man occupied a central position. Guetzkow and Simon (1955) secured similar results, as did Mohanna and Argyle (1960). Hierarchical structures appear to be superior for simple tasks, since an operator at a central position possesses an independence that permits him to organize the system more efficiently.

However centralized structures do not induce performance at a level superior to that of decentralized structures in all tasks (Heise and Miller, 1951). If the central man is not given or does not assume authority, a central network may not be superior (Shaw, 1959). The operator at the central position may become saturated or overloaded by work (Case 7), and his commanding position will therefore degrade system performance (Shaw, 1954).

System performance is not determined by structure alone, and in fact structure may be less important than other task and procedural variables. Nevertheless, Luce et al. (1953) point out that three factors favor a centralized structure: (1) on the average, it is more reliable in its operations; (2) it requires less training time to achieve proficiency; and (3) it appears to be more suitable for various system requirements than decentralized structures. Briggs and Johnston (1967) suggest that most real world systems "are designed as hierarchical structures with clear, centralized control" and this results from "the high regard our culture holds for efficiency (reliability) and for low overhead (training time)."

Conclusions. Interaction between operators should be minimized. Interaction at either the input or output positions in the system causes each operator to devote less than his full capacity to task-specific activities.

Illustrative Studies

In the air intercept controller experiments by Briggs and Naylor (1964) and Briggs and Johnston (1965), the amount of interaction between team members was varied. Half the teams could operate as teams by interacting over a voice channel; the other half had to operate as individuals. Under certain conditions, whenever a target passed out of one controller's zone of responsibility to another, interaction was desirable because one controller could alert the other to the handoff.

In each of these experiments superior performance was achieved by teams in which controllers acted as individuals. Even when a high degree of coordination was required (Briggs and Johnston, 1965), teams trained as teams were not superior to those trained as individuals. The results of these studies suggest that the task should be organized to permit (as much as possible) individual operator functioning; verbal interaction between team members should be minimized if possible.

The interaction in the studies just mentioned was at the output (response) level. Roby and Lanzetta (1956, 1957a) studied interaction at the input level. Three-man teams worked at an analogue version of a bomber crew station. Each man observed two displays and was required periodically to set two switches depending on inputs. Interaction was varied from low to extremely high, determined by the amount of information the operator had to gather from a team member to adjust his switches. Results indicated a progressive increase in team errors as the amount of input interaction required was increased (see also Case 68, Chapter 3). Kidd (1961) observed the same phenomenon. Interaction apparently reduces the amount of attention the crew member can apply to task-specific activities.

Conclusions. Team performance under load is superior when interaction requirements are minimized. However if operator independence is carried too far, a load imbalance may result and performance will deteriorate.

Illustrative Studies

Additional support for the desirability of minimizing interaction is evident in the study by Lanzetta and Roby (Case 47) in which system load was varied systematically (6, 10, and 15 seconds between display changes), as were the low and maximum interaction conditions just described. Both load and interaction significantly affected team performance. As one would expect, the higher the load (the shorter the interval between display changes), the poorer the performance; but, more important, performance under load was superior in the low interaction condition compared to load performance in high interaction.

Roby and Lanzetta (Case 48) performed an experiment in their aircrew station simulation to illustrate the conflict between independence and load balancing. They used three stations: the first had a single switch; the second, two switches; the third, three switches.

Subjects sat in enclosed booths that prevented direct communication. Each station contained a control box with three levers and signal lights, each corresponding to one of the three stations. Thus to call station 2, station 1 depressed his number 2 lever, which lighted the number 2 lights in all three stations. To respond, station 2 depressed his own number 2 lever. Displays were voltmeters, and on each one an instrument dial was simulated from an experimenter console. Instruction cards indicated what correct switch settings should be, depending on instrument readings. Because of the unequal distribution of displays per station, communication with other stations that possessed the required information was necessary.

Load balancing (equal distribution of work) was defined by the assignment of displays providing information for switch activation to stations (Table 5-3). Under condition I observation responsibilities were equally divided. In condition II the heaviest observation load was assigned to the station that had the most control responsibility. Condition III counterbalanced control responsibility by assigning the heaviest observation load to station 1, which had fewest controls.

Conditions I and II were considered by the experimenters to be approximately equal in terms of independence because the display/switch ratio was roughly equal; conditon III required more interaction. How-

Table 5-3 Number of Displays per Load Balancing Condition[a]

	Number of Controls per Station		
Condition	1	2	3
I	2	2	2
II	1	2	3
III	3	2	1

[a] From Roby and Lanzetta, 1957, Case 48 (as described in Briggs and Johnston, Case 148).

ever in terms of load balancing, which is determined by both input and output load, condition III was hypothesized to be better than I, and both were believed to be superior to II. Results indicated that performance was best under condition I (low interaction, medium load imbalance), next best under II (low interaction, high load imbalance), and poorest under III (high interaction, low load imbalance). Both variables—independence and load balancing—had a significant effect on performance, although independence was apparently the stronger of the two variables. See also Case 42.

Conclusions. Assuming sufficient flexibility in operating procedures (e.g., the opportunity to reassign functions and make other load-balancing adjustments), teams will adjust to a variety of system loads. Chapter 2 has also provided a number of illustrations of operator adjustment to system load.

Illustrative Studies

In Case 1, which involved two-man teams in an air traffic control task, there were two load levels (an approach every minute or every two minutes) and two team arrangements, one in which each team member could compensate for early or late arrivals by aircraft under control of his partner, and another in which a team member could not compensate for partner-generated arrival time errors.

During initial training superior performance was obtained by the teams working under low load; by the end of training, however, teams working under high load in the compensatory arrangement performed better than both non-compensatory teams. This suggests that if teams are permitted to adjust to high work loads, they are capable over time of developing compensatory procedures that can result in performance matching that of teams assigned relatively low work loads. This, of course, assumes flexibility in the operating procedures of the system. The deterioration of team performance under increasing system load may be particularly troublesome in inflexible structures.

Adjustment to load may have negative as well as positive effects. Behan (Case 7) used a serial structure in the five-man simulated air defense teams. The man at the input position became highly overloaded. Behan noted that with practice, the skill of an operator at one position in the serial structure increased sufficiently for him to process more data and thus to raise the load on succeeding positions. This in turn caused a temporary performance loss at the succeeding positions. Training for teams in a serial structure should anticipate this.

Conclusions. Adding redundant personnel to a team may have both positive and negative effects.

Illustrative Studies

When in adequate proficiency is a serious problem, another operator may be added to a team in parallel with an existing operator. For example, if prompt detection in a radar monitoring situation is critically important, several operators may be assigned to the activity simultaneously (see also Cases 117 and 118).

Six teams participated in one of the series of studies by Klaus and Glaser (1968). Following individual training two two-man teams were formed: an operator and the first monitor, and an operator and a second monitor. The teams were given team training independently until criterion had been reached, whereupon they were combined into a single three-man team with *parallel monitors* and practice was continued.

Five of the six three-man teams thus formed showed an initial improvement with the addition of a redundant operator as compared with the final performance of the two-man teams from which each three-man team was formed. However four of the three-man teams deteriorated in performance over time following the initial improvement achieved by the addition of the redundant operator. In two cases this deterioration resulted in a final level of performance below the final level of its original two-man teams, even though individual operator proficiency, as measured by a warmup period at the beginning of each day's session, remained high throughout the experiment.

TEAM TRAINING

Several questions underlie our interest in team training.

1. What is being learned when the team is being trained as a team? The desirability and utility of team training is based on the assumption that some distinctive elements determine the effectiveness of the team output; it is these elements that must be learned. The plain fact is that it is difficult to specify what should be taught in "teamwork." Perhaps as a consequence of this, teamwork is ordinarily taught not in terms of the acquisition of specific qualities or skills but by providing the operator with an opportunity to practice individual skills in a team context. But if one does not require something special of the team member (beyond what he would do as a single operator), can it be said that a team training requirement exists? If one cannot specify what is being learned, can training be controlled? Without such control, team training effectiveness is likely to be fortuitous.

2. Does team training exhibit the same characteristics as individual operator training? If the skills to be learned as a team are team specific, team training effects may differ from those of operator training. [In this connection, however, little attention has been paid to the development

and application of group learning curves, in part because of methodological problems in determining what a group learning curve as a combination of individual curves really means (Baloff and Becker, 1968)].

3. What is the temporal relationship between operator and team training? Is it most effective to train operators in individual skills first and follow with team training, to train all skills from the beginning in a team context, or to train individual skills only? Assuming that both operator and team training can be provided, is there some general point at which the transition should occur?

4. Is team training effective in increasing system output? This after all is the critical point: if team training does not significantly improve performance beyond that secured with individual training, it is not needed. The question may appear heretical—of course operators must be trained as teams if they are to perform as teams. If, however, system output does not improve significantly after scheduled team training, one could perhaps just as well insert operators into existent teams as replacements to develop their skills by on-the-job (OJT) training.

5. Are the major variables that affect (interact with) team performance training the same ones that influence operator training (e.g., type of task, operator proficiency at start of training, operator learning ability, team structure, feedback provided)? If these are influential, they must be considered in developing a team training curriculum.

Task characteristics are probably significant for team training, but little is known about this variable. It seems unlikely that team training of highly determinate and indeterminate tasks would involve the same mechanisms. The feedback variable is also of special interest in team training. The team information to be fed back must be examined in terms of exactly what is being taught (and learned) in the team context. Since feedback is experienced by the individual operator and referred back to his own responses, team feedback may be erroneous if it is accepted by the operator as correctly describing the consequences of his individual performance (Chapter 6 gives more detail on this point).

Conclusions. On the surface team learning exhibits the same performance curve displayed by the individual operator.

Illustrative Studies

Case 138. The training situation in this study was the operant conditioning task employed by Glaser and his colleagues in cases cited previously. Three team members were individually trained to at least a 63% level of response accuracy, and then team training began.

Individual learning curves were contrasted with those of the three-man team (Figure 5-4). Team response acquisition obviously is almost identical

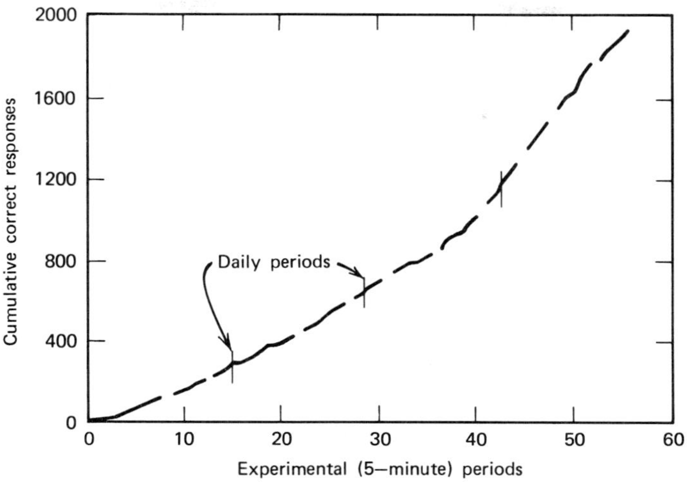

(a)

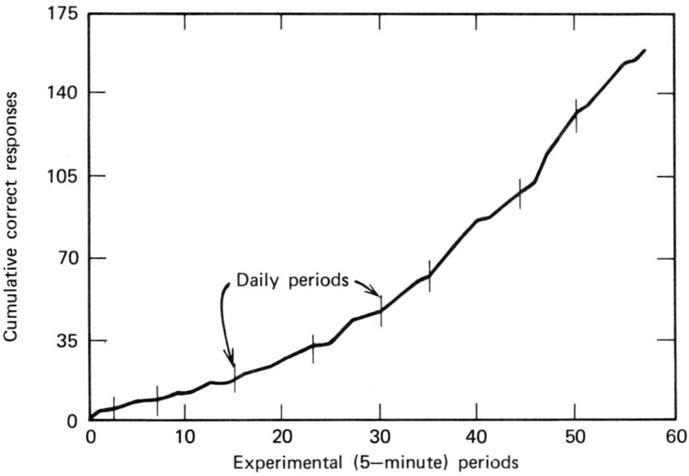

(b)

Figure 5-4 Comparison of individual and team learning curves. *(a)* Individual learning curve (single subject); *(b)* team learning curve (three-man team). With permission, from Klaus and Glaser, 1968, Case 138.

with that of the individual operator. If the team is considered as a series of independent components, each having a probability of successful performance, the multiplicative rule in probability applies and the team should be successful on .63 × .63 × .63 or 25% of their trials. This is essentially what occurred.

It seems to the author that this study oversimplifies the relationship between team and individual learning, first because exclusively series operations are not indicative of the majority of team situations, and second because the only essential change in the transfer from individual to team operations was a shift from individual feedback to team feedback (i.e., team members did not interact with each other more or differently when they were placed in the team situation). If the operator's task requirements are not changed because of the team context, can one say that the team is being trained for anything team specific? This type of study needs to be repeated with different types of tasks. (But see also the shape of the learning curve in Figure 5-1.)

Conclusions. Individual training is superior to team training, at least for simple or only moderately complex tasks. Team training appears to be relatively ineffective in producing performance beyond that resulting from individual operator training.

Illustrative Studies

Horrocks et al. (1960) found team training no more efficient than individual training in their six-man decoding task. In another experiment (1961), they designed a three-man task requiring members to memorize the location of alphabet letters. There were two groups: an experimental group that had just finished serving in a 5-day team experiment and presumably had learned from their team experiences, and a control group consisting of teams whose members had never worked together. There were no significant differences between teams on the new task.

As one of a series of studies in team variables (certain of these have already been described as Cases 71 and 75), Briggs and Naylor (1964) trained two members in a three-man team individually or as teams on a checkerboard replica of an air defense (intercept) task. Upon transfer, half of each group of teams worked on an independently organized task where interaction was not possible; the other half worked in a true team context (interaction possible).

Highly superior performance on the transfer task was achieved by teams whose members had been trained previously as individuals (Table 5-4). As was indicated in Cases 71 and 75, the independent transfer organization was superior to the interaction organization. The measure was fuel consumed per intercept (the more consumed, the poorer the performance).

In another experiment (Briggs and Johnston, 1965), teams were made up of members trained as individuals while other team members were

Table 5-4 Performance as a Function of Training Conditions and Transfer Task Organization[a]

Transfer Task Organization	Training Mode	
	Team	Individual
Interaction	18,444	10,067
Independent	9,761	10,575

[a] From Briggs and Naylor 1964, Case 75.

trained only in the team context. There were no significant performance differences between the two types of teams.

Conclusions. High fidelity training conditions are important determiners of operational task performance, whether fidelity refers to equipment, procedures, or skill requirements. Input fidelity is more crucial than output fidelity.

Illustrative Studies

The physical and functional similarity of training to operational tasks has received considerable attention in laboratory research, particularly with regard to aircraft simulators, although little has been done on training task fidelity in the team training context. However Briggs and Naylor (1964) and Briggs and Johnston (1965) studied fidelity (similarity) of training to transfer (operational) tasks in terms of equipment and procedures. A second experiment described in Briggs and Johnson (1965) investigated task fidelity in terms of skill requirements.

The task performed was aerial intercept control via simulated radar displays. Two of the team's three members, the radar controllers (RCs), issued heading and speed commands over simulated radio links to simulated interceptor pilots. The third crew member was a supervisor-coordinator.

These studies featured a training and transfer task. The training task consisted of two checkerboards having 2304 squares. There were eight checkers on each board: four targets and four interceptors. During training, the experimenter moved the targets from square to square, while RCs moved each interceptor checker concurrently. The goal was to make each interceptor land on or block a target, thus registering a "hit."

In the transfer task all targets and interceptors appeared as radar returns moving in real time on a CRT. Thus the transfer task closely simulated operational equipment.

In the first study (Briggs and Naylor, 1964), team response (R) fidelity during training was high for half the teams and low for the other half. Stimulus (S) fidelity was relatively low for all teams.

Under law fidelity training, RCs moved their own checkers and merely called out heading and speed changes. Under high fidelity conditions the situation more closely resembled the transfer (operational) task, because experimenter assistants moved interceptor checkers and repeated orders from RCs over simulated radio links.

In the second study (Briggs and Johnston, 1965, also reported in Briggs and Johnston, 1966), both S and R fidelity were varied. High S fidelity was characterized by the use of two 5-inch CRTs, identical save in size to transfer displays. Low S fidelity was represented by the checkerboard game described previously. In the high R fidelity situation RCs issued commands to pilots over simulated radio links. In the low R fidelity situation RCs issued commands by turning speed and heading dials to appropriate positions, and pilots received commands from their display boards.

In both studies transfer performance was superior following R fidelity training (which permitted team members to learn efficient verbal commands to interceptor pilots). In both experiments, however, this superiority was relatively short-lived; by the second transfer session there were no differences between teams. Yet since the superiority of teams trained under high S fidelity conditions was maintained over all four transfer sessions, S fidelity appears to be more important.

Nevertheless it is important to note that in both experiments, training task fidelity influenced performance only in interaction with other variables (e.g., transfer task organization, sessions). When fidelity was extracted as an independent variable, differences between low and high fidelity were not statistically significant. This tends somewhat to reduce the importance of the fidelity variable.

In a second experiment described in Briggs and Johnston (1965), the transfer task required a higher degree of interaction between the two men in each team. During training, there were five groups of teams. Groups 1 and 2 were not required to attempt coordinated intercepts during training, but coordination was required of groups 3, 4, and 5.

From the standpoint of training fidelity, groups 1 and 2 received relatively low fidelity training experience (no coordination required), whereas the other three groups received relatively high fidelity training. Performance during transfer was in line with expectation: groups 3, 4, and 5 were statistically superior to groups 1 and 2 in terms of coordination. It is possible therefore that training fidelity is more important in complex tasks that require the operator to time-share not only among controls and displays but also with operators. Or so Briggs and Wiener (1966) suggest.

Conclusions

Given two incompatible criteria of system performance, such as speed and accuracy, teams will emphasize that aspect of performance about which they receive knowledge of results (KOR), even though other aspects of system performance deteriorate as a consequence. If criteria are changed, team performance will be modified quickly to emphasize that aspect of performance which was previously neglected. (However, see Adam, 1972.) Concurrently, team performance on the other aspect will deteriorate.

However if the team changes from a single criterion to several criteria, teams will continue to stress that aspect of performance about which they had previously received KOR.

Illustrative Studies

Case 139. The same aerial interceptor task used in Case 133 was employed here. Each radar controller viewed his own CRT, on which returns from both interceptor and target aircraft were displayed. Eight groups were trained, four in terms of a time criterion (time required to make an interception), four on a coordination criterion (degree of coordination). Groups 1 to 4 transferred to a simple criterion (*either* time or coordination), while groups 5 to 8 transferred to a complex criterion (coordination *plus* time). Within the first four groups, groups 1 and 4 experienced a change in criterion; the others did not.

Equipment and procedures in both training and transfer tasks were essentially the same. Verbal feedback consistent with the criterion employed was provided. Investigators measured performance in terms of the amount of separation between the nonintercepted target and its interceptor, and the time required to make an interception.

Training results were quite straightforward: the groups that used the coordination criterion (C) obtained better coordination (smaller separation error) than time criterion (T) groups; the reverse occurred with the speed criterion.

Performance levels during transfer for groups 1 to 4 appear in Figures 5-5 and 5-6; data for groups 5 to 8 are in Figures 5-7 and 5-8. If transfer was to a simple criterion, teams adapted to the new work situation quickly; there was little residual effect of the previously experienced criterion except for the speed performance of group 4. When transfer was to the complex criterion (involving incompatible criteria), however, teams continued to emphasize the aspect of performance that had been reinforced

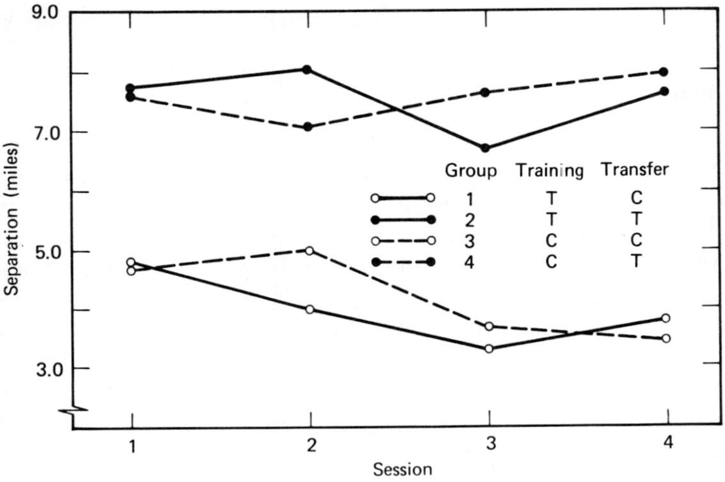

Figure 5-5 Coordination performance during transfer for groups 1–4. With permission, from Briggs and Johnston, 1965, and Case 139.

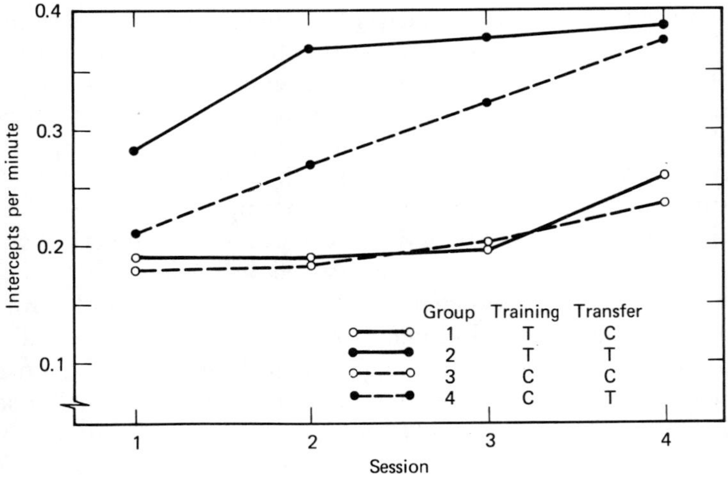

Figure 5-6 Speed performance during transfer for groups 1–4. With permission, from Briggs and Johnson, 1965, and Case 139.

under the previous simple criterion. This makes sense intuitively: teams operate as instructed except when instructions are internally incompatible, in which case the logical approach is to continue emphasizing that aspect of performance with which one has had more experience. For

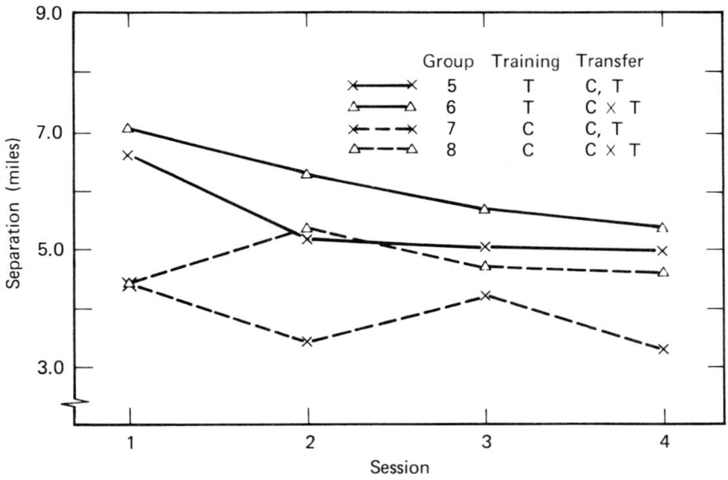

Figure 5-7 Coordination performance during transfer for groups 5–8. With permission, from Briggs and Johnston, 1965, and Case 139.

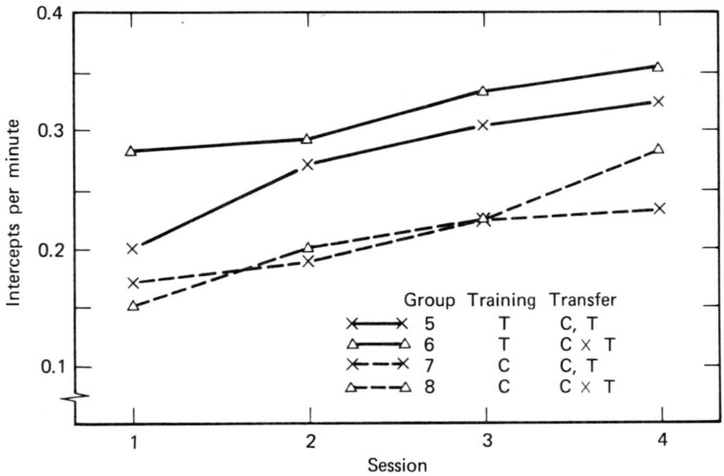

Figure 5-8 Speed performance during transfer for groups 5–8. With permission, from Briggs and Johnston, 1965, and Case 139.

example, groups 7 and 8 both trained on the coordination criterion and they were better coordinated during transfer (Figure 5-7). Yet groups 5 and 6, which trained under the time criterion, were superior to groups 7 and 8 in terms of time (Figure 5-8).

It appears also to make little difference whether the complex criteria are expressed to teams as specific feedback (groups 5 and 7) or as general feedback (groups 6 and 8). In terms of coordination performance (Figure 5-7), for example, groups 5 and 6 are not significantly different on any transfer session, even though the teams in group 5 received specific scores for time and coordination separately, whereas group 6 received only a single figure of merit (time multiplied by separation).

Conclusions. Team training, like individual training, is affected by task complexity. Training is, of course, easier and performance superior on less complex tasks. Task organization in team training is a less significant variable than task complexity.

Illustrative Studies. See Case 71 in Chapter 3.

Conclusions. Individual reinforcement during early team training overcomes the characteristic decrement in proficiency that occurs during the transfer from individual to team training. However this reinforcement does not reduce the amount of team practice required.

Illustrative Studies

Case 140. This study was another in the series by Glaser and his colleagues (1965) in which subjects had to learn an operant conditioning task as individuals and then as three-man teams. The task was to estimate the duration of stimulus lights presented on a panel and to depress the correct switch.

In previous studies in this series, a marked reduction in individual team member proficiencies had been noted when individual training was terminated and team training began. This decrement was hypothetically related to the reduction in the ratio of reinforcement that occurs when the change is made in training conditions. Under team training conditions it is obvious that individual subjects were reinforced for their responses only at a reduced rate, since the reinforcement of each member depended on the overall team product; and in a series structure, any member who performs correctly while the team as a whole is incorrect, will not be appropriately reinforced (see also the question of confounded feedback discussed in Chapter 6).

It was hypothesized that if during team training, the individual team member was reinforced individually for his correct responses, even if the team as a whole performed incorrectly, this would help to maintain individual team member proficiency at the same level attained when he entered the team situation.

The experimental results indicated that team reinforcement combined with individual reinforcement produced more rapid acquisition than team reinforcement alone. However this facilitating effect of individual reinforcement disappeared when the added feedback was removed; the experimental teams required almost as many trials to achieve criterion a second time with team reinforcement alone as did the control teams. The overall number of team training trials was not affected materially by the experimental condition.

Individual reinforcement in team training seems to be advantageous only when it can be maintained indefinitely and then primarily for low proficiency individuals. (The effects of individual reinforcement were most striking for teams formed of low proficiency members.)

Conclusions. During training, teams in highly uncertain situations should be provided with opportunities outside system operation to establish operating procedures and principles.

Illustrative Studies

Case 141. Shure et al. (1962) used a five-man group in a communications network in which the task was to transmit information among the group in order to identify the item that was missing. There were three conditions. In separate planning (SP) the group was allowed time between work periods during which members could transmit written messages on matters such as group organization; in cotemporal planning (CT) messages during team operation could include planning as well as task-specific information; and in no planning (NP) only task-specific information could be transmitted.

By the end of the experiment all 10 SP groups had set up hierarchical organizations (see previous conclusions relative to hierarchical team structures), whereas only 2 of the 10 CT groups and none of the NP groups had done so. Predictably, the SP groups were significantly superior in performance to the other groups, which did not differ from each other. See also Guetzkow and Dill (1957). Self-organization by teams is probably most required in indeterminate systems.

TEAM PERFORMANCE AND COMMUNICATIONS

Since communications are an essential part of team operations, researchers seek to answer the following questions:

1. To what extent do non-task-required communications influence team performance and in what way (positively or negatively)?
2. Will training in task-required communications improve team performance?

Conclusions. Non-required (irrelevant) communications have a negative effect on team performance. Team communications should be minimized except when required information can be secured only in this way.

Illustrative Studies

Analysis of air defense operator performance (Cohen and McKelvey, 1954) indicated that between 3 and 6% of operator activity was spent in general social talk that contributed nothing to overall task performance except possibly by improving morale.

Case 142. The task was the aerial intercept task also described in Case 139. One condition included specific training in team coordination, including the opportunity to communicate; in another condition communication during training was prohibited. Intrateam communications retarded performance by increasing irrelevant information transmission or information that could have been obtained more readily direct from displays. Personnel permitted to communicate tended to ignore their visual display in order to communicate. Prohibiting such communications during training did not lessen their disruptive effect in transfer trials. When the team exhibiting superior performance was permitted to communicate, their performance quickly dropped to the level of the least capable team.

Conclusions. Visual methods of communicating information are more effective than verbal methods. The addition of verbal communications to a visual mode adds nothing to the efficiency of the visual mode.

Illustrative Studies

Case 143. In another Ohio State University study of aerial interception, two-man radar controller teams were trained and then tested under alternative information channel conditions: verbal alone (Ve), visual alone (Vi), and verbal combined with visual (Ve-Vi). Figure 5-9 indicates that training performance was superior in the Ve-Vi condition. Teams transferred to the Ve-Vi condition exhibited performance superior in terms of coordination (separation between the target and the interceptor) to the Ve condition. Teams trained with the visual channel alone performed about as well as those trained with the combined channels. Teams transferring to the visual channel (regardless of how they had previously been trained) performed better than those transferred to the verbal mode. This suggests that visual superiority is not a matter of training alone but is inherent in the nature of the channel. Although the verbal channel has been and obviously can be used for the transmission of

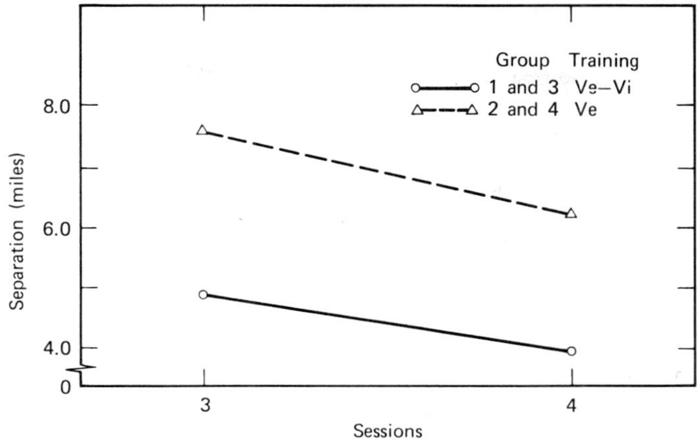

Figure 5-9 Effects of training with verbal-visual (Ve-Vi) and verbal (Ve) channels. With permission, from Briggs and Johnston, 1965, and Case 143.

complex coordination information, it is not as effective as the visual channel. One reason for the relative ineffectiveness of verbal communications may have been the failure of controllers to recognize the types of communications that could help system performance, or their lack of the verbal skills to make use of this knowledge if they had it.

Conclusions. Differences in performance result from differences in communication structure. The more direct the transmission of information permitted by a structure, the better the team performance.

Illustrative Studies

Research on communications structures (networks) in small groups dates back to the pioneering work of Bavelas (1950), who wanted to know what effect group structure had on group efficiency. The focus of the work stemming from his efforts has been on comparison of centralized (wheel) with decentralized (circle or completely connected) networks (see Chapter 3 for a more detailed description of the network situation). The tasks performed with these networks have generally involved problem solving. Results from early studies indicated that centralized networks were more effective with simple problems, decentralized ones with complex problems.

The rather extensive network studies performed (summarized in Glanzer and Glaser, 1961; Shaw, 1964) have not been very successful in answering the question raised by Bavelas, partly because of the artificial laboratory situation developed to model communications processes and partly because the most effective structure appears to be determined chiefly by task characteristics. Differences between centralized and decentralized networks diminish in importance once decentralized groups proceed to organize themselves (often along the lines of the wheel network); thereafter they communicate in much the same manner as do centralized groups and with about as much speed. When communications restrictions affect the task performance of groups, the effect appears to be exerted indirectly, by influencing the ability of group members to organize themselves, perhaps by reducing the number of suggestions made for organization (Guetzkow and Simon, 1955; Guetzkow and Dill, 1957). More recent work has emphasized small communications groups embedded in larger groups (Cohen et al., 1969). This research suggests that member responses are not only a function of position in one's own group but also of the member's relationship to members of other groups within the same organization, i.e., the extended team.

Porter and Roberts (1972) have pointed out that network studies "seem largely to have run their course, with little really new or exciting evolving from them in recent years." Missing from all these studies have been data concerning the relationship of communications patterns and behavior to overall system performance. Little is known about how high performing personnel differ in communications behavior from low performers. The following study (which is similar to Case 48) is interesting because it suggests that communication structures do have an effect on performance.

Case 144. Airmen in groups of three were required to perform a laboratory task that simulated aircrew operations. Each operator sat in a booth that restricted direct communication. Each booth contained two switches. Slides projected to the operators information that required them to change switch positions based on information that had to be secured in part from other operators. Operator errors were recorded for performance of the task under four communications structures. The communications structures were as follows: *(a)* operators had no access to information needed to operate; consequently they had to obtain four units of information from other group members; *(b)* operators had access only to one information unit and had to get three from other sources; *(c)* there were two "accessible" and two "nonaccessible" units; and *(d)* operators had direct access to three units and got one from other sources.

Differences in performance were associated with differences in the structures employed. The more direct the transmission of information permitted by a structure, the better the team's performance. (See also Case 52 in Chapter 2).

Conclusions. Variations in verbal communication patterns are related in an obscure manner to variations in team performance. However training in team communications has not been found to significantly improve that performance.

Illustrative Studies

Efforts to discover a relationship between communications variables and team performance have made use of several research strategies. One strategy has been to perform content analyses of communications in team performance to find variables that correlate with that performance. In another strategy team members have been trained in communications presumably required for team performance to determine whether performance improves (as a result of the training). Actually both types of research are required, since the communications variables to be acquired by training must first be discovered through some form of analysis.

Case 145. This study utilized experienced Air Force personnel (four-man teams) on simulated B-52 missions. An experimental group experienced more team training in a simulator; a control group received more individual training. In terms of several measures (simulator flight check, airborne flight check, navigational and bomb accuracy scores, and peer group assessment), there were no real differences between the groups. To revert to a previous theme, there seemed to be little to recommend team-specific training at the expense of individual training.

Differential experience did, however, seem to result in differences in communications (see Chapter 3, p. 107 for the communications taxonomy developed by the experimenters). Increasing experience in team exercises was accompanied by an increase in the number of volunteered messages. If voluntary messages indicated increased coordination among crew members, we would have support for the validity of team training procedures. The pattern of communications varied between teams that engaged in a high volume of verbal interaction and those that transmitted relatively few messages. Communications judged (on the basis of ratings) by the experimenters to be necessary for team coordination were associated directly with team proficiency. The upshot of Case 145 is that differences in communication patterns exist between teams, but the significance of these differences is obscure.

A later study by Siskel et al. (1965) compared within-crew communications in B-52 and KC-135 aircraft as a function of crew experience and selected mission segments. In line with the results of Case 145, it was hypothesized that as a result of their lower level of coordination skill, the less experienced student crews would have a higher rate of communications than the more experienced combat crews. No significant differences were found as a function of experience level.

The author has observed comparable communications incongruities in Army helicopter pilot-navigator teams (Fineberg, 1974; Sanders et al., 1975) flying at extremely low altitudes. The copilot/navigator guides the pilot by reporting significant terrain checkpoints and advising the pilot when to turn and in which direction. Since this is a condition of high uncertainty (the cues used are extremely fine-grained), it might be assumed that the more information the navigator transmitted to the pilot, the more effective navigation would be. Major differences in communications volume were discerned among navigators (some being silent for long periods of time, others communicating constantly), but these differences were not correlated with significant differences in performance.

Case 146. Having analyzed communications in performance of three-man helicopter crews, the investigators derived a set of communications categories different from and much more elaborate than those obtained in Case 145. One hypothesis examined was that at mission phases in which uncertainty is greatest, communications volume is increased to reduce the uncertainty and develop a course of action. As uncertainty is reduced through evaluation of the task situation, however, the appropriate course of action becomes more obvious and restrictions are imposed by the team commander on communications channels. Both hypotheses seem reasonable; unfortunately only 2 of the 26 communications categories developed around these concepts were significantly correlated with team performance.

Case 147. In this study the antisubmarine warfare helicopter teams of Case 146 were again used as a vehicle for examining communications behavior. Based on a content analysis of team communications, using the 27 categories of Case 146, 3 factors were derived by factor analysis. A 16-hour training course based on the three factors involved group discussion by students of the communication behaviors observed when other students were flying simulated missions.

Sixteen two-man crews participated in the training. The criteria employed in evaluating the effectiveness of the communications training courses were (1) number of attacks on the correct one of the two submarines presented, (2) percentage of hits regardless of submarine attacked, and (3) miss distance from target. Table 5-5 compares 16 trained and 16 untrained crews on the three criteria.

Although every measure was in favor of the trained subjects, none was

Table 5-5 Effects of Communications Training on Performance[a]

Criterion	Trained	Untrained
Percentage of crews attacking correct sub	88	69
Percentage of hits	75	69
Miss distance	400 yards	1194 yards

[a] From Siegel and Federman, 1973, Case 147.

statistically significant. It was also learned that the trained group communicated more frequently in the same length of time.

Williges et al. (1966) found that verbal training did not transfer to a test situation, nor did it have a significant effect on teamwork.

Case 148. Intrateam communications were analyzed in the same series of air intercept tasks reported in Cases 75, 133, 139, 142, and 143. Differences in amount and type of communications were reported in various training/transfer sessions. Teams modified their communications as mission demands changed. Under time stress, for example, teams communicated less and restricted messages to objective data. This is entirely logical and is one of the mechanisms described in Chapter 2 for operators who must deal with overload conditions. Teams transferred from the training condition to a different "operational" condition were reluctant to change their patterns of communications. When demands were relaxed, teams relaxed their communication discipline quickly.

The relationship between communications variables and team performance was negative in the five studies described in Case 148. Not only were there few significant correlations between communications categories and performance, but often these were negative. These negative correlations are in agreement with previous conclusions that personnel interaction (primarily through communication) tends to degrade system performance.

Briggs and Johnston (1967) derive the following conclusions from their studies in Case 148:

1. The volume of communications reduces with training. The pattern of messages changes as a function of both training and task characteristics.

2. Communications between team members have the following characteristics:

(a) One type of message represents attempts to reduce input uncertainty (see the hypotheses addressed in Case 146).

(b) A second message type represents an attempt to evaluate what is known, a step preceding the formulation of alternative courses of action.

(c) Following data evaluation, another message type describes possible courses of action.

(d) Once a course of action is decided on, a fourth message type (commands) occurs as action is implemented.

3. Time stress will result in fewer communications. Under time stress the pattern of team communications will involve more objective information messages than tactical, evaluative, or opinion-type messages. The opposite occurs when teams work under low time stress and when they are encouraged to develop highly coordinated performance. Time stress, therefore, fosters communications discipline. Although teams will maintain discipline when changing from high to low time stress conditions, interaction among team members fostered under low time stress persists when that stress is increased. Therefore instructors should attempt during and between training problems to encourage greater self-discipline and control in the team as time stress increases.

4. The availability of information channels in a system markedly influences communications content. Team members can utilize less efficient verbal channels to transmit objective information when machine channels (such as a radar display) suffer partial failure. Especially in less capable teams, however, objective information transmissions may occur in such volume that they exclude other necessary message types. The transmission of objective data may become (particularly in less skilled teams) an end in itself, with the result that messages necessary to utilize these data are not transmitted.

5. The volume and content of communications between team members can be controlled by the use of immediate feedback that reinforces one message type and extinguishes another. However feedback can produce unusual effects on performance; no clear conclusion is yet available on the value of specific training for verbal communications. The acquisition of communication skills is a rather lengthy process, despite the tremendous overlearning that occurs in communications usage.

6. In general, laboratory research on team communications indicates that the less interoperator interaction, the better.

ATTITUDES AND MOTIVATION

It is reasonable to assume that just as the individual operator is more or less motivated, so is the team, and this motivation in both cases affects system output. To present a comprehensive picture of what is known about the MMS, one should therefore include motivational variables.

Unfortunately, much of the motivational research describes the individual (as distinct from the operator or multiman system) and is derived from studies of ad hoc groups rather than task-oriented teams. Even when studies deal with operators, primarily those on the production line, the measures reported emphasize feelings of job satisfaction. These may influence system output (particularly in indeterminate systems), but it is impossible to discover how, since the degree of job satisfaction is almost never related to an output. One of the exceptions is a study by Berkowitz (1956), an attempt to relate aircrew attitudes to combat effectiveness. The relationships revealed in that study were difficult to interpret. Many of the variables investigated in these studies are tenuous and hard to define operationally, at least in real world terms (e.g., group atmosphere, described in Frank and Anderson, 1971).

Rather than engage in aimless theorizing, the author prefers not to present any conclusions. However, some hypotheses (but only hypotheses) can be suggested.

1. The effect of operator/team motivation on system output is greatest in indeterminate systems and tasks in which operators are most free to determine their outputs. Motivational effects are accomplished by changes in operator goals, which in turn influence productivity (quantity) and quality.

2. The effect of motivation on output is generally less than that of other more structural system variables, such as the nature of the task.

This topic is explored in studies in the "new" field of organizational development (e.g., Friedlander and Brown, 1974; Miner and Dachler, 1973; Hinrichs, 1970) which emphasizes job satisfaction and motivation in industrial organizations.

SUMMARY OF CONCLUSIONS

1. In problem-solving tasks group performance is usually superior to individual performance, although groups may not be cost effective because of the number of men they require.

2. The superiority of group performance in such tasks is manifested primarily in very small groups (i.e., as one proceeds from the individual to the two-man group). Groups of larger size are not usually significantly better than the two-man group.

3. The addition of an observer in monitoring improves detection if the two monitor independently. Monitoring teams adopt various strategies for combining their performance, but these strategies are rarely optimal.

4. In industrial systems there appears to be a curvilinear relationship between organization size, work output, and work quality. Organization size is positively related to absence rate.

5. Crew composition significantly affects crew behavior; evidence relating to effects on output, however, is somewhat inconclusive.

6. Little disruption of team performance results from turnover as long as only one team member is replaced (at one time) and the replacement is as skilled as the person he replaces. Disruption is increased, however, if additional team members are replaced.

7. There is some evidence that operator intelligence and personality influence crew performance, perhaps in terms of leader behavior.

8. Team performance variance is only partially determined by individual member performance.

9. Differences in team learning can be specifically related to the proficiency exhibited by team members as individuals when entering the team situation.

10. Team organization in part determines team output. Teams are able to adapt to a variety of system loads. A hierarchical team structure is a favorable arrangement provided it is not a strictly serial one. Team performance in a serial structure is determined by the weakest link in the chain, and overloading is a distinct possibility.

11. Team training effectiveness is partly a function of the type of organization for which the team is being trained.

12. A limiting factor in teams working in highly indeterminate organizations is the difficulty they have in organizing themselves to come to grips with their task. Groups performing information processing tasks are also limited by the difficulty they experience in gathering information from various sources and communicating it to others.

13. The effectiveness of team organization varies as a function of task characteristics.

14. System performance is generally superior when tasks are divided among team members.

15. Procedural flexibility is highly desirable; under emergency conditions, however, a more restricted range of options is preferable.

16. Both in training and in actual performance, independence (noninteraction) of operators is highly desirable. Interaction at either the input or output level distracts the operator from task-specific duties.

17. This independence of team members must not be achieved at the cost of overloading the individual operator. Both independence of function and load balancing are needed for superior team performance.

18. Adding redundant personnel to a team may have both positive and negative effects.

19. Team training exhibits some of the characteristics of individual training. Individual training is, however, superior to team training, at least for simple or moderately complex tasks. Team training appears to be relatively ineffective in producing significantly better performance, as opposed to individual training.

20. High fidelity training conditions are important determiners of operational task performance. Input fidelity is more crucial than output fidelity.

21. Individual reinforcement during early team training overcomes the characteristic initial loss of proficiency that occurs in the transfer from individual to team training. However it does not reduce the total amount of team training required. During training, particularly in highly uncertain situations, teams should be given the opportunity to discover operating principles.

22. Operators will attempt to maximize the aspect of performance about which they receive specific feedback, even at the expense of other team performance characteristics. Teams experiencing a change in feedback criteria will adjust their outputs to maximize the aspect of performance being emphasized by the feedback. But if a change is made from single to compound criteria (feedback), teams will persist in emphasizing the single criterion they had used successfully before.

23. The effect of non-task-required communications during crew performance is to reduce the effectiveness of that performance. Interactions among operators should be minimized.

24. Visual methods of communicating information within the team are superior to verbal methods. The addition of a verbal channel to a visual one does not produce superior performance.

25. Differences in performance result from differences in communications structure. The more direct the transmission of information permitted by a structure, the better.

26. Variations in communications patterns are related to performance variations but not in any one-to-one way. Training in team communications does not significantly improve team performance.

27. The amount of information on the effects of motivation or other attitudinal variables on crew performance is too sparse to permit us to draw any meaningful conclusions.

DEVELOPMENTAL IMPLICATIONS

Most of the recommendations made were taken from Briggs and Johnston (1967), and refer primarily to the design of the training of prospective system operators.

1. In developing the system, create a hierarchical structure that has a certain degree of centralization. Develop parallel subsystems; avoid developing a series structure.

2. In major systems, avoid developing excessively large teams of operators.

3. Insofar as the system mission permits, opportunities for interaction between team members, particularly those requiring verbal communication, should be avoided in developing the system. Provide visual displays for communication of information rather than verbal channels.

4. In developing operating procedures, emphasize procedural flexibility as much as possible.

5. The following conditions should be emphasized in developing operator/team training:

(a) individual is preferable to interdependent (team-specific) training except when specific procedures for coordination must be taught;

(b) emphasize fidelity (similarity) of the training to the operational situation, concentrating on input fidelity;

(c) provide as much direct, individual feedback as possible;

(d) provide debriefing sessions following team training sessions to allow teams to explore alternative ways of organizing team tasks;

(e) emphasize in training the criteria that will determine success or failure in the operational situation;

(f) create teams out of operators who have achieved greatest individual proficiency before team training begins.

6. In system operations, do not be concerned about a moderate degree of personnel turnover. Ensure, however, that replacements are trained at least to the same skill level as the team to which the replacement is assigned.

7. In indeterminate systems requiring the use of teams, greater success will be achieved with teams structured along project lines rather than on a specialization basis.

8. Addition of a second watchkeeper in a monitoring situation often increases monitoring performance sizably. However this applies only to the monitoring situation, and in any case the addition of more than one watchkeeper is usually not warranted.

GENERAL REFERENCES

Acton Society Trust. *Size and Morale.* London: Acton Society Trust, 1953.

Adam, E. E. An analysis of changes in performance quality with operant conditioning procedures. *J. Appl. Psychol.,* 1972, **56,** 480–486.

Adams, S. Status congruency as a variable in small group performance. *Social Forces,* 1953, **32,** 16–22.

Anderson, L. R. Small group behavior. In S. Seashore and R. McNeil, (Eds.), *The Management of Crisis: Metropolitan Government and the Behavioral Sciences.* New York: Free Press, 1970, pp. 204–256.

Argyle, M. et al. Supervisory methods related to productivity, absenteeism and labor turnover. *Hum. Relat.,* 1958, **11,** 23–40.

Baloff, N. and Becker, S. W. A model of group adaptation to problem-solving tasks. *Organ. Behav. Hum. Performance,* 1968, **3,** 217–238.

Baumgartel, H. and Sobol, R. Background and organizational factors in absenteeism. *Pers. Psychol.,* 1959, **12,** 431–443.

Bavelas, A. Communication patterns in task oriented groups. *J. Acoust. Soc. Am.,* 1950, **22,** 725–730.

Behan, R. A. An experimental investigation of the interaction between problem load and level of training. *Hum. Factors,* 1961, **3,** 53–59.

Bergum, B. O. and Lehr, O. J. Vigilance performance as a function of paired monitoring. *J. Appl. Psychol.,* 1962, **46,** 341–343.

Berkowitz, L. Group norms among bomber crews: Patterns of perceived crew attitudes, "actual" crew attitudes, and crew liking related to aircrew effectiveness in far eastern combat. *Sociometry,* 1956, **19,** 141–153.

Berry, P. C. *An exploration of the interrelation among some non-intellectual predictors of achievement in problem solving.* Technical report, Contract Nonr 609–20, Dept. of Psychology, Yale University, New Haven, Conn., 1958.

Briggs, G. E. and Johnson, W. A. *Team training research.* Report NAVTRADEVCEN 1327–2, Naval Training Device Center, Port Washington, N.Y. November 1965. (AD 477–963.)

Briggs, G. E. and Johnston, W. A. Stimulus and response fidelity in team training. *J. Appl Psychol.,* 1966, **50,** 114–117.

Briggs, G. E. and Johnston, W. A. *Team training.* Report NAVTRADEVCEN 1327–4, Naval Training Device Center, Orlando, Fla., June 1967. (AD 660–019.)

Briggs, G. E. and Naylor, J. C. *Experiments on team training in a CIC-type task environment,* Report NAVTRADEVCEN 1327–1, Naval Training Device Center, Port Washington, N.Y., June 1964. (AD 608–309.)

Briggs, G. E. and Wiener, E. L. Influence of time sharing and control loading on transfer of training. *J. Appl. Psychol.,* 1966, **50,** 201–203.

Campbell, J. P. Individual versus group problem solving in an industrial sample. *J. Appl. Psychol.,* 1968, **52,** 205–210.

Clement, D. E. Group performance in a visual signal-detection task. *Hum. Factors,* 1973, **15,** 43–52.

Clement, D. E. and Schiereck, J. J. Sex composition and group performance in a visual signal detection task. *Mem. Cognition,* 1973, **1,** 251–255.

Cohen, A. M. et al. The effects of changes in communications networks on the behaviors of problem-solving groups. *Sociometry,* 1962, **25,** 177–196.

Cohen, A. M. et al. Experiments in organizational embeddedness. *Adm. Sci. Qt.,* 1969, **14,** 208–221.

Cohen, J. D. and McKelvey, R. K. *The behavior of individuals and personnel systems in the surveillance functions of an air defense direction center. III. Distribution of responses with respect to job functions.* Report AFPTRC-TR-54-100, Crew Research Lab., Air Force Personnel and Training Research Center, Randolph AFB, Texas, December 1954. (AD 64482.)

Davis, J. H. Group decision and social interaction: A theory of social decision schemes. *Psychol. Rev.,* 1973, **80,** 97–125.

Davis, J. H. *Group Performance.* Reading, Mass.: Addison-Wesley, 1969.

Davis, J. H. and Restle, F. The analysis of problems and predictions of group problem solving. *J. Abnorm. Soc. Psychol.,* 1963, **66,** 103–116.

Dunnette, M. D. et al. The effect of group participation on brainstorming effectiveness for two industrial samples. *J. Appl. Psychol.,* 1963, **47,** 30–37.

Egerman, K. et al. *Increasing team proficiency through training. 3. Detrimental effects of reinforcement in teams with redundant members.* Technical report AIR-B64-6/62-TR, American Institute for Research, Pittsburgh, June 1962. (AD 276-428.)

Faust, W. L. Group versus individual problem solving. *J. Abnorm. Soc. Psychol.,* 1959, **59,** 68–72.

Fineberg, M. L. Navigation and flight proficiency under nap of the earth conditions as a function of aviator training and experience. *Proceedings,* Human Factors Society Meeting, October 1974, pp. 249–254.

Fox, D. J. and Lorge, I. The relative quality of decisions written by individuals and by groups as the available time for problem solving is increased. *J. Soc. Psychol.,* 1962, **57,** 227–242.

Frank, F. and Anderson, L. R. Effects of task and group size upon group productivity and member satisfaction. *Sociometry,* 1971, **34,** 135–149.

Friedlander, F. and Brown, L. D. Organization development. *Ann. Rev. Psychol.,* 1974, **25,** 313–341.

Gibb, J. R. The effects of group size and of threat reduction upon creativity in a problem-solving situation. *Am. Psychol.,* 1951, **6,** 324 (abstract).

Glanzer, M. and Glaser, R. *A review of team training problems.* Technical report Contract N7onr-37008, American Institute for Research, Pittsburgh, September 1955.

Glanzer, M. and Glaser, R. Techniques for the study of group structure and behavior: II. Empirical studies of the effects of structure in small groups. *Psychol. Bull.,* 1961, **58,** 1–27.

Glaser, R. et al., *A study of some dimensions of team performance.* Tech. Rpt. Contract N7onr-37008, American Institute for Research, Pittsburgh, September 1955.

Glaser, R. et al. *Increasing team proficiency through training.* 2. *The acquisition and extinction of a team response.* Technical report AIR-B64-5/62-TR, American Institute for Research, Pittsburgh, May 1962. (AD 276-429.)

Greer, F. L. et al. Interpersonal knowledge and individual and group effectiveness. *J. Abnorm. Soc. Psychol.,* 1954, **49,** 411–414.

Guetzkow, H. and Dill, W. Factors in the organizational development of task oriented groups. *Sociometry,* 1957, **20,** 175–204.

Guetzkow, H. and Simon, H. The impact of certain communication nets upon organization and performance in task-oriented groups. *Manage. Sci.,* 1955, **1,** 233–250.

Hackman, J. R. and Morris, C. G. *Group tasks, group interaction process, and group performance effectiveness: A review and proposed integration.* Technical report 7, School of Organization and Management, Yale University, 1974. To be published in L. Berkowitz (Ed.), *Advances in Experimental Social Psychology,* New York: Academic Press.

Hall, E. J. et al. Group problem solving effectiveness under conditions of pooling vs. interaction. *J. Soc. Psychol.,* 1963, **59,** 147–157.

Havron, M. D. et al. *The effectiveness of small military units.* Personnel Research Section Report No. 980, Adjutant General's Office, Washington, D.C., 1952.

Havron, M. D. and McGrath, J. E. The contribution of the leader to the effectiveness of small military groups. In L. Petrullo and B. M. Bass (Eds.), *Leadership and Interpersonnel Behavior.* New York: Holt, Rinehart & Winston, 1961.

Haythorn, W. W. The composition of groups: A review of the literature. *Acta Psychol.,* 1968, **28,** 97–128.

Heise, G. A. and Miller, G. A. Problem solving by small groups using various communication nets. *J. Abnorm. Soc. Psychol.,* 1951, **46,** 327–336.

Hewitt, D. and Parfitt, J. A note on working morale and size of group. *Occup. Psychol.,* 1953, **27,** 38–42.

Hinrichs, J. R. Psychology of men at work. *Ann. Rev., Psychol.,* 1970, **21,** 519–554.

Hoffman, L. R. Similarity of personality: A basis for interpersonal attraction. *Sociometry,* 1955, **21,** 300–308.

Hoffman, L. R. Homogeneity of member personality and its effects on group problem-solving. *J. Abnorm. Soc. Psychol.,* 1959, **58,** 27–32.

Hoffman, L. R. and Maier, N. R. F. Quality and acceptance of problem solutions by members of homogenous and heterogeneous groups. *J. Abnorm. Soc. Psychol.,* 1961, **62,** 401–407.

Hoffman, L. R. and Maier, N. R. F. An experimental re-examination of the similarity-attraction hypothesis. *J. Pers. Soc. Psychol.,* 1966, **3,** 145–152.

Hoffman, L. R. and Smith, C. G. Some factors affecting the behavior of members of problem-solving groups. *Sociometry,* 1960, **23,** 273–291.

Hornseth, J. P. and Davis, J. H. Individual and two-man team target finding performance. *Hum. Factors,* 1967, **9,** 39–43.

Horrocks, J. E. et al. *Team training II: Individual learning and team performance.* Report NAVTRADEVCEN 198-2, Naval Training Device Center, Port Washington, N.Y. August 1960. (AD 247–147.)

Horrocks, J. E. et al. *Team training III: An approach to optimum methods and procedures.* Report NAVTRADEVCEN 198-3, Naval Training Device Center, Port Washington, N.Y. August 1961. (AD 267–666.)

Howell, W. C. et al. *System performance following radar failure in a simulated air traffic control situation.* Report WADC-TR-59-573, Wright Air Development Center, Wright Patterson AFB, Ohio, September 1959. (AD 234–011.)

Husband, R. W. Cooperative versus solitary problem solution. *J. Soc. Psychol.,* 1940, **11,** 405–409.

Indik, B. P. Some effects of organization size on member attitudes and behavior. *Hum. Relat.,* 1963, **16,** 369–384.

Indik, B. and Seashore, S. *Effects of organization size on member attitudes and behavior.* Survey Research Center of the Institute for Social Research, University of Michigan, Ann Arbor, 1961.

Ingham, G. *Size of Industrial Organization and Worker Behavior.* Cambridge: Cambridge University Press, 1970.

Jones, M. B. Regressing group on individual effectiveness. *Organ Behav. Hum. Performance,* 1974, **11,** 426–451.

Kerr, W. A. et al. Absenteeism, turnover and morale in a metals fabrication factory. *Occup. Psychol.,* 1951, **25,** 50–55.

Kidd, J. S. Social influence phenomena in a task-oriented group situation. *J. Abnorm. Soc. Psychol.,* 1958, **56,** 13–17.

Kidd, J. S. *A summary of research methods, operator characteristics, and system design specifications based on the study of a simulated radar air traffic control system.* Report WADC-TR-59-236, Wright Air Development Center, Wright-Patterson AFB, Ohio, July 1959. (AD 235-126.)

Kidd, J. S. A comparison of one-, two- and three-man work units under various conditions of work load. *J. Appl. Psychol.,* 1961, **45,** 195–200.

Kidd, J. S. and Hooper, J. J. *Division of responsibility between two controllers and load balancing flexibility in a radar approach control team.* Report WADC-TR-58-473, Wright Air Development Center, Wright-Patterson AFB, Ohio, April 1959. (AD 214–616.)

Kidd, J. S. and Kinkade, R. G. *Operator changeover effects in a complex task.* WADC 59–235, Wright Air Development Center, Wright-Patterson AFB, Ohio, August 1959. (Also *J. Engin. Psychol.,* 1962, **1,** 82–91.)

Kinkade, R. G. and Kidd, J. S. *The effect of team size and intermember communication on decision-making performance.* Report WADC-TR-58-474, Wright Air Development Center, Wright-Patterson AFB, Ohio, April 1959. (AD 215–621.)

Klaus, D. J. and Glaser, R. *Increasing team proficiency through training. 5. Team learning as a function of member learning characteristics and practice conditions.* Report AIR-E1-4/65-TR, American Institute for Research, Pittsburgh, May 1965. (AD 471-469.)

Klaus, D. J. and Glaser, R. *Increasing team proficiency through training.* 8. *Final summary report.* Report AIR-E1-6/68-TR, American Institutes for Research, Pittsburgh, May 1968. (AD 669-688.)

Leavitt, H. J. Some effects of certain communication patterns on group performance. *J. Abnorm. Soc. Psychol.,* 1951, **46,** 38–50.

Lefkowitz, J. Effect of training on the productivity and tenure of sewing machine operators. *J. Appl. Psychol.,* 1970, **54,** 81–86.

Lorge, I. and Solomon, H. Two models of group behavior in the solution of Eureka-type problems. *Psychometrika,* 1955, **20,** 139–148.

Lorge, I. et al. Solutions by teams and by individuals to a field problem at different levels of reality. *J. Educ. Psychol.,* 1955, **46,** 17–24.

Lorge, I. et al. A survey of studies contrasting the quality of group performance and individual performance, 1920–1957. *Psychol. Bull.,* 1958, **55,** 337–372.

Lorge, I. et al. Evaluation of decisions written by ad hoc groups and simulated commanders. In A. H. Rebenstein and C. J. Haberstroh, (Eds.), *Some Theories of Organization.* Homewood, Ill.: Irwin, 1960, pp. 448–451.

Luce, R. D. et al. *Information flow in task-oriented groups.* Technical report 264, Research Laboratory of Electronics, Massachusetts Institute of Technology, Cambridge, Mass., August 1953.

Mandell, M. *Recruiting and selecting office employees.* New York: American Management Association, 1956.

Marquardt, D. I. Group problem solving. *J. Soc. Psychol.,* 1955, **41,** 103–113.

McGrath, J. E. and Altman, I. *Small Group Research.* New York: Holt, Rinehart & Winston, 1966.

Meadow, A. et al. Influence of brainstorming instructions and problem sequence on a creative problem-solving test. *J. Appl. Psychol.,* 1959, **43,** 413–416.

Meister, D. A critical review of human performance reliability predictive methods. *IEEE Trans. Reliab.,* 1973, **R-22,** 116–123.

Metzner, H. and Mann, F. Employee attitudes and absences. *Pers. Psychol.,* 1953, **6,** 467–485.

Miner, J. B. and Dachler, H. P. Personnel attitudes and motivation. *Ann. Rev. Psychol.,* 1973, **24,** 379–402.

Mohanna, A. I. and Argyle, M. A. A cross-cultural study of structured groups with unpopular central members. *J. Abnorm. Soc. Psychol.,* 1960, **60,** 139–140.

Murphy, G. and Newcomb, T. M. *Experimental Social Psychology.* New York: Harper & Row, 1937.

O'Brien, G. E. and Owens, A. G. Effects of organizational structure on correlations between member abilities and group productivity. *J. Appl. Psychol.,* 1969, **53,** 525–530.

Osborn, A. F. *Applied Imagination.* New York: Scribner's, 1957.

Pelz, D. C. Some social factors related to performance in a research organization. *Adm. Sci. Qt.,* 1956, **1,** 310–325.

Porter, L. W. and Lawler, E. E. Properties of organization structure in relation to job attitudes and job behavior. *Psychol., Bull.,* 1965, **64,** 23–51.

Porter, L. W. and Roberts, K. H. Communication in organizations. Technical report 12, Contract N00014-69--A-0200-9001, University of California, Irvine, July 1972. Also published in M. D. Dunnette (Ed.), *Handbook of Industrial and Organizational Psychology.* Skokie, Ill.: Rand McNally, 1973.

Porter, L. W. and Steers, R. M. Organizational, work, and personal factors in employee turnover and absenteeism. *Psychol. Bull.,* 1973, **80,** 151–176.

Revans, R. W. Human relations, management and size. In E. M. Hugh-Jones (Ed.), *Human Relations and Modern Management.* Amsterdam: North Holland, 1958 (pp. 177–220).

Roby, T. B. and Lanzetta, J. T. *An investigation of task performance as a function of certain aspects of work-group structure.* Report AFPTRC-TN-56-74, Air Force Personnel and Training Research Center, Lackland AFB, Texas, June 1956. (AD 109-174.)

Roby, T. B. and Lanzetta, J. T. *A replication study of work group structure and task performance.* Report AFPTRC-TN-57-85, Air Force Personnel Training and Research Center, Lackland AFB, Texas, June 1957. (a) (AD 134-205.)

Roby, T. B. and Lanzetta, J. T. Conflicting principles in man-machine system design. *J. Appl. Psychol.,* 1957, **41,** 170–178.(b)

Rohde, K. J. Theoretical and experimental analysis of leadership ability. *Psychol. Rep.,* 1958, **4,** 243–278.

Rosenberg, S. and Hall, R. L. The effects of different social feedback conditions upon performance in dyadic teams. *J. Abnorm. Soc. Psychol.,* 1958, **57,** 271–277.

Rosenberg, S. et al. Some effects of varying combinations of group members on group performance measures and leadership behaviors. *J. Abnorm. Soc. Psychol.,* 1955, **51,** 195–203.

Rotter, G. S. and Portugal, S. M. Group and individual effects in problem solving. *J. Appl. Psychol.,* 1969, **53,** 338–341.

Sanders, M. G. et al. Communication during terrain flight. Report 75–13, US Army Aeromedical Research Laboratory, Ft. Rucker, Al., March 1975. (AD A009336.)

Schafer, T. H. *Detection of a signal by several observers.* Research report 101, U.S. Navy Electronics Laboratory, San Diego, Calif., 1949.

Schoner, B. et al. Quality of decisions: Individuals versus real and synthetic groups. *J. Appl. Psychol.,* 1974, **59,** 424–432.

Shaw, M. E. A comparison of individuals and small groups in the rational solution of complex problems. *Am. J. Psychol.,* 1932, **44,** 491–504.

Shaw, M. E. Some effects of problem complexity upon problem solution efficiency in different communication nets. *J. Exp. Psychol.,* 1954, **48,** 211–217.

Shaw, M. E. Acceptance of authority, group structure, and the effectiveness of small groups. *J. Pers.*, 1959, **27**, 196–210.

Shaw, M. E. Communication networks. In L. Berkowitz (Ed.), *Advances in Experimental Social Psychology*. New York: Academic Press, 1964, pp. 111–147.

Short, J. G. et al. *Increasing team proficiency through training. 7. The simulation of team environments.* Report AIR-E1-5/68-TR, American Institute for Research, Pittsburgh, May 1968. (AD 669-687.)

Shure, G. H. et al. Group planning and task effectiveness. *Sociometry*, 1962, **25**, 263–282.

Siskel, M. et al. *Intra-crew communication of B-52 and KC-135 student and combat crews during selected mission segments.* Report AMRL-TR-65-18, *Aerospace Medical Research Laboratories*, Wright-Patterson AFB, Ohio, May 1965. (AD 617-598.)

Spector, P. and Suttell, B. J. *An experimental comparison of the effectiveness of three patterns of leadership behavior.* Report AIR-196-57-FR-164, American Institute for Research, Washington, D.C., 1957.

Stafford, A. R. et al. *The effects of choice of working partner on student achievement and attitudes.* Report AFPTRC-TN-55-61, Air Force Personnel and Training Research Center, Lackland AFB, Texas, 1955.

Steiner, I. D. Models for inferring relationships between group size and potential group productivity. *Behav. Sci.*, 1966, **11**, 273–283.

Sweeney, E. J. *Sex differences in problem solving.* Technical report 1, Contract Nonr 251–25 Department of Psychology, Stanford University, Stanford, Calif., 1953. (AD 27343.)

Taylor, D. W. and Faust, W. L. Twenty questions: Efficiency in problem solving as a function of size of group. *J. Exp. Psychol.*, 1952, **44**, 360–368.

Taylor, D. W. et al. Does group participation when using brainstorming facilitate or inhibit creative thinking? *Adm. Sci. Qt.*, 1958, **3**, 23–47.

Thomas, E. J. and Fink, C. F. Effects of group size. *Psychol. Bull.*, 1963, **60**, 371–384.

Triandis, H. C. et al. Member heterogeneity and dyadic creativity. *Hum. Relat.*, 1965, **18**, 33–55.

Tuckman, B. W. Group composition and group performance of structured and unstructured tasks. *J. Exp. Soc. Psychol.*, 1967, **3**, 25–40.

Waag, W. L. and Halcomb, C. G. Team size and decision rule in the performance of simulated monitoring teams. *Hum. Factors*, 1972, **14**, 309–314.

Williges, R. C. et al. Role of verbal communication in teamwork. *J. Appl. Psychol.*, 1966, **50**, 473–478.

Van Beek, H. G. The influence of assembly line organization on output, quality and morale. *Occup. Psychol.*, 1964, **38**, 161–172.

Van Zelst, R. H. Validation of sociometric regrouping procedure. *J. Abnorm. Soc. Psychol. (Morton Prince Suppl.)*, 1952, **47**, 299–301.

Zajonc, R. B. and Taylor, J. J. The effect of two methods of varying group task difficulty on individual and group performance. *Hum. Relat.*, 1963, **16**, 359–368.

CASE REFERENCES

Case 115. Taylor, D. W. and Faust, W. L. Twenty questions: Efficiency in problem solving as a function of size of group. *J. Exp. Psychol.*, 1952, **44**, 360–368.

Case 116. Bouchard, T. J. and Hare, M. Size, performance and potential in brainstorming groups. *J. Appl. Psychol.*, 1970, **54**, 51–55.

Case 117. Morgan, B. B. and Alluisi, E. A. On the inferred independence of paired watchkeepers. *Psychonomic Sci.,* 1965, **2,** 161–162.

Case 118. Weiner, E. L. The performance of multi-man monitoring teams. *Hum. Factors,* 1964, **6,** 179–184.

Case 119. Clement, D. E. Group performance in a visual signal-detection task. *Hum. Factors,* 1973, **15,** 43–52.

Case 120. Hornseth, J. P. and Davis, J. H. Individual and two-man team target finding performance. *Hum. Factors;* 1967, **9,** 39–43.

Case 121. Marriott, R. Size of working group and output. *Occup. Psychol.,* 1949, **23,** 47–57.

Case 122. Kennedy, J. L. The system approach: A preliminary exploratory study of the relation between team composition and financial performance in business games. *J. Appl. Psychol.,* 1971, **55,** 46–49.

Case 123. Roby, T. B. and Lanzetta, J. T. A study of an "assembly effect" in small group task performance. *J. Soc. Psychol.,* 1961, **53,** 53–68.

Case 124. Hoffman, L. R. and Maier, N. R. F. Sex differences, sex composition, and group problem solving. *J. Abnorm. Soc. Psychol.,* 1961, **63,** 453–456.

Case 125. Trow, D. B. *Team work under turnover and succession.* Technical report 2, Contract Nonr 3679(00), Harpur College, June 1964. (AD 601–816.)

Case 126. Forgays, D. G. and Levy, B. I. *Combat performance characteristics associated with changes in the membership of medium-bomber crews.* Report AFPTRC-TN-57-140, Air Force Personnel and Training Research Center, Lackland AFB, Texas, 1957. (AD 146–414.)

Case 127. Horrocks, J. E. et al. *Team training III: An approach to optimum methods and procedures.* Report NAVTRADEVCEN 198-3, Naval Training Device Center, Port Washington, N.Y., August 1961. (AD 267–666.)

Case 128. Comrey, A. L. Group performance in a manual dexterity task. *J. Appl. Psychol.,* 1953, **37,** 207–210.

Case 129. Gilster, H. L. *A combat crew production function.* USAF Academy, Colorado, Technical report 69-1, SAC-ORI Personnel Study, September 1969. (AD 698-336.)

Case 130. Klaus, D. J. and Glaser, R. *Increasing team proficiency through training. 5. Team learning as a function of member learning characteristics and practice conditions.* Report AIR-E1-4/65-TR, American Institute for Research, Pittsburgh, May 1965. (AD 471-469.)

Case 131. Egerman, K. Effects of team arrangement on team performance: A learning-theoretic analysis. *J. Pers. Soc. Psychol.,* 1966, **3,** 541–550.

Case 132. Doten, G. W. et al. The use of teams in image interpretation: information exchange, confidence and resolving disagreements. *Hum. Factors,* 1968, **10,** 107–116.

Case 133. Briggs, G. E. and Naylor, J. C. Team versus individual training, training task fidelity, and task organization effects on transfer performance by three-man teams. *J. Appl. Psychol.,* 1965, **49,** 387–392.

Case 134. Vroom, V. H. et al. The consequences of social interaction in group problem solving. *Organ. Behav. Hum. Performance,* 1969, **4,** 77–95.

Case 135. Shifflett, S. C. Performance effectiveness and efficiency under different dyadic work strategies. *J. Appl. Psychol.,* 1973, **57,** 257–263.

Case 136. Stone, T. H. Effects of mode of organization and feedback level on creative task groups. *J. Appl. Psychol.,* 1971, **55,** 324–330.

Case 137. Kidd, J. S. and Kinkade, R. G. *Air traffic control system effectiveness as a function of the division of responsibility between pilots and ground controllers: A study in human engineering aspects of radar air traffic control.* Report WADC-TR-58-113, Wright Air Development Center, Wright-Patterson AFB, Ohio, June 1958. (AD 151-082.)

Case 138. Klaus, D. J. and Glaser, R. *Increasing team proficiency through training. 8. Final summary report.* Report AIR-E1-6/68-FR, American Institute for Research, Pittsburgh, May 1968. (AD 669-688.)

Case 139. Briggs, G. E. and Johnston, W. A. Influence of a change in system criteria on team performance. *J. Appl. Psychol.,* 1966, **50,** 467–472.

Case 140. Klaus, D. J. et al. *Increasing team proficiency through training. 6. Supervisory furnished reinforcement in team training.* Report AIR-E1-5/65-TR, American Institute for Research, Pittsburgh, May 1965. (AD 471-470.)

Case 141. Shure, G. H. et al. Group planning and task effectiveness. *Sociometry,* 1962, **25,** 263–282.

Case 142. Johnston, W. A. Transfer of team skills as a function of type of training. *J. Appl. Psychol.,* 1966, **50,** 102–108.

Case 143. Briggs, G. E. and Johnston, W. A. *Laboratory research on team training.* Report NAVTRADEVCEN 1327-3, Naval Training Device Center, Port Washington, N.Y., May 1966. (AD 485-636.)

Case 144. Roby, T. B. and Lanzetta, J. T. *An investigation of task performance as a function of certain aspects of work-group structure.* Report AFPTRC-TN-56-74, Air Force Personnel and Training Research Center, Lackland AFB, Texas, June 1956. (AD 109-174.)

Case 145. Krumm, R. L. and Farina, A J. *Effectiveness of integrated flight simulator training in promoting B-52 crew coordination* Report AMRL-TDR 62-1, Aerospace Medical Research Laboratories, Wright-Patterson AFB, Ohio, February 1962. (AD 273-891.)

Case 146. Federman, P. J. and Siegel, A. I. *Communications as a measurable index of team behavior.* Report NAVTRADEVCEN 1537-1, Naval Training Device Center, Port Washington, N.Y., October 1965. (AD 623-135.)

Case 147. Siegel, A. I. and Federman, P. J. Communications content training as an ingredient in effective team performance. *Ergonomics,* 1973, **16,** 403–416.

Case 148. Briggs, G. E. and Johnston, W. A. *Team training.* Report NAVTRADEVCEN 1327-4, Naval Training Device Center, Orlando, Fla., June 1967. (AD 660–019.)

VI
FEEDBACK

SUMMARY. Feedback, which permits the system to regulate itself, may provide various types of information (response confirmation, information about response correctness, and consequences, and system status data). It can guide, motivate, and reinforce the operator. The temporal, quantitative, and qualitative variables that modify its effectiveness are discussed. Although feedback is a powerful tool in both learning and operational situations (any feedback is better than none at all), the operator is somewhat insensitive to minor variations in feedback structure. Within the multiman system, feedback may be presented at various levels of remoteness from the operator, and this creates problems of data confounding. Operator goals are complexly related to feedback effects and may mediate these effects. The evidence for feedback transfer to the operational situation is equivocal.

INTRODUCTION

Feedback (F) is essential to regulation of the system, just as it is to the regulation of the individual. In both cases performance must be evaluated and then modified to bring it into accord with criteria. This cycle is repeated continuously. Without F, evaluation is difficult and modification is imprecise.

For both the individual and the system, the events to which F is compared are always one's own response or those of one's team or the outputs of one's own system. Within a single-operator system the operator perceives F in terms of his own responses; in a multiman system he may also be concerned about how his crew is performing. It is easier, however, for the operator to regulate his own behavior by means of F than that of his team or the system as a whole.

Like load or stress or drive, F is a theoretical construct. The operator who receives F does not observe F directly but infers it from the stimuli that present F (e.g., a display or a test grade) and from the effects of those stimuli. The researcher in the area of feedback cannot be sure that what he presents is in fact perceived and accepted by the operator as F. He must rely on the effect of F on operator responses and system outputs to determine whether what he presented was indeed effective as F.

Nevertheless, there can be no question that certain phenomena that we call F occur in almost all situations. Feedback as commonly studied by psychologists is F employed to assist learning, where it is usually highly effective in modifying performance. Our primary interest is in F as it affects the system and the operator, and, particularly, the system functioning operationally with trained personnel rather than students. The system developer wants to determine when and how to build F into a system. This is not an easy task, because an operational system does not function with anything as obvious as an instructor calling out correct answers. Since most of our data describe F as it functions in relation to the individual and in the learning context, it is necessary to extrapolate these data to the system. The same mechanisms may or may not be involved in both cases, however.

Definitions

In the most general sense, F in the system context is information provided to the operator or the crew concerning the events that follow upon the actions that have been performed.*

These events are then referred back (in the operator's memory or by means of a display/hard copy printout) to the original action to determine the appropriateness of that action. Concurrently the operator uses this evaluation to guide the performance of his next action. Therefore F has effects in both feeding *back* and feeding *forward.*

The effect of F is to modify performance in various ways: (1) by *directing* or guiding changes in the operator's procedures and performance to satisfy system requirements, (2) by *motivating* the operator or the crew to adjust work output to other (presumably higher) goals, and (3) by *reinforcing* certain response patterns that are seen by the operator/crew as being correct or successful.

These functions need not be mutually exclusive; the same F may have all these effects, although depending on the task and the task conditions, one or the other may be primary. The question of which function is most important is also essentially irrelevant, since all depend on the information provided by the F. It would in any event be difficult to differentiate these functions operationally.

Each function may, however, be affected somewhat differently by F parameters. It seems reasonable, as Goldstein et al. (1968) suggest, that

*The expression "knowledge of results" (KOR) is often used interchangeably with F.

the percentage of F required to maintain high motivation is relatively fixed and rather small, whereas that necessary for proper direction varies with task complexity, specificity of instruction, and availability of other sources of guidance.

These functions may also be used as criteria of effectiveness of F information, which presumably is effective to the extent that it guides and reinforces correct responses and motivates the operator to work harder. Feedback may provide different types of information relating to the confirmation, correctness, and consequences of the response, and to system status.

***Response Correctness* (Rcr)**

Information provided to the operator (or others who need to know about his performance) that he has in fact performed an operation is designated Rcf. For example, an indicator illuminates after the operator has thrown a power switch to the ON position. Since Rcf says nothing about the longer range consequences of the action taken, it has extremely limited utility except when the operator must make many discrete responses in a fixed sequence and Rcf helps him to order these, and when the consequences of an incorrect act or failure to act would be very serious.

***Research Correctness* (Rcr)**

Information about the correctness or incorrectness of the operator's response (Rcr) is required when several response alternatives are possible and the correct choice is not known to the operator in advance; thus response performance in the learning situation usually is followed by an indication of the correctness (or incorrectness) of the response (or a series of responses). This type of F (commonly called *augmented F*) is found only in a learning situation (whether it be in a classroom or in the workplace). In a nonlearning (mission performance) situation, knowledge of response correctness is determined by means of response consequences (see below). Rcr has little applicability to operational systems in which the judgment of response correctness is mediated by the determination of response consequences. (Situations such as supervisor monitoring/correction or computer checking of operator responses involve tutoring, even though such tutoring may be incorporated into normal system operations.)

***Response Consequences* (Rcn)**

Information about the consequences of the action taken (Rcn) is supplied, for example, when turning the ignition key in an automobile is followed by the sound of the engine turning over. Rcn subsumes the previous functions. First, it confirms the response made by the operator, which is trivial. Much more important in operational situations that do not include tutoring, the correctness of a response (Rcr) can be inferred only from its consequences. For example, in a battle situation the success or failure of an action is determined by its effect on the enemy. Rcn may also serve to cue the operator to perform the next response in sequence. The sound of the automobile engine causes the driver to depress the gas pedal. (This is not, however, peculiar to Rcn. All F can perform this function.)

System Status (Rss)

Information about the condition of one's own or another system or the external environment, on the basis of which an operator or crew must act is called Rss. For example, when a voltmeter that displays generator output presents an out-of-tolerance indication, it keys the operator to some maintenance action. Rss differs from other F functions in that this information is not necessarily (or even frequently) the *immediate* consequence of or follow-on to a specific operator/crew action; it may reflect system events that have been put in motion by much earlier operator actions. It does, however, provide information that regulates operator and system actions in the sense that when a particular status condition occurs, the operator must often take action to maintain the integrity of his system.

Just as general F functions are not mutually exclusive, an item of F may provide several types of information. For example, Rss may combine with Rcr, as when a change in the generator reading indicates that a previous maintenance action was the correct one to have taken.

It should be apparent that the concept of F addressed in this chapter is more inclusive than that ordinarily implied by the common term KOR, as it describes individual actions. It includes any representation of operator, system, or environmental status resulting from own or other system actions on the basis of which regulatory action can be taken.

Feedback is probably more important in indeterminate systems than ir structured ones. When operating procedures are preprogrammed, the program includes at least some of the information (e.g., cueing) F would have provided. In probabilistic (uncertain) situations, on the other hand, F may be the only basis for guiding the operator to select among alternatives.

Intrinsic and Extrinsic F

The four types of information do not automatically trigger a specific function. Feedback information does not automatically guide, motivate, or reinforce—these functions are determined by the operator's interpretation of the information.

It is also necessary to distinguish between *intrinsic* and *extrinsic* F. Intrinsic F is commonly thought of as being inherent in operator actions. This simply means that the information or cues on the basis of which the operator interprets F information are assumed to be part of or correlated with the action that produced the F.

This statement is admittedly quite vague. By one interpretation intrinsic F exists only as a physiological correlate of the individual response and, therefore, would have no exact counterpart on the system level. On the other hand, if F is intrinsic when it is part of the total system, the MMS does contain intrinsic F. For example, information describing the status of one's own system can be considered intrinsic because it is inherent in the situation (i.e., is displayed by system displays used routinely); it is F because it describes the effects of system (including personnel) operations. The same might be said about information describing external systems or environments that are routinely monitored by own system displays.

Extrinsic F is not inherent in the operator action or system operations but is supplied by an external source. For example, information describing the correct response in an instructional sequence is extrinsic because it is provided by an instructor or an answer sheet. All four types of F information may be either intrinsic or extrinsic, depending on whether their source is inherent in the system or outside it.

All F is veridical; it describes what has happened within its capability of providing information. Extrinsic F is commonly assumed to be correct because the external (tutorial) source should not deliberately provide misinformation. Intrinsic F also should be completely truthful because it is not under the operator's conscious control. In both cases, however, the F information provided may be incomplete or imprecise, or the operator's perception and comprehension of the F may not be completely accurate. For example, an engine fire warning indicator that illuminates is assumed to be correct by the pilot, although it may be based on a display malfunction. The indicator reported what it was supposed to report; the pilot drew the wrong conclusion because the F information was incomplete. Feedback may not be veridical in the sense that the operator's interpretation of its meaning may not correctly mirror "objective truth" (whatever that is). This is the case even when the F supposedly describes the operator's own responses and his own system.

Structural Variables

Structural variables apply to all F information. The most important describe the manner in which F can be presented.

Time Factors

Feedback can vary in terms of the rapidity with which F stimuli appear following completion of an operator action or the occurrence of a system state that is to be displayed.

It is logical to assume (but remains to be demonstrated) that F is more effective in cueing the correct next response, the closer it occurs in time to the operator's previous response or the system state that elicits Rss. When a prolonged interval elapses between an action and its F, the operator has the additional task of *remembering* the relationship between the two. The interval, which may be filled by other required actions or system events, may distort the operator's interpretation of the relationship between his or the system's response and its F.

Feedback delay may vary from a fraction of a second in the physiological concomitants of an individual response to many minutes in systems; in the latter case, F describes other systems in the environment (e.g., an enemy force) or its own long-duration system events. Whoever controls extrinsic F can decide on the duration of that interval; ordinarily it is presented reasonably quickly after the occurrence of an action or event. Sometimes this is not possible, though, and the question then arises: how long can F be delayed before it loses effectiveness? In intrinsic F, the nature of the action, the task, or the environment ordinarily determines the F interval; and then we must determine what system factors increase F delay. In indeterminate systems, difficulty in recognizing the occurrence of F stimuli (what constitutes the F) may make it appear that F is unduly delayed. The HF researcher is interested in determining the maximally permissable interval and what the effects of F delay are.

Quantitative Factors

Feedback also can vary in terms of the *frequency* of its occurrence and *amount* of information it provides. In the real world, F may not be available for every response; it may be prohibitively expensive to provide it for every system event. When F is available, all the information it should or could provide may not be supplied. On the other hand, if all available information were provided, it might overload or confuse the operator. What then is the minimal frequency and amount of F to permit effective system operation?

This question is critical both in the learning situation and in real world system operations. What one can do about frequency and amount depends on the degree of control one has over F parameters. In the learning situation one can control Rcr; in the real world system, Rss. First, however, we must determine the extent to which F must and can be controlled.

Quality Factors

Quality factors include specificity and relevance.

Specificity is defined in terms of two factors. (1) The distinction between quantitative and qualitative (outcome) information. For example, in a battlefield situation, battalion G3 may be told by one of his attacking units that "almost all ammunition is expended" (qualitative F) or "we have 20 rounds per man left" (quantitative). (2) More detailed categories of information, which overlap with amount of information because additional F dimensions may provide additional information.

Logically, the more specific the F, the more effective its guidance should be. The research question is: how much is F effectiveness improved by additional specificity?

Relevance is determined by the logic of the relationship (as perceived by the operator) between F and the action that elicited F. The clearer that relationship is, the more correctly F will guide further action. Ordinarily relevance would pose little problem to the operator, but this factor can be distorted by many other factors, such as F delays, multiple-F dimensions and alternative evaluative criteria. All of these make it difficult to tie F stimuli to the response that elicited them.

Irrelevant F is less important in learning situations because the instructor can exert some control over the relevance of the F he provides. In indeterminate system operations, however, where the system has less control over the F provided to the operator, the logical relationship that determines relevance may become obscure. Relevance is difficult to handle experimentally because it is highly subjective, being tied in also with value considerations. The literature as a consequence tells us little about relevance as a whole, although refer to Chapter 5 for certain relevance effects. Quality factors are hard to deal with because their component parts resist precise definition.

F content

Time, quantity, and quality are subordinate to and interact with the more fundamental problem of *what F information* should be provided. Once the latter is known, the former can be inferred without too much difficulty. The nature of the F to be provided cannot be determined experimentally, however, since it is specific to a particular system and task; the answer can be supplied only by analysis of the given system and task. Although the task description (Chapter 3) should specify the required F, it does so inadequately. One can, however, discover experimentally how important quantitative and qualitative factors are *generally* (over a variety of task situations) by examining how F effectiveness varies with different factor values.

Interactive Variables

The context in which F is provided also influences its effectiveness.

Medium of Presentation

Feedback may be presented by means of displays, by direct (face to face) or indirect verbal or by written contact. The research question is: which medium is most effective?

The System Context

Ideally, each operator should receive only that F information which he can utilize most directly. Under certain system-specific circumstances, however, F describing team, system, or subsystem operation may also be necessary. How can the operator make the most effective use of team and system F, and how effective for operator and system performance is that F, when F is ordinarily interpreted in relation to individual operator actions?

Task Characteristics

The nature of the task and the way in which subtasks are organized must also be considered. Frequency, specificity, and amount of F may become increasingly important as the task grows in complexity or as one moves from highly structured to highly probabilistic tasks. Unfortunately, there are little data on this point.

In the system context F creates special problems for investigation. The F data available in the literature relate mostly to individual molecular responses. In the system, however, F describing more molar task and system events is likely to occur or be required at least as frequently as F for molecular responses. Will the mechanisms postulated for molecular F function in the same way for tasks and events? What, for example, is the impact on F effectiveness of having multiple-F events occurring concurrently? What is the relationship between F for the individual operator and team/system output? Extrapolations from available laboratory data must be made cautiously.

GENERAL FINDINGS

Conclusions. Often (but not invariably) F is effective in improving operator performance. However this effectiveness is described most often in psychomotor (e.g., tracking) and relatively simple tasks, and in learning situations. Attempts to incorporate F in real world tasks frequently produce conflicting results. Moreover, the measurement of F effectiveness is often specific to the task and the measure of performance efficiency employed.

Illustrative Studies

In studies that go back 50 years or more (e.g., Judd, 1905; Arps, 1920; Thorndike, 1927), it has been determined that F often (but not invariably) increases the level of performance in tasks that are highly overlearned (e.g., Arps, 1920; Johansen, 1922; Manzer, 1935); increases the rate of improvement early in practice on a new task (e.g., Bilodeau, 1952; Lindahl, 1945; MacPherson et al., 1948; Pressey, 1950); and increases the frequency of reports that tasks are more interesting and less fatiguing when F is present (e.g., Ammons, 1956; Arps, 1920; Ellwell and Grindley, 1938; MacPherson et al., 1948; Pressey, 1950).

In vigilance situations F often (but again not invariably, see Colquohon, 1966) produces an increase in detection frequency (Adams and Humes, 1963; Baker, 1958, 1960; Garvey et al., 1959; Sipowicz et al., 1962; Wiener, 1963, 1974).

In experiments by Wallis and Samuel (1961) and Wiener (1968) immediate KOR may have prevented the decrement over time that characteristically occurs in vigilance tasks. Reaction time (RT) to visual stimuli tends to increase linearly with time on task (McCormack, 1962); when KOR is provided, RT does not increase (McCormack et al., 1962).

On the other hand, the utility of the F provided is often specific (at least in vigilance tasks) to the measure used. For example, Chinn and Alluisi (1964), found that KOR for missed signals reduced false responses and KOR for correct detections decreased missed signals.

These studies almost always describe the effect of extrinsic, augmented F. Yet a review of F studies by McKeachie (1974) suggests that F in the instructional situation does not always produce improved learning. (See also a review by Higgins, 1972, which also has a pessimistic tone.) The more complex the task to be learned, the less likely it is that F will be effective, possibly because it is difficult for the instructor to provide the required F information and for the student to extract it. Gagne and Rohwer (1969) also suggest that F effects in instructional situations are highly variable, depending on learning characteristics, type and timing of F, and type of task.

For a general review of F effects and theoretical approaches to F, see Annett (1969). For a review of F effects and variables specifically in relation to psychomotor tasks, see Bilodeau (1966).

TIME FACTORS

In studying temporal effects it is necessary to distinguish between (1) the pre-F interval (between the antecedent response and the F for that response), and (2) the post-F interval (between the occurrence of F and the presentation of the next stimulus).

In both cases one would hypothesize on the basis of reinforcement theory that the longer the interval, the less effective operator performance would be. However the mechanisms involved may differ, the pre-F interval affecting the reinforcement of the previous response, the post-F interval influencing the cueing effect of F on the next response.

An instructor will attempt to supply F as quickly as he reasonably can (with minimal pre-F interval, although not necessarily with minimal post-F interval). Since speed is not always possible, we must ask whether there is a significant loss in learning as a function of these F intervals.

In contrast to the instructor, the system designer is constrained by the nature of the situation and task, which very largely determines whether he will be able to manipulate F intervals. For example, in designing a roving lunar vehicle to be remotely controlled from earth, the distance between the two areas produces an unavoidable delay to the operator in receiving visual F following each control action.

Conclusions. The literature on pre- and post-F intervals is confusing and contradictory. In cognitive tasks the pre-F interval appears to have relatively little effect on error. In some studies the longer the post-F interval, the better performance seems to be, up to a point that is yet unknown.

Illustrative Studies

Many studies with lower animals have shown the detrimental effect of reinforcement delay (pre-F interval) in simple learning situations. There are, however, few such studies with human subjects.

One of the early studies with humans involved concept identification and employed pre-F delays of 0, 0.5, 1, 2, 4, and 8 seconds (Bourne, 1957). As delay increased, performance decreased, and task complexity was related positively to delay effects.

The work was apparently flawed, however, because pre-F and post-F intervals were confounded. The defects in this study were remedied by Bourne and Bunderson (1963). Their experiment varied the pre-F interval while keeping the post-F interval constant; concurrently they varied the post-F interval, simultaneously keeping a constant pre-F interval. The results appear in Figure 6-1.

Obviously pre-F intervals have little effect on error production; post-F intervals have some effect, but it is difficult to interpret. The longer the post-F interval, the better performance seems to be. This may mean that subjects used the post-F interval to rehearse, memorize, or assimilate the information they had been given by F. Bourne et al. (1965) threw some light on this problem by varying the post-F interval from 1 through 25

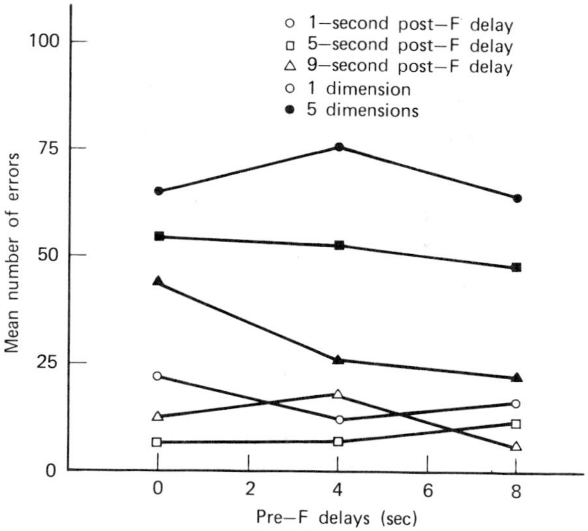

Figure 6-1 Number of errors to solution as a function of pre-F and post-F delays and number of irrelevant dimensions. With permission, from Bourne and Bunderson, *J. Exp. Psychol.*, 1963.

seconds. They found a facilitating effect up to 9 seconds for easy tasks and 17 seconds for difficult ones.

On the other hand, Roweton and Davis (1968) reported that varying the pre-F interval in a concept formation task significantly improved performance, whereas post-F intervals did not. Contrary effects were found by White and Schmidt (1972). Matthews (1972) and Namikas and Carey (1971) observed that post-F interval variations had no effect.

If one can extrapolate to the system designer, the results suggest that it may be useful to allow the operator some time to assimilate the information provided by F before the next input arrives (see Chapter 2 for studies of input frequency). This assumes, of course, that the system designer has some control over the rate at which inputs reach the operator, and often he does not. The intervals studied are of course unrealistically short.

FREQUENCY AND AMOUNT OF F

In many systems F cannot be provided after each operator response or system output, and the amount of F provided may vary. This raises two questions. (1) What is the minimum F frequency and amount that the operator must have? (2) What is the effect on operator/system output of reducing F frequency and amount below that mimimum?

The answers to these questions may be situation specific. An F-relevant response must be defined in the context of a specific task. Discrete, molecular control actions have little meaning except in the context of the total task. Building in F for responses of this type might provide only trivial information (e.g., a series of indicators to confirm that individual switches had been thrown); this would be inordinately expensive from an equipment standpoint and might overload the operator, as well.

The amount of F given for a specific action may depend on the number of dimensions that action possesses. Hence any quantitative value we might pull out of the literature could be quite inaccurate for an individual task.

For both F frequency and amount, the essential factor is the nature of the response for which one wishes to supply F. Unfortunately the literature on this subject has been largely confined to relatively simple laboratory tasks; there are few cross-task comparisons.

This does not mean that no applicable data exist, only that in the design situation they can serve merely as guidelines.

Frequency

Conclusions. Any frequency of F is more effective than no F at all. However the frequency required for effective operator performance need not be very great.

Illustrative Studies

The results in the literature are somewhat ambiguous. Kimble (1961), in reviewing several studies describing percentage of F, indicated that performance is usually but not always poorer for conditions of less than 100% F. On the other hand, others have reported that performance is not seriously impaired on simple conditioning and RT tasks until F is withheld on 70 to 80% of the trials (McCormack et al., 1963; McCormack and McElheran, 1963). Goldstein et al. (1967; see Chapter 4) found similar results for a complex *DM* task.

Case 149. In this *vigilance* situation, subjects observed an oscilloscope for an hour, during which eight signals appeared in each 15-minute period. KOR was given after 0, 25, 50, 75, or 100% of the signals. Significant differences occurred between the number of targets detected by the 0 and 25% groups and the 25 and 50% groups, but none between the 50, 75, and 100% groups. Figure 6-2 suggests only that performance with F is significantly better than performance with no KOR; varying percentages of KOR were not significantly different from each other. Vigilance decrement over time was not significantly affected by percent of KOR.

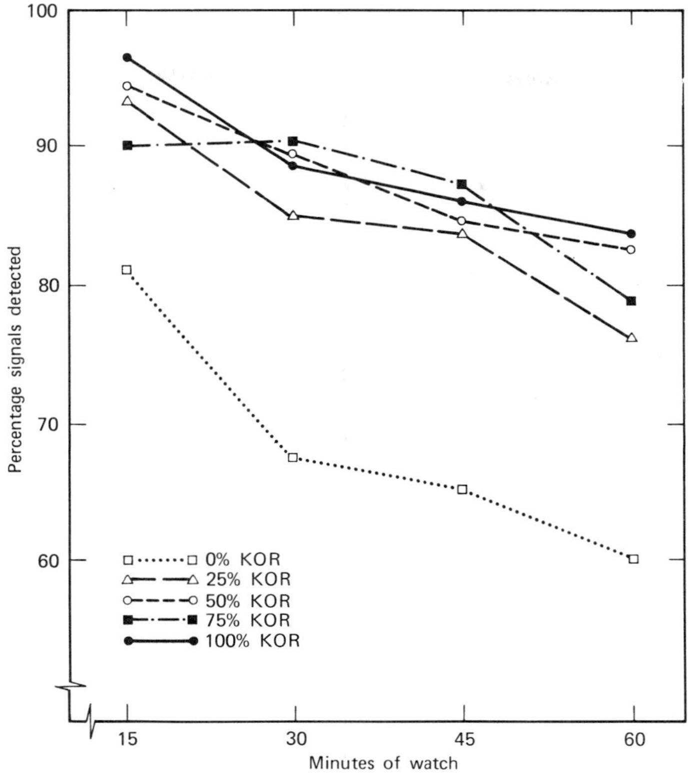

Figure 6-2 Detection of signals as a function of percentage KOR. With permission, from Johnson and Payne, *J. Appl. Psychol.,* 1966, Case 149.

Case 150. In another vigilance task involving a circular display of 20 red lamps that illuminated in sequence, the signal was failure of a lamp to light in normal sequence. KOR was provided by a red light to indicate when the signal was missed, a green one for a correct response. KOR frequencies were 100, 50, and 0%, and the order of effectiveness, although not statistically significant, was the same.

The results in *reaction time* studies have been very similar. In a series of such studies (McCormack, 1962), it was determined that RT increases linearly with time on task. When KOR was provided, however, RT did not change significantly (McCormack et al., 1962).

Case 151. The subject's task was to depress a switch as fast as possible when a light appeared. Individual groups received 0, 30, 50, 70, and

100% KOR, which was provided by illuminating either a red or a green pilot light immediately following the response. The red light indicated RT slower than the previous one, and the green light signaled a faster response.

The only reliable change in RT occurred in the 0 KOR group, in which RT became longer after time (traditional vigilance decrement). All other groups showed no difference in performance as a function of KOR (see Figure 6-3). Note that Figures 6-2 and 6-3 are almost mirror images of each other.

Case 152. In a follow-on study using the same task, 0, 10, 20, 30, and 40% KOR frequencies were employed to explore the frequency range

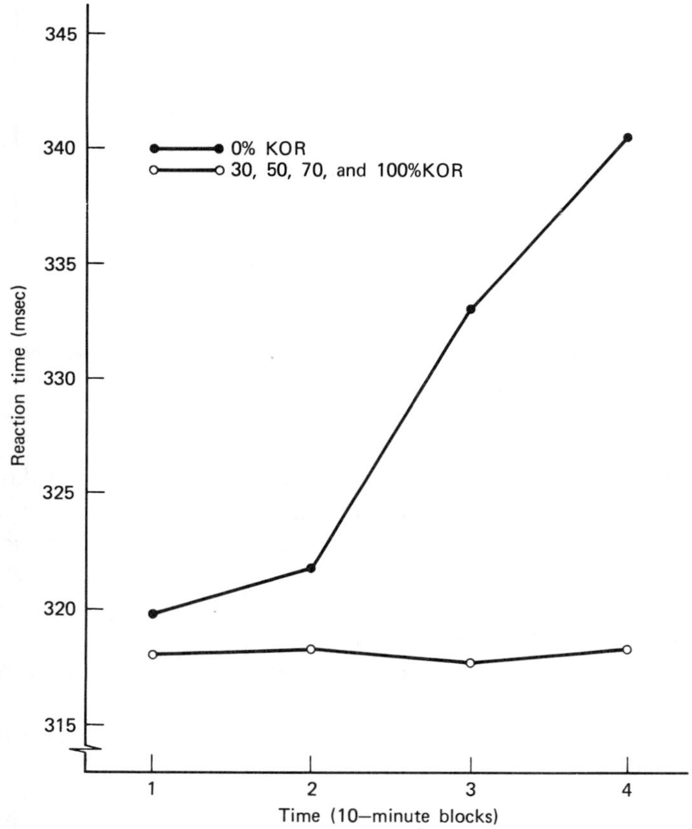

Figure 6-3 Reaction time as a function of percentage KOR. With permission, from McCormack et al., *Percept. Mot. Skills,* 1963, Case 151.

0–30%. There was a sizable performance decrement under the 0 KOR condition but no reliable performance changes in the 30–40% KOR range. The experimenters concluded that the critical point along the KOR continuum lies between 20 and 30%.

Case 153. The question of F frequency has also been investigated in the context of a *probabilistic* task. The operator's task was to guess whether a displayed meter value was higher than that sampled from the distribution of values presented in a second (not shown) meter. As more and more stimulus values were presented, the subject could gain increasing familiarity with the distribution of numbers. Feedback was presented on 0, 30, 67, and 100% of the trials. Only trivial differences among the four frequencies were found.

One would not suggest that no F is needed, but apparently the frequency required is very small, at least in the tasks used. Unfortunately, with the exception of Case 153, the tasks employed have been very simple. We do not know what would happen if more complex tasks were selected.

Amount

Conclusions. Little reliable information is available about the effects of amount of F on performance, and the results are conflicting. It is hypothesized that amount of F can have a significant effect on performance, but only when the operator can make use of the increased amount of information presented.

Illustrative Studies

Whereas F frequency involves only one dimension (occurrence or nonoccurrence), the effect of F amount depends on the number of dimensions the F and the antecedent response possess.

In Case 153 amount of F was varied in terms of correctness of choice versus choice correctness plus actual correct reading. This variable was insignificant.

Case 154. In contrast, Bourne and Pendleton (1958) compared two conditions of F completeness in concept identification problems that required subjects to sort geometric patterns into four categories. In one condition the subject was told on each trial whether his response was correct or incorrect; in the other he was shown not only the correct/incorrect information but the correct response whenever he made an error. These two degrees of completeness were combined with three levels of task complexity. Incomplete F was significantly inferior to complete F (twice as many errors). (Having incomplete data is apparently not the same as having incomplete F. In Chapter 4, Case 82, we saw that for

DM at any rate, the operator was not significantly affected by 10% missing data.) As one would expect, errors in this task increased almost linearly with increasing task complexity.

Case 155. Amount of information was also studied in the context of a multiman information processing threat evaluation system (Case 113 also used this system). Two conditions were varied: frequency (percentage of trials receiving F: 0, 33, 67, and 100%); and F amount (accuracy of the operator's judgments vs. his accuracy plus judgments based on Bayesian processing).

Results of the study confirmed previous conclusions that DM does not need 100% KOR. The effects of adding F based on Bayesian judgments did not improve subjects' scores, although the Bayesian model did produce superior evaluations. Subjects may not have been able to use the extra information provided; the number of discriminable response categories provided by the KOR may not have matched the subject's ability to discriminate.

Case 156. In another study the task was compensatory tracking of needle deflections on a zero-center voltmeter. In the high level F condition four groups received information during trials by means of counters indicating time-on-target (TOT), plus a series of clicks over earphones, indicating cumulative TOT. In the low level F condition, four groups received TOT information only at the end of each trial, together with noninformative broadband noise over earphones. Thus the high level F groups received substantially more F information than did the low level groups, and there was less delay associated with the former condition. In a transfer phase half the subjects continued on their original F schedule, while the other half transferred to the alternative schedule. High level F was found, as one would expect, to be highly superior (by about 25% TOT) to low level F during both training and transfer; in addition, high F subjects indicated more interest in their task. It should be noted that these results confound amount of F with pre-F delay.

SPECIFICITY

Logically, the more specific the information provided by F, the more effective performance should be. The degree of specificity is, of course, the crucial matter to the system designer. In some cases a simple indication of whether a system parameter is within limits (go/no-go displays) is enough. Other situations require information not only about whether the parameter has exceeded limits, but direction and amount of excess.

To make use of F, the operator must be able to discriminate the various categories of information it contains. Annett (1961) has pointed out that increased specificity of KOR involves two components: one, a directional component that states a relationship between the subject's response and the desired correct response; the other, the number of discriminable response categories available to the subject. Increases in specificity that involve the second component do not always result in superior performance (Eriksen, 1958;Green et al., 1955) and may, in fact, degrade performance through the confusion generated by unnecessary detail. There is apparently little benefit from using a wider range of KOR categories than the subject has available in his potential response repertoire.

The studies performed on this topic have dealt only with the rather commonplace question of whether increased specificity produces improved performance. The more critical questions of degree of specificity and when highly specific F should be provided apparently can be dealt with only by analysis of specific task requirements.

In reviewing the illustrative studies, the distinction between F specificity and amount must be kept in mind. Specificity takes information and fractionates it. Amount adds information. This distinction has often been blurred in experimental tests. In the real world, too, it is often difficult to differentiate between the two variables, since fractionation into more detailed categories may appear to add an increment of information.

Conclusions. Performance tends to improve as F information becomes more specific, as long as the operator needs and can use more detailed information. However, the effects depend in part on task difficulty and operator proficiency.

Illustrative Studies

Case 157. This study involved a compensatory tracking task performed on a CRT. The variables included (1) an easy or difficult target course, and (2) four degrees of F specificity (i.e., 3 information categories vs. 7, vs. 13 vs. continuous information). Measures were tracking error and control motion.

Increases in the precision of displayed error information (increase in the number of information categories) resulted in a negatively accelerated reduction in tracking error (Figure 6-4). Most of the improvement was in the range of three to seven categories. As the operator became more proficient, the superiority resulting from increased precision decreased. The effects of the increased precision did not depend on the difficulty of the task, except when control motion was the criterion measure.

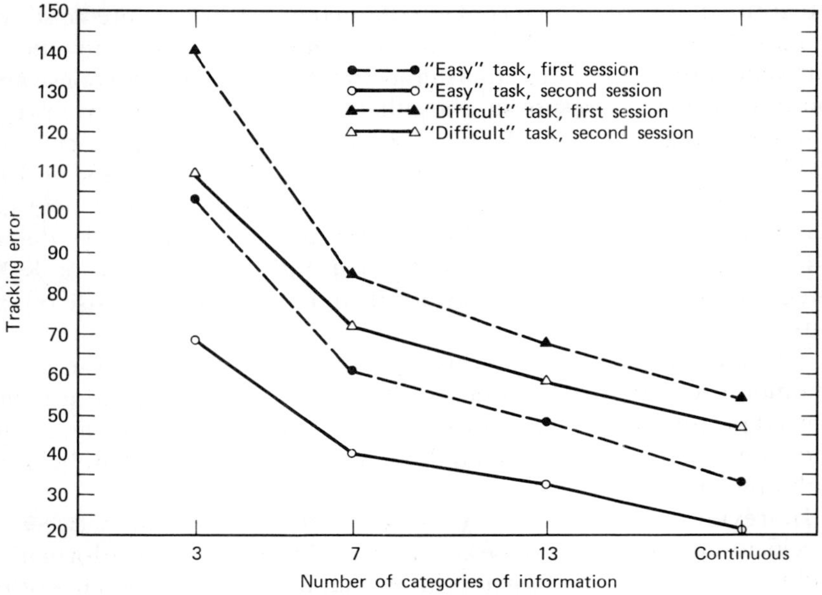

Figure 6-4 Tracking performance as a function of F specificity. With permission, from Hunt, *Hum. Factors,* 1961, Case 157.

Case 158. Subjects in another study were required to perform two simultaneous tasks that simulated the operation of the Mercury capsule: a procedural task and a three-dimensional tracking task with the added complication of fuel monitoring.

Feedback was presented in terms of varying noise intensities. Specificity was manipulated by varying the width of five tracking error zones, each one associated with a different noise intensity. The poorer the performance, the louder the noise became. Thus the "tight noise" condition meant that error zones were very narrow and only a small amount of error was needed to trigger the loudest noise. In the "gross noise" condition, a much larger amount of error was required to trigger the loudest noise.

During training, groups receiving augmented noise F had lower tracking error scores than did control groups, although the means were not significantly different. However F specificity was significantly related to performance at transfer. The medium noise group was significantly inferior to the gross and tight noise conditions.

Case 159. The task was to estimate from 120 photos how many degrees a 1/4-inch arrow would have to be turned to parallel an adjacent arrow-headed line. KOR was given orally and varied as follows: (1) no information, (2) "right" (correct) if no more than one degree in error, otherwise wrong, (3) "right" if correct to the nearest degree, otherwise wrong, (4) "over" (or under) *X* degrees, and (5) correct answer to nearest degree. All groups except the control group decreased their error over trials, thus demonstrating once again the importance of KOR. Learning was generally greater, the more specific the KOR. However there was considerable variation between groups as a function of training sessions. Although adjacent treatment groups (those having similar amounts of specificity) did not differ significantly from each other, the more extreme ones did. It is possible therefore that specificity produces significant effects only when its increments are large.

FEEDBACK CORRECTNESS

Feedback correctness involves two components: the validity of the F report and the operator's interpretation of that report. In the learning situation F should always be correct, but its presentation to the operator may be less than adequate and may not be correctly perceived by him. In the operational system Rss should correctly mirror the status of the system, but the operator's interpretation of the information (e.g., malfunction symptoms) may be lacking. In a highly indeterminate situation Rcn (e.g., battlefield reports) may only partially represent what has actually occurred as a consequence of the action that elicited the feedback, its presentation may be faulty, and the operator may fail to understand its meaning. Misinterpretation of F may occur for a number of reasons.

Stimulus Characteristics

The essential information in the F may not be discriminated because of its stimulus characteristics; for example, it may be difficult to differentiate F from non-F information ("noise") if the two are tightly "packaged" together. The input characteristics described in Chapter 2 will impact on the operator's ability to analyze the F stimulus.

Inconsistent Environmental Events

If one's own system has only partial or no control over events in the external environment, the F information provided as a consequence of the actions of one's own system in that environment may appear to be inconsistent or unpredictable. This is characteristic of indeterminate adversary type (e.g., military) systems, where the adversary's actions in response to one's own may be difficult to interpret.

Delayed or Incomplete F

Sometimes F is so delayed that it is difficult for the operator to identify the action producing the F. (What did I do that produced this effect?) Or, in a probabilistic situation in which the operator develops his response on the basis of multiple sequential inputs, F may occur at such low frequencies or irregular intervals that it is difficult to relate the F to the eliciting action. Perhaps only some of the information that should be contained within the F input will reach the recipient. The information may be relatively nonspecific or associated with relatively unimportant or even irrelevant aspects of the situation. The problem of F correctness occurs most often and most seriously in the context of the indeterminate system and in systems that attempt to control a stochastic environment or other systems within that environment. Feedback correctness may also be a problem in multiman systems, when the F provided on the basis of the crew response does not relate appropriately to the operator's own performance (more on this in a later section).

Problems of interpretation, inconsistency, and incomplete data occur much less often in highly structured systems because such systems usually contain far fewer F dimensions and the relationship between F and the action leading to the F is relatively easy to understand. When problems of F correctness arise in structured systems, it is usually because one of its subsystems is indeterminate (e.g., malfunction symptoms in the maintenance subsystem). Problems of F correctness arise primarily in operational systems, although oddly enough most F research is performed in a learning context.

From the standpoint of the system designer, even though he deliberately tries to make his F completely veridical, he may fail partially either because he has not analyzed correctly what F information should be supplied and to whom, and/or because in complex systems it is difficult to supply information that describes all relevant subsystem and system events. In addition to the illustrative studies described below, the problem of misinformation has been investigated by Bourne (1963), Johannson (1962), and Pishkin (1960).

Conclusions. Operator performance is in inverse relation to amount of F misinformation. The adequacy with which operators can filter out incorrect F is a decreasing function of the percentage of incorrect F. Awareness that incorrect F is being presented helps in maintaining correct performance. Even erroneous quantitative F is more effective than correct qualitative (nonspecific, outcome) F.

Illustrative Studies

Case 160. This task involved analysis of a historical pattern: to learn to depress one of two keys when one of a series of lights illuminated. Feedback was provided by separate red and green lights. Misinformation was varied in terms of the percentage of total trials in which incorrect information (right-wrong) was provided (0, 12, 20, 30, 40, and 50%). Certain subjects were also made aware that they were to receive misinformation.

As the amount of misinformation increased, learning decreased. As subjects' awareness of the incorrect information increased, however, the decrement resulting from the misinformation was reduced. In fact, subjects receiving complete instructions about the misinformation performed almost as effectively as those receiving zero misinformation.

Saunders and Neimark (1967) also found that concept formation subjects, informed of misinformation, performed more effectively than those not so advised. Mackworth (1964) found that detection performance with false KOR was intermediate between that with veridical KOR and no KOR. Loeb and Schmidt (1960), Weidenfeller et al. (1962), and Antonelli and Karas (1967) showed that false KOR was as effective as KOR in vigilance performance. In a reaction time task Peretti (1970) reported that subjects receiving false KOR performed more effectively than those who received no KOR.

Case 161. The tactical *DM* situation utilized was that described in Case 98, and it is of special interest because much of the information received in advanced systems is probabilistic. The experimental variables were outcome F, in nonspecific (success/failure) terms; correct F, in quantitative (P_s) terms; F deviating from correct F by σ 5%; F deviating from correct F by σ 10%. F was supplied after each alternative had been selected by the operator. The percentages of correct choices made over 126 problems were true P_s, 73.4%; σ 5% error, 67.8%; σ 10% error, 66.1%; outcome F, 59.5%.

Although erroneous F reduced *DM* performance, there was no significant difference between the two groups receiving the two amounts of misinformation. Operators may be relatively insensitive to small increments of incorrect F. The system designer therefore need not be overly concerned (from the standpoint of the operator only, of course) about small amounts of error in his system. Further research is required to determine the critical level of incorrect F that leads to significant breakdown of performance.

With reference to F specificity, it is important to note that slightly erroneous F facilitated performance more than outcome (or nonspecific) F. This suggests that *DM* (and perhaps other types of tasks) could be improved by aids (such as a computer) that provide quantitative F, even if that F is somewhat incorrect.

Case 162. In earlier research Taubman (1944) had reported that KOR seemed to increase counting accuracy. Case 162 involved 40 subjects required to count the number of rapid auditory signals. The variables were number of signals (sequences of 1, 2, 4, 6, 8, 10, 15, 20, 30, and 40 tones) and correct/incorrect F (actual number of tones vs. misinformation). With error as a criterion, operator accuracy was increased by knowledge of the number of tones presented; this was more effective, as one would imagine, for the longer sequences. The effect of misinformation was to increase error, and this effect also appeared to generalize (see phase III results, Figure 6-5).

Conclusions. Misinformation based on *transformations* of F data does not reduce performance as much as F that is grossly at variance with the "truth."

Illustrative Studies

Several studies (Bilodeau, 1953; Case 161) have investigated the effect of F transformations in which the true error value was reduced, amplified, or otherwise transformed. This can be considered misinformation, since

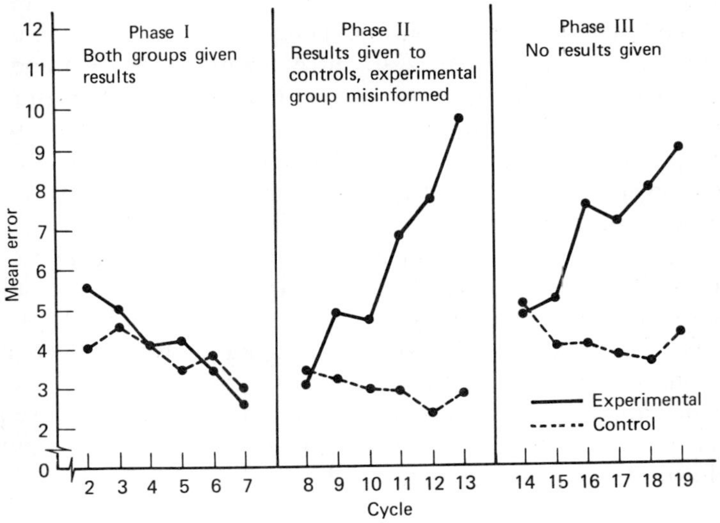

Figure 6-5 Counting performance as a function of misinformation. With permission, from Monty, *Hum. Factors,* 1962, Case 162.

F is distorted by some amount. The effect of these transformations has often been insignificant. For example, Sulzer and Levy (1966) found that inflating F by multiplying the basic F scale value by various powers of 10 was ineffective in the acquisition and extinction of a discrete knob-positioning response. Conceivably, the insensitivity that prevents DM from being as efficient as a Bayesian processor (a negative effect) also prevents him from being overly influenced by random error in the F information he receives from his environment (a positive effect).

Case 163. This study utilized a task similar to that of Sulzer and Levy, 1966, that involved moving a lever a given amount blindly (i.e., without being able to use a scale for guidance). Thus the only cues the subject had came from the F provided by the experimenter. The F was arbitrarily inflated or reduced by a specified amount, and the effect of misinformation on performance was completely negative.

Case 164. In this task blindfolded subjects tried to achieve a specified score by turning an exposed thimble on a micrometer. The accuracy of a subject's response on a given trial was contingent on the experimenter's report of his performance on preceding trials. The KOR variations were (1) no score transformation, (2) scores transformed by a negatively accelerated transformation equation, and (3) transformation by a positively accelerated transformation equation. The true KOR group showed superior accuracy in initial practice trials, although after about eight trials all groups performed equally well. This suggests that when misinformation is provided and intrinsic F is available in the subject's response (even if it is not initially used), the subject will eventually learn to use that intrinsic F to counteract error effects.

It appears from these studies that F transformations do not seriously impair performance. However the tasks employed were extremely simple, and misinformation in more complex (e.g., probabilistic) tasks might have greater impact. On the other hand, it is possible that misinformation has significant effects only when F is grossly at variance with "truth."

MEDIUM OF PRESENTATION

Feedback can be presented visually and aurally (by means of displays), by face-to-face communication, by intercom or radio link, and in written form. One could, of course, present F by tactile, kinesthetic, or even olfactory means, and studies have been performed to investigate these possibilities (e.g., Geldard, 1974); but these more exotic media are experimental only and would rarely be used in operational systems.

Most F in complex systems is presented visually or by verbal means, although a good deal also occurs in written form. When a choice supposedly exists among media, the question naturally arises: which one produces the most effective performance?

In approaching this question, the results of Chapter 5 must be remembered—namely, that verbal communication was found to be less effective than visual displays (Case 143).

The designer does not have complete freedom to select his F medium, the choice often being dictated by other system considerations. For example, if F information must be transmitted any appreciable physical distance, it cannot be presented by direct verbal means, although telephonic communication might be a satisfactory substitute. When there are multiple F dimensions, one would be more likely to code the F information (see Chapter 2) and transmit it visually, because aural F has very few dimensions.

One must also consider the particular function F is supposed to play. For example, direct verbal F may be considered more motivating than F presented automatically, on the assumption that the presence of a human is inherently more motivating to another human than a display. Of course the utilization of direct verbal F for motivation is almost always secondary to F for informational purposes.

It is possible that each presentation medium is best for a particular type of task situation. If this were so, studying the question of presentation format in only one task situation would lead to erroneous conclusions.

Conclusions. Studies of this variable have usually involved a vigilance task in which the effectiveness of the medium of F presentation seems to depend largely on the measure of performance utilized. Feedback presented directly by operators may improve performance more than F presented automatically.

Illustrative Studies

Case 165. In this study the task was to detect signals (periodic interruptions of a lamp) as soon as each was presented. The variables investigated were verbal versus nonverbal KOR, where the latter were presented either visually or aurally, and two types of KOR, for signals detected and for signals missed.

When the measure was percentage of signals detected, verbal KOR was found to be superior to nonverbal, auditory KOR was superior to visual, and KOR for misses was superior to KOR for detections (Figure 6-6). There was no difference between KOR for misses and for detections when F was verbal; but KOR for misses was superior to KOR for detections when F was nonverbal.

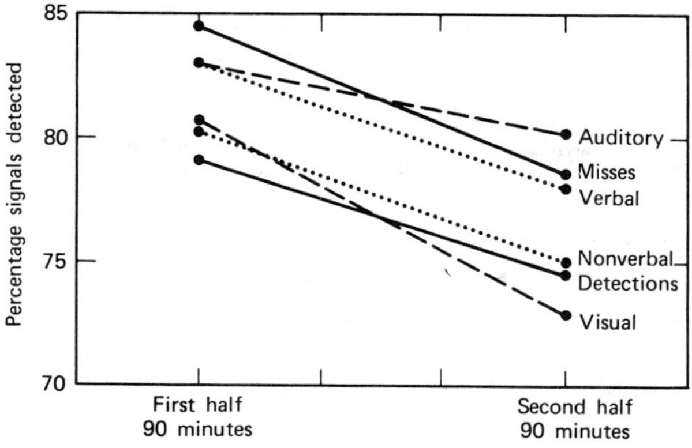

Figure 6-6 Signal detection as a function of F media. With permission, from Ware and Baker, *J. Eng. Psychol.*, 1964, Case 165.

Case 166. This study attempted to differentiate the roles of informational and motivational F by identifying them with different modes of presentation. The hypothesis was that mechanical presentation of KOR (i.e., by signal lights) would affect the informational input, whereas a personal verbal presentation by an observer would add a motivational aspect to the F. The Mackworth clock situation (detection of double unit jumps of a moving clock hand) was presented. The variables were no KOR; verbal (observer) KOR; and automatic KOR.

Results indicated that the type of KOR presented has significant effects on performance; verbal KOR provided about 10 to 15% more detections than did the signal lights. There was little difference between automatic and no-KOR conditions. The superiority of verbal KOR persisted throughout the test period until the eighth day, when all groups were tested without KOR. It appears that KOR effects persisted even when KOR was removed; the varying groups maintained their relative standing over the 8 days.

FEEDBACK SOURCE LEVEL

Feedback in a multiman system can be considerably more complex than F in a single-operator system. In the latter, only two types of F measures are ordinarily available: those describing the operator's performance, and the system output (which, because of the close link between the operator and the system in single-operator systems, is often identical with his performance). In the multiman system F can describe performance at various levels: *(a)* the operator, *(b)* his teammates, *(c)* subsystems other than those in which the operator functions, and *(d)* the system as a whole. Feedback in the multiman system may therefore permit the operator to evaluate his performance in terms of criteria specific to his own job and in relation to that of other crew members, subsystems, and the system as a whole.

It is evident that F is most appropriate when it relates to the operator's own responses or task performance, less so when it describes a higher level performance to which the operator's own performance has contributed only partly. Such F has been termed in the literature "confounded" because it reflects intermingled contributions.

It must not be thought that confounded F is erroneous information. Higher-order F is ordinarily entirely veridical, and it is incorporated in the system for specific, presumably required, purposes. (In system design every effort is made to provide only that minimal amount of F absolutely required to perform system functions at a specified level of accomplishment; anything beyond that is expensive and wasteful. Moreover, every effort is made to assign to particular operator positions only the F which that operator needs.) However, higher-order F may provide misleading cues to the operator because its relation to his responses is obscure. To the extent that team, subsystem, and system F are only partially relevant to the operator's own performance, one would expect some loss of effectiveness in regulating that performance.

There are three pertinent research questions.

1. How do the various types of F available in the multiman system affect operator and overall system performance? This relates to one of the central research questions the behavioral scientist must address: the relation of the operator's performance to total system output (Chapter 1).

It is possible that the more complex the multiman system, the less effect any individual operator's performance has on total system output (except where the operator is in a central "key man" position). The effectiveness of F in improving operator performance may also decrease with increasing system complexity.

2. How effectively does F influence the individual operator's performance as a function of F source level (i.e., the level at which it originates)?

It is possible that the more remote the operator is from the source of the F he receives, the less effect that F will have. A corollary possibility is that F describing other subsystem or higher system level performance may confuse the operator, since he interprets F primarily in terms of his own performance.

3. How do we determine how much of each type of F should be provided in response to system requirements?

From the system designer's standpoint this is the essential question. Aside from certain F that is explicitly *required* for system operation (e.g., emergency warnings, status of critical instruments), the system developer has some control over the amount and type of F he may provide by means of status displays. Manifestly he cannot provide F for every operator action or system event, and therefore he must select among his alternatives. Selection among the *desirable* F stimuli he can supply becomes the system developer's primary problem in designing F into his system. In the specific system context this question can be answered only by analysis of system tasks, although F studies may yield useful principles.

The previous paragraphs describe F *during* system operation. Yet F may be given *following* system operation in the form of a team *debriefing* which is specifically instructional, pertains to the team as a whole rather than to any single operator, and does not usually deal with specific discrete responses but with the task as a whole.

The amount of data available to answer the preceding questions is small. Almost all the studies reported describe individual F, a few, team F; still fewer are concerned with subsystem and system F.

Conclusions. Operator performance differs significantly as a function of F source level. The operator's performance is influenced most by F when its source is closest to him.

Illustrative Studies

Case 167. This study is described in some detail because it is a model of what a multiman system study should be, partly because of its emphasis on system as well as on individual operator measures.

The system was a complex information processing system, already described in Case 42. Its mission was to gather information by directing aerial reconnaissance flights, to process that information, to determine the value of the information for predicting target locations and to use the predicted locations in directing the search.

Performance was studied as a function of four levels of F remoteness. The following conditions were compared: *(a)* the output of an operator and his partner at processing stage F-1, *(b)* the output at stage F-2 of another processor, who treated the output from the F-1 operators, *(c)* the output of all operators at stage F-2, and *(d)* the output of a third operator at stage F-3. These were compared to a no-F condition. Four reconnaissance operators (RO) formed the first system stage; two sector commanders (SC) and assistants were the second stage, and an area commander (AC) was the third stage.

Since only the ROs experienced variations in F conditions, performance was measured in terms of RO performance. Some measures were directly related to RO performance (output, operating costs, and efficiency); other measures were indirectly affected by RO adjustments (measures of performance at the SC and AC level). The results of the study are described in Table 6-1.

Although most measures failed to demonstrate significant differences for the F conditions, there appeared to be a consistent tendency for performance to improve when F was based on the F-1 output. When total scores were ranked by F conditions for the various measures, performance was highest for six of the nine measures at the F-1 stage.

Because of the somewhat disappointing results of this initial study, a second study was performed, utilizing roughly the same task and environment.

In contrast to the previous study in which F was available only to ROs (first stage), in the second study F was given to both the SC and AC (second and third stages). Two different kinds of F were provided: F-P (power feedback: number and value of cells included in the AC's flight plans) and F-E (efficiency feedback: a ratio based on number of areas recommended for surveillance divided by total number of areas included in the flight plans). Under a third experimental condition, both F-P and F-E were given. Performance measures were available at the individual RO level (F-1 stage) and at the team/system level (F-2, F-3 levels).

Measures of RO performance were largely insensitive to manipulation

Table 6-1 Measures of F Effectiveness as a Function of Source Level[a]

Measures of RO costs and efficiency

1. No difference among F conditions for total number of flights.
2. Significant differences among F conditions for aircraft use time, primarily at the first (nearest) F source level (F-1); other F conditions did not differ.
3. Although differences were not significant, performance was greater for all three measures under the first F source condition.

Measures of RO output

1. No significant F differences in terms of these measures.
2. Teams of different composition reacted differently to various conditions.

System measures

1. Total value of recommendations made by SC under F-1 condition was significantly greater than those made under other conditions.
2. No significant differences for number of recommendations made by SC to AC.

[a] From Feallock and Briggs, 1963, Case 167.

of F at higher system levels. The investigators considered this somewhat surprising, since the ROs were controlled by higher system levels, but it lends support to the previous hypothesis that F effects on the operator tend to decrease with increasing system complexity. The lack of significant effects may also have been due to the ability of ROs to provide the recommendations needed for flexibility at higher system levels without changing strategies.

At the crew/system level (F-2, F-3), differences between the F conditions were significant for number of cells overflown and reconnoitered by SCs. The largest number of cells was overflown under the power (more specific) condition, and the smallest number when efficiency (less specific) information was supplied. When both types of F were provided, an intermediate number of cells was overflown.

The analysis for the power and efficiency of SC flights also showed significant F differences. For both measures, maximum performance occurred when both power and efficiency F were given. The response at the team level to each F condition apparently was one in which operators adopted strategies that would maximize all F received.

It seems clear then that the operator's performance is different when he receives F based on his immediate output and when the F he is given is based on his output after further processing by other operators.

In contexts in which performance is determined by several factors, it seems highly probable that the most important function of F is to furnish response guidance. In the first study, the various types of F failed to give a basis for response selection. Under these circumstances, the response to F, if any, was simply an increased level of activity for whatever adjustment operators made (a motivational effect?).

In the second study F was expressly defined to be instructional. To perform adequately operators had to adopt different strategies. Operators were responsive to F under these conditons and adjusted their performance in a way that tended to maximize performance in terms of all the information they received. These results support the conclusion that F can have a marked effect on operators performing complex system tasks when it guides operators in selection of response alternatives.

"Where tasks are multidimensional in nature, the efficacy of feedback depends upon the extent to which the feedback provides information for evaluating alternative approaches. When feedback is not of this type, differential performance related to the manipulation of feedback is not expected to occur" (Feallock and Briggs, 1963, p. 110).

Conclusions. Generally the operator in team tasks learns more rapidly and reaches a higher level of proficiency when F is direct and individual specific rather than confounded, especially when it is not possible for one man to compensate for the deficiencies of his teammates. The utility of team F is questionable.

Illustrative Studies

Case 168. The original study in confounded F was performed by Hall (1957). The task was to adjust a micrometer knob to a desired setting when the scale could not be seen and the operator had to guess at the setting. The two subjects who made up a team could not see each other and were not allowed to communicate. Each subject had control over his own knob, but both subjects had to adjust the knob at the same time, and the F provided to each by the experimenter was a team score consisting of a weighted average of their two responses. (Note the similarity of the situation to that involving transformations, Cases 163 and 164).

The variables studied were the type of F presented during the pretraining period. In one condition each team member was pretrained with F describing his own response; in another he was given team scores in which individual operator weights were varied systematically (equal or 3–1). In the testing period all subjects received the same feedback, team scores in which teammate scores were weighted equally or 3-1.

The results suggest that during the testing period teams trained on confounded F improved significantly faster than teams pretrained on individual F, but performance differences, at least by the conclusion of testing, were not large. There were no differences between the two weight conditions.

It is important to note that in contrast to other studies to be reviewed, the confounded F did not have any negative effect.

This study illustrates the use of F in a compensatory task.

"With the feedback functions used in this study, it was possible for one teammate to compensate for another's errors. That is, for example, if one team member was responding too little, the other could increase his response, and both would obtain a satisfactory team score. . . . One property of feedback functions appears to be the possibility for one teammate to compensate for another's errors" (Hall, 1957, p. 303).

Case 169. In a follow-on study using the same task, the investigators varied F after each trial to each member of the two-man team in the following ways: *(a)* F to one group described the results of the operator's own responses (direct F), *(b)* a second F condition averaged the responses of the two operators and presented that average to each operator (confounded F). and *(c)* a third condition provided F to each team member on the adequacy of his partner's responses ("others" F). Figure 6-7 shows the

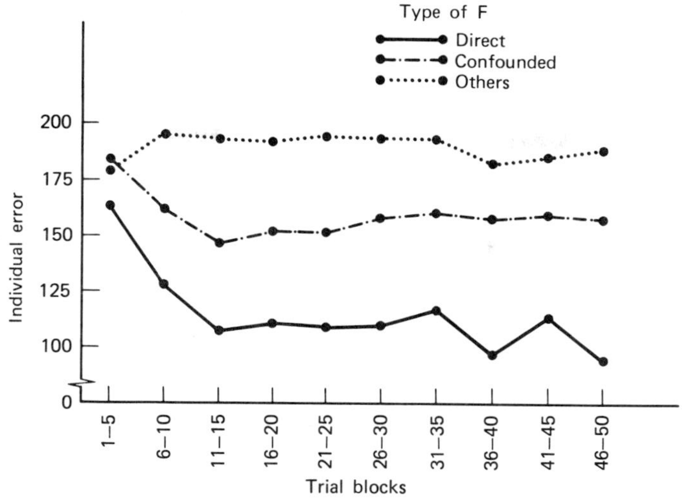

Figure 6-7 Effects of confounded F on individual accuracy. With permission, from Rosenberg and Hall, *J. Abnorm. Soc. Psychol.*, 1957, Case 169.

effect of the various kinds of F on *individual operator* accuracy. The operator learned more rapidly and reached a higher level of proficiency when F was direct rather than confounded. There was only slight improvement with confounded F. No improvement was noted with "others" F, which can be considered as either no-F or even as a misinformation condition.

The effects of the various types of F on *team performance* are somewhat different. Team accuracy is apparently improved by confounded F in much the same way as individual accuracy is improved by direct F (Figure 6-8). Confounded F to an individual did not cause him to make a poorer contribution to the team product than did direct F. The two curves are almost identical. On the other hand, it is difficult to see why team performance was not improved more by team ("confounded") F than by direct F.

It is important to note that these results derive from a task having quite clearly compensatory characteristics, in which the two team members can learn to adjust to each other. The effect of confounded F might be different if the task demanded that each partner perform equally or if team members' contributions to performance were weighted unequally.

A follow-on experiment (Rosenberg, 1959) asked whether individual accuracy once acquired can be retained under confounded F. All subjects were given individual direct F training on the knob-turning task until they

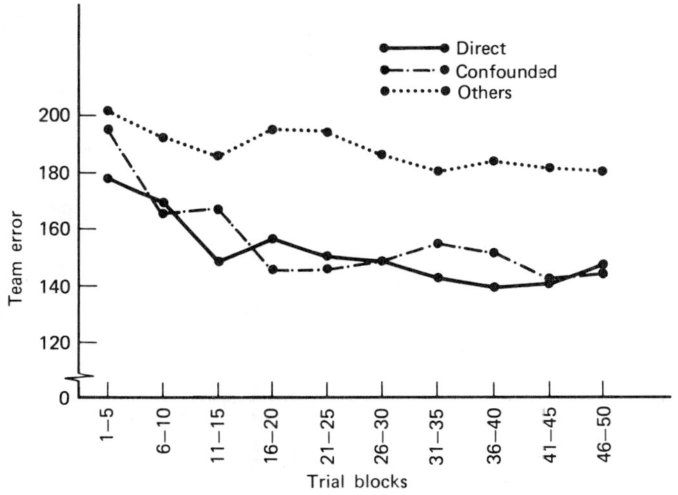

Figure 6-8 Effects of confounded F on team accuracy. With permission, from Rosenberg and Hall, *J. Abnorm. Soc. Psychol.*, 1957, Case 169.

had learned it. From these individuals, pairs of subjects were assembled and given direct, confounded, or "others" F. The results demonstrated a rapid deterioration of performance when "others" F and confounded F were provided, although performance with confounded F was not as seriously affected as was performance by "others" F.

Case 170. Another effort at determining the mechanisms involved in individual versus team F was made by Zajonc (1962). He felt that the results of Case 169 could be explained by taking into account that the two team members could compensate for each other's performance. Therefore he established a seven-man reaction time task in which it was not possible for one team member to "make up" for inappropriate behavior by other team members.

The task was to press a key as quickly as possible, thus turning off a stimulus light. Half the teams experienced confounded F (information on total team performance), while the other half received direct F (each man was informed of his own performance, that of other team members, and that of the team as a whole). This condition seems to indicate that different amounts of information were also involved in the study.

As we would expect, improvement in individual RT was substantially greater for direct F. In terms of group performance, teams working under direct F conditions gave clear evidence of improvement, whereas teams working under confounded F showed practically no improvement.

Improvement resulted primarily from the individual component of the direct F. Relatively poor performers improved in their performance under either confounded or direct F conditions, a finding comparable in some respects to the results of Case 75, where poorer subjects improved under either independent or interactive training but better subjects improved only under independent training.

For additional research related to F in the team context, the reader is referred to the series of studies by Glaser and his colleagues (Cases 130, 131, and 138 in Chapter 5), employing an operant conditioning reinforcement model. In these studies team performance was a direct function of two elements: the reinforcement (F) schedules established and the team structures used. Egerman et al. (1962, 1963) found that team performance deteriorated in disjunctive parallel team structures (correct KOR provided whenever one of the two team members was correct, even though the other may have been wrong—in other words, confounded F).

The effect of team F on individual operator performance, however, is by no means a simple matter of differential reinforcement. Johnston and Nawrocki (1967) found that individual performance in a team setting is at least partially a function of criterion difficulty, which in turn is affected by the performance of the other team members. Johnston (1967) points out that the behavioral effects of teammate ability (Goldman, 1965) and replacement (Ziller, 1965) may be mediated by team F.

Conclusions. Postperformance (debriefing) F can be highly effective in improving team performance in complex tasks.

Illustrative Studies

Case 171. Team F need not be as discrete and specific as the reinforcement schedules used in Egerman et al (1962, 1963) or that provided by Rosenberg and Hall (1958). Feedback may be supplied in the form of postperformance (debriefing) group discussion, organized around team functions performed and problems encountered in the immediately previous team exercise. The question is whether such relatively diffuse and delayed F improves performance.

The task studied was an air defense problem requiring detection, recognition, and tracking of displayed enemy aircraft, the assignment of weapons, and so on. Half the group of operators held debriefing discussions following the work shift.

In 13 measurable functions (e.g., detection, plotting, track establishment and maintenance), the debriefed crews improved their performance significantly (from pre- to posttest) whereas in 12 the performance of the non-debriefed crews either remained the same or decreased slightly. The average performance gains for the debriefed crews over all functions were 35.1% for crew A and 49.6% for crew C. The gains of the non-debriefed crews were 0.1% for crew B and 4.6% for crew D.

Certain team functions improved significantly more than others. The explanation given is very suggestive. The data point to an inverse relationship between the visibility of a function and the amount that that function improves as a function of debriefing. High visibility may correspond to high intrinsic F, and vice versa. The functions that yield less information on performance adequacy *during* operations (are less visible) show greater improvement as a result of postexercise F than do those that show more. This is demonstrated by Table 6-2, in which five functions are ordered from high to low visibility.

The authors of this study make certain suggestions about the identification of high visibility functions in team situations. A highly visible function is one in which the following conditions obtain:

1. The results of the operator's actions are directly displayed to him and to his supervisor.
2. The operator who is affected by another's actions can communicate with the latter directly.
3. The operator can observe the activities of other operators who may be affected by his actions.
4. The operator receives information about other inputs that the recipient of his actions may be obtaining.
5. The operator is aware of the kind and distribution of information needed by the recipient.

"If a function is visible, the operator performing it is able to see the consequences of his actions upon the performance of another crew member. Therefore, it is possible for him to learn to adjust his behavior in

Table 6-2 Visibility Related to Average Performance Gain[a]

		Average % Gain from Pre- to Posttest	
Function	Visibility	Briefing	No Briefing
Tactical action	High	14.0	11.5
Track establishment		46.5	1.5
Input processing	Medium	35.5	5.5
Track maintenance		49.0	−6.5
Lateral tell-out	Low	62.0	−9.5

[a] From Alexander et al., 1962, Case 171.

accordance with the task requirements of the other man. If a function has low visibility, the operator will not receive information about the adequacy of his performance from the operational situation itself. If he is to improve at all, it must be on the basis of knowledge of results provided in some other way" (Alexander et al., 1962, p. 210).

In other words, augmented F substitutes for and increases the visibility of functions with low intrinsic F.

In determining the system F to be designed into the system, the system developer should first decide which of the tasks for which he is designing possesses intrinsic F (high visibility). He should then concentrate his design efforts on the others that need "boosting" by extrinsic F provided through displays.

Conclusions. Another type of F that can be useful when veridical F is not available is consensus F, which is F based on the agreement of a majority of team members. Consensus F is not as effective as F presumably based on veridical sources; however, it is superior to the no-F condition.

Illustrative Studies

Case 172. Image interpreters working in teams can produce more complete and accurate intelligence information from aerial reconnaissance imagery than interpreters working alone (Doten et al., 1968; Bolin et al., 1965; Sadacca et al., 1962). Presumably the consensual judgment of team members reduces the number of identification errors made by single interpreters. The purpose of this study was to investigate the F mechanisms involved.

The experiment consisted of pre- and posttest comparison, and an intervening practice phase in which various types of F were provided. These were consensus F (target identifications agreed upon by at least two interpreters in a three-man team) and precise F (identifications of known targets made by experts).

Scores were in the predicted direction: precise F methods showed the highest overall performance (120.4), the consensus F method produced intermediate performance (107.4), and the no-F method was least effective (88.0).

Consensus F has much in common with the F provided by debriefing. There is, however, nothing inherently veridical in this type of F, and it is of course possible for consensus F to furnish incorrect information.

Case 173. In follow-on study to the one just reported, the experimental conditions were *(a)* three-man versus two-man team, *(b)* discussion versus no discussion. *(c)* heterogeneous teams (in terms of initial proficiency) versus homogenous teams, *(d)* interpreters of high, medium, and low initial detection proficiency, and *(e)* no-F (interpreters who practiced alone).

Interpreters working in teams with consensus F exhibited greater improvement than interpreters working alone. These results are in agreement with those of the preceding study.

Interpreters working in heterogeneous teams improved more on all measures than did interpreters in homogeneous teams. The poorer interpreters in heterogeneous teams probably learned more from their interactions with better interpreters than did interpreters in homogeneous teams. (See comparable results described in the section on team composition, Chapter 5.) There were no significant differences between the discussion versus no discussion conditions and three-man teams versus two-man teams.

Conclusions. As in the case of F for the individual operator, team F is more effective when it is specific. However it is possible to overload the team by providing more detailed F than the personnel can use.

Illustrative Studies

Case 174. In the very simple knob-turning task habitually used by Rosenberg and his co-workers (see Cases 168 and 169), information specificity was varied. In one condition of Case 174 a team was informed only whether the total response was good or poor; thus if a team error occurred, the members were informed of the event but did not know the direction of error. In another condition team F contained specific directional information; poor—too little; good; or poor—too much. Cooperative performance was more nearly optimal in the teams that received the more specific F.

Case 175. One might conclude from the preceding case that it is desirable to increase the informational content of F to secure greater specificity. Although this conclusion seems intuitively sensible, the results of an experiment by Horrocks et al. (1960) make it necessary to be somewhat cautious. Four-man teams made range, elevation, and azimuth estimates and used these data to locate the angular direction and ground distance of a target. Since this was a more complex task than the simple Rosenberg knob-turning situation, it was possible to feed back more specific information than could have been supplied for the simple task.

The investigators provided either crude directional error information or more precise and voluminous F information. Surprisingly, performance was superior under the less precise F conditions. Apparently it is possible to supply more detailed F than is necessary or desirable, which can lead to unrealistic attempts by the operator to use the superfluous information. Therefore, although specific F is desirable, one should not overload a team with the detailed F available in complex tasks. All this emphasizes the necessity for an intensive task requirements analysis before F is provided. There may also be an interaction between increasing F specificity and skill level. As the team becomes more skilled, more detailed F may become appropriate, whereas the same team earlier may not have been able to make use of the higher-quality F.

MOTIVATIONAL FACTORS

At the start of this chapter we pointed out that F can perform a motivational function. For example, Clarke (1972) found that F was highly significant in promoting persistence when subjects were presented with an unsolvable finger-tracing task.

Some researchers (e.g., Payne and Hauty, 1955) have said that a dichotomy exists between directive and motivational functions in F and have sought to determine that one or the other is primarily or exclusively responsible for F effectiveness. The work in favor of motivational factors has been largely directed by Locke and his colleagues (Locke and Bryan, 1966a,b, 1967; Locke, 1968).

Locke and Bryan (1969) have defined motivational F as having few or no cue properties (i.e., F that neither informs the operator of the nature or locus of his errors nor suggests how these might be corrected). Examples of motivational F might include knowledge of total score on a task summed over number of trials (imprecise F) and F on simple psychomotor tasks the subject already knows how to perform (overlearned F). Although F of these types could be interpreted by the subject as a signal to change his method of performing a task, it would not tell him what changes to make or how to go about correcting his error. Since purely motivational F lacks cue value, it also can be considered to be less adequate than informational F.

The argument is made that when F lacks informational value, the operator uses that F to adjust his performance goals. If he interprets F as signifying inadequate performance, he will ordinarily set a goal to improve his performance. If the operator interprets F as indicating adequate or superior performance, he will set a goal to maintain or reduce his level of effort. (But if F contains no information relating to how to improve performance, how will the improvement be brought about? Perhaps motivational F works only on overlearned or very simple tasks in which performance quantity and not quality is affected.)

The systems in which such goal adjustments can occur are those in which specific output criteria are not given and the operator is merely told (explicitly or implicitly) to maximize his performance. Could goal adjustment occur in structured tasks in which each discrete response must lead to another discrete response (e.g., operating a console in strictly prescribed sequence)?

The studies of motivational F suggest that in predicting performance, it is not sufficient to know that an operator was given F, one must also know what he has decided to do about it—that is, those goals he has set in response to the F.

The question of motivational versus informational F has been studied by developing experimental situations in which F can have little or no directive qualities and noting whether under these circumstances F can be effective in improving performance. If performance improves, one can begin to explore the possibility F can have motivational properties as well as directional qualities.

It is possible, however, that F ordinarily has both motivational and guidance functions and which one is activated depends on the situation the operator faces. The characteristics of the system to be designed will determine whether motivational F aspects can be exploited. For example, the motivational function may be most useful in dull, repetitive tasks. The motivational dimensions of F may also be effective in systems that do not have fixed outputs and in which the operator can improve his performance. When an increase in response output (e.g., console operation) is not possible, the motivational aspect may remain dormant. If the operator can derive guidance from F, he will do so, but if he cannot, he will make use of its motivational properties.

One can therefore accept the fact that the operator's goals play an important role in regulating task performance (see Locke, 1968, for a summary, as well as studies by Howell and Kreidler, 1963, 1964) without rejecting the directional function of F (confirmed by Kim and Hamner, 1976).

The important thing is to determine the conditions that permit F to exercise a positive effect on performance, whether that opportunity derives from the information F provides or the goals the subject sets for himself on the basis of F.

From a system design standpoint, it is easier to work with the informational properties of F, because these can be modified more readily than can the operator's motivation.

Conclusions. One of the effects of F is to cause the operator to adjust his performance goals, and this in turn produces an improvement in performance. However goal adjustment requires that F be meaningful and useable as a basis for the goal adjustment.

Illustrative Studies

Case 176. The classic work from which the others stem is that of Gibbs and Brown (1955). They set subjects to perform the uninteresting, unskilled, and repetitive task of copying pages from difficult scientific reports and similar material, using a document copying machine. In half the trials subjects could see a counter that tallied each page as it was copied; in the other half the counter was covered. Subjects were given no daily quotas to meet, they were paid uniformly, they were never supervised or monitored during their work, and the experimenter never commented on daily production figures. Presumably a subject's output was determined entirely by self-competition and the individual's job satisfaction. The results showed that when the men could see the counter, their output was significantly higher than when they could not. The increase was approximately equal to one extra day's production in every four.

Case 177. Chapanis (1964) attempted to replicate the findings of Case 176 by requiring subjects to punch sequences of digits into a teletype tape by using a perforator; F was indicated by a counter mounted on top of the perforator. The experimental conditions differed somewhat from those employed by Gibbs and Brown. [Brown (1966) discusses these and their possible effects.]

Group I subjects never received any information about their output; the counter was never operated. The counter was operative for group II subjects but was never reset to zero at the start of the work session; thus the subject, if he wished, could subtract his daily readings to get a measure of his performance. For group III subjects the counter was always operative and was always set to zero at the start of a work period. Group IV subjects, like group III, were presented with a counter always reset to zero, and they also had to jot down on a card the counter readings and sign their names and dates; however no comment was made by the research assistant who collected the cards.

No difference even approaching significance was found among the four groups, although performance increased over the 24 days of the experiment, suggesting that subjects adjusted their goals but that this goal adjustment did not depend on extrinsic F.

Case 178. Results similar to those of Chapanis were reported in another study designed to assess the purely motivational effect of KOR in a repetitive industrial task (grinding a piece of metal to specified size and shape). Indian workers in one group were permitted no opportunity to secure information about their daily output, another group could only roughly infer what that output was, and it was possible for a third group to infer their output from an index number (which, however, was not explained to them). There were no meaningful differences among the groups.

Case 179. Locke and Bryan (1969) obtained different results, The task was simple addition, a highly overlearned task. The conditions were knowledge of raw score (KOR) versus no knowledge (no KOR), and hard and easy goals. Scores in the KOR condition were given in a form in which they could not be used to set goals. The hard-and easy-goal subjects, on the other hand, were informed only of their progress in relation to a standard set by the experimenter.

Hard-goal subjects worked significantly faster than the easy-goal subjects, but the KOR and no-KOR groups did not differ in performance. Obviously F would have little utility in such a simple task. The results are indicated by Figure 6-9. These findings, suggest, as one would expect, that performance increases when operators adopt goal setting procedures (level of aspiration). Cummings et al. (1971) also found that KOR directly influences goal levels, but the situation is not quite so simple, since previous performance on the task also modifies goals.

Such studies suggest that in the operational situation, it is desirable not only to provide F but also to ensure that the F is relevant to the operator in terms of some standard that is given to him.

In designing F into the system, it is necessary to ask several questions. Does the operator need the F? Do the system and the task permit him to modify his performance to ensure that F will be useful? How will the operator perceive the F? What effect will it have on his performance?

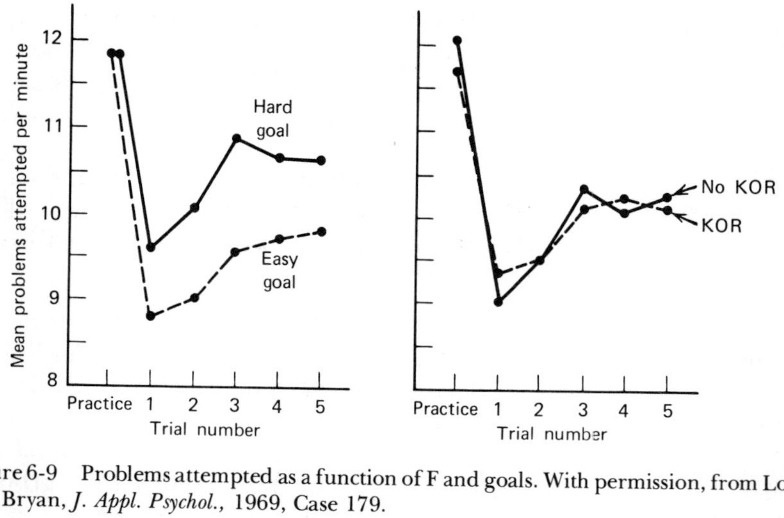

Figure 6-9 Problems attempted as a function of F and goals. With permission, from Locke and Bryan, *J. Appl. Psychol.*, 1969, Case 179.

Case 180. The obscurity inherent in the question of motivational factors in KOR is revealed by a recent study by Leamon (1974) attempting to explicate the differences between Gibbs and Brown (Case 176) and Chapanis (Case 177).

The task was an industrial, shop floor operation. Seven female workers had to collect a box of components, select a number, align the items, count them into bundles of varying number, examine the pack for faulty edges, and package the items.

The KOR consisted of a scale of "ergonomics work units" based on synthetic predetermined motion times—MTM. A graphic record for each operator was prepared, showing hourly performance rates at half-hourly intervals.

The three treatments (no KOR; KOR; and withdrawal of KOR, all presented in that order to each subject) produced significant differences in performance: KOR was superior to no KOR, and withdrawal of KOR was superior to KOR.

Leamon considered that the effect of KOR was produced by reinforcement of pacing changes. In other words, the pace of worker effort was improved by the KOR, which functioned by transferring information; "the effectiveness of this functioning is determined by motivation levels."

One feels after reading the report of this study that even in a simple industrial situation it is extremely difficult to determine precisely what has been transmitted to workers by KOR.

It is significant that all studies of this question have been performed with simple, gross, routine, repetitive tasks in which the amount of work performed was the criterion. The precision of the information that can be provided for such tasks is probably equally gross; if KOR is to have an effect in such tasks, it will be largely motivational. More complex MMS tasks call for more precise KOR, and often such tasks require only a specific output that cannot be increased; hence the motivational factor is irrelevant. The important point from a system design standpoint is not motivational versus informational KOR, but the amount and type of KOR required by the task itself.

Conclusions. Incentives interact with informational F to improve performance. The effect of incentives alone seems, however, to be weak and transient.

Illustrative Studies

Case 181. This was a categorization task time-shared with a memory task. Three F conditions were presented: *(a)* no F and no incentive *(b)* F alone, and *(c)* F plus incentive. Each of five stimuli contained two shapes, two colors, and one or two figures; the stimuli varied in position. Subjects had to categorize each stimulus as "+" or "−." In the simple condition subjects always classified in the same way; in the complex condition they alternated classifications between "+" and "−." Feedback was provided by indicators labeled right and wrong.

Feedback conditions were significant for both tasks (Table 6-3). For the memory task type of task and F interacted significantly; F and task complexity also interacted significantly.

One of the studies performed as part of Case 150 (also described in Bergum and Lehr, 1964) was a vigilance task involving the monitoring of circular moving light displays in which the signal was failure of a light to illuminate. Monetary rewards were given instead of KOR (20¢ gained for each correct detection, 20¢ lost for each failure). The control group received no rewards. The reward group performed significantly better in the first vigilance period but no better in the last two periods (Figure 6-10). As is usual in vigilance tasks, there was an overall performance decrement regardless of treatment. Interestingly, in the transfer condition the controls were significantly better than the reward group. These results suggest that the facilitating effect of monetary rewards is short-lived. The depressing effect on performance when rewards are removed is similar to that often found when KOR is removed. (See also Smith et al., 1967.)

Montague and Webber (1965) found that the combination of monetary reward plus KOR in a vigilance task raised the overall level of performance without eliminating the decrement over time. In this study the combined reward/KOR condition produced significantly superior performance over KOR alone, which was no more effective than a control condition. Somewhat similar results were obtained by Sipowicz et al. (1962) in a simple vigilance task; monetary reward was more effective than KOR in maintaining a higher detection level, and the combination of reward plus KOR was more effective than KOR alone.

Table 6-3 Mean Percent Correct Responses as a Function of Task and F Condition[a]

Task	F Condition		
	No F	F Alone	F + Incentive
Categorization	77.8%	83.9%	85.6%
Memory	55.8%	65.4%	74.3%

[a] From Denny and Gruemmer, 1970, Case 181.

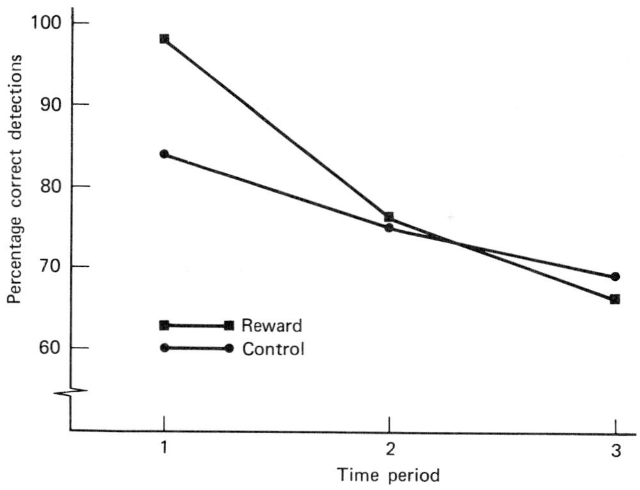

Figure 6-10 Effect of reward on vigilance performance. With permission, from Bergum and Lehr, 1963, Case 150.

However monetary rewards in a series of cognitive and psychomotor tasks were effective only when these modified operators' goals (Locke et al., 1968). When the effect of goal change was partialled out, reward effects were considerably reduced.

Case 182. The trickiness of KOR as a mechanism for regulating performance was demonstrated in a study of "keeping track" performance. Subjects had to tally mentally the number of occurrences of each of four letters presented visually and sequentially. KOR was presented by illuminating various indicators when a correct tally was reported.

The purpose of the study was to employ KOR and financial incentives containing "motivational" but not "directive" properties to manipulate the subject's strategies. The experimental conditions were *(a)* no KOR, *(b)* KOR alone, *(c)* KOR plus a 10¢ reward for a correct tally, and *(d)* KOR plus a 15¢ reward.

Neither KOR nor the presence of financial incentives was effective in bringing about performance improvement.

Conclusions. The effect of F on a stressful task appears to be positive, although the mechanisms are obscure.

Illustrative Studies

Case 183. It seems reasonable to assume that in a highly stressful situation, F might help to reduce the stress, thus improving performance.

Evidence tending to confirm this principle was found in one study that involved a 5-minute, self-paced, four-choice discrimination task. When one of four possible colors appeared in a display, the correct response was to press the key associated with that color.

The stress simulated a mission in which operators had to drop a bomb on a target at time zero. Failure would result in damage to the aircraft, simulated by a severe electric shock. Feedback was presented in score form at the end of each successive 10-second period of the test. The F variables were F and no F and relevant and non-revelant performance. Although differences between F and no-F groups were insignificant, operators receiving relevant F performed at a higher rate in the last two blocks of trials prior to time zero.

AUGMENTED F

Feedback in the learning situation is extrinsic (augmented) F that is supplied to the student by an instructor or instructor surrogate (e.g., computer assisted instruction). Augmented F represents an explicit evaluation of operator/system performance, a comparison of performance against requirements.

Intrinsic F, which is inherent in the action preceding F or in the task situation as a whole, does not include an explicit evaluation. It is assumed that the operator evaluates performance unconsciously.

Of course it is only a presumption, based on our knowledge that it is possible to learn even without extrinsic F, that such a thing as intrinsic F exists; it is difficult to specify the stimulus characteristics comprising intrinsic F (although some of the conditions under which it is found are described in Case 171). All F, including augmented F, is of course a construct; one can see the F stimulus and note the result in performance, but F itself is invisible.

Augmented F is commonly considered as F for training purposes, since the training situation is one in which this type of F plays a very dramatic part. However many system operations that do not involve training employ a type of augmented F (system F) in the form of system status displays. The system status F described at the beginning of this chapter has characteristics of both intrinsic and augmented F. A special meter to display power output is a form of augmented F because the system could function––albeit less efficiently perhaps—without it, and the meter and the information it provides are not available to the operator without special design provisions. Once system status F has been designed, however, it is inherent in system operations and can be considered intrinsic F at the system level.

We can consider F in terms of two phases that overlap in real life but can be separated, somewhat artificially, for analytic purposes. In the first phase the operator is learning a new task, and augmented F has the responsibility for aiding that learning. Formal training is complete in the second phase, and F has the responsibility of regulating performance (maintaining its adequacy and consistency).

During learning, whatever intrinsic F cues there are coexist with extrinsic ones. If the operator does not make use of intrinsic F, it is probably because he has not learned how to use it and he must rely on more obvious augmentation. After the operator has learned his task, he should no longer need extrinsic F and available intrinsic F can be substituted.

Kinkade (1963) has demonstrated that removal of augmented F (upon graduation from training to operational tasks) can result in performance deterioration if, after training, the operator attempts to continue use of augmented F instead of replacing it with intrinsic F. Such behavior is highly undesirable because in an operational task augmented F occurs quite infrequently. The operator is particularly prone to this error when intrinsic F cues are obscure.

System F (i.e., status displays) in the MMS supplements other intrinsic cues available to the operator. But whereas individual intrinsic F regulates only operator performance, system F regulates both operator and system performance. Since system F is deliberately designed into the system, its intensity can be set at a high level and *it cannot be removed.*

Unfortunately, little or no attention has been paid (except perhaps for human engineering studies of display parameters) to the mechanisms of incorporating system F into the system and to the most effective means of enabling the operator to make use of such F. Perhaps this is because the methods required to determine what system F should be provided are analytic rather than experimental.

It is hardly necessary to ask whether augmented F is useful in the learning situation. The reader's own experience will testify to that. A much more significant question is: do augmented F effects transfer to the operational (nontraining) situation, and under what conditions? If the answer is no, augmented F during training serves only to shorten the training period (which, however, is important in itself).

Conclusions. The effect of augmented F on learning is almost always positive. However the implications of the evidence for the usefulness of agumented F when the operator transfers to the operational situation and the augmented F is withdrawn, are unclear. One group of studies suggests that performance deteriorates when F is withdrawn from a tracking task (e.g., Bilodeau, 1952). Another group of studies found positive transfer effects after F is removed (e.g., Smode, 1958; Kinkade, 1959; Karlin, 1960; Williams and Briggs, 1962).

Illustrative Studies

Case 184. An experiment typical of the studies showing retention of performance effects when augmented F is removed was a vigilance task using the Mackworth Clock Test (detection of double jumps of a moving clock hand). Here KOR was presented by three lights (for each of three types of errors) mounted above the clock. The KOR conditions were (1) no KOR, (2) knowledge furnished only upon the subject's response (i.e., correct detections and commisive errors), and (3) KOR for correct detections, commissive errors, and omissive errors.

Full KOR (group III) was superior to partial KOR (group II), which in turn was superior to no KOR (group I). When KOR was withdrawn on the second day, the groups formerly receiving KOR retained their superiority (indeed, in the case of group II, the partial KOR group, performance was improved). (Figure 6-11 presents these trends.)

Case 185. It has been suggested that the discrepant results noted in the Conclusions may result from the relative usefulness of the augmenting F during acquisition trials (Annett and Kay, 1957; Annett, 1961; Karlin and Mortimer, 1963). The hypothesis was that for tasks in which the usefulness of the intrinsic F is large relative to the usefulness of the augmenting F, the decrement during transfer trials with the augmenting

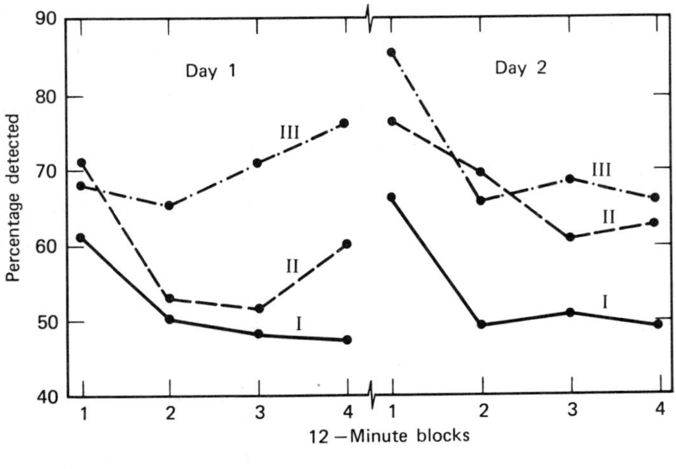

Figure 6-11 Retention of KOR effectiveness after withdrawal. With permission, from Wiener, *J. Appl. Psychol.*, 1963, Case 184.

F removed will be smaller. For tasks in which the usefulness of the intrinsic F is slight relative to the removed augmenting F, the performance deterioration should be greater (see also Case 171).

Case 185 was a compensatory tracking task. Augmenting F was presented either by visual blip or 2000 Hz tone. There were three levels of augmented F: (1) continuous F, indicating the direction and extent of tracking error at any moment, (2) 13 discrete categories of F, with tracking error presented in only one of the 13 categories, and (3) one F category only (on/off target).

Results indicated that increasing the amount of augmenting F information for intrinsic F conditions that presented little information resulted in positive transfer to the operational task when augmented F was removed; this is contrary to the initial hypothesis. The use of F apparently interacted with task difficulty, which may have confounded the study. It is also possible that when a loss results after augmenting F is removed, the effect of the F has been partly motivational.

Case 186. Further evidence for positive F transfer to the operational situation is provided by a study in which the task was a simple compensatory tracking task. Feedback was provided by auditory clicks that informed the operator of his errors during training. The operator was then transferred to a non-F condition. The results (Figure 6-12) indicate that various forms of F had some value in training, and the F effects persisted into the operational task. The relative ranking of the four groups remained the same in the transfer situation as in the training trials.

The manner in which augmented F during training is eliminated during operations is of interest also. A second experiment investigated three schedules for withdrawing augmented F—abruptly, decreasing the number of trials in which F was given, and increasing the augmented F tolerance limits. During the transition period, increasing the tolerance limits for error was found to be the most beneficial schedule for withdrawal. Withdrawal of augmented F was most disruptive when the augmented F was inconsistently available during withdrawal (decreasing the number of trials during which augmented F was available). This is entirely logical in terms of reinforcement theory. However no difference was found between the withdrawal groups and a control group during transfer.

To study the role of complex criteria in activating augmented F, two criteria were compared: amplitude plus rate for two groups, amplitude plus acceleration for two others. The use of augmented F that was based on complex criteria did not improve performance when compared with a control group. This may be another example of overwhelming the operator with more information than he can use.

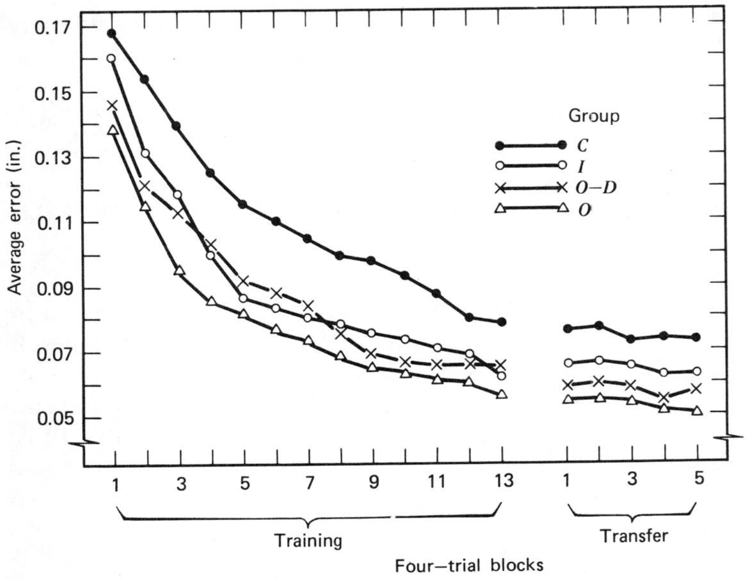

Figure 6-12 Maintenance of KOR effects following withdrawal in a tracking task. *C* = No F; *I* = inside tolerance F; *O* = outside tolerance F; *O–D* = outside tolerance plus directional F. With permission, from Briggs, 1962, Case 186.

Case 187. Three other experiments involved the same compensatory tracking task as in Case 186. In the first study the operator tracked first with a completely aided display (a cursor whose position was dependent on error amplitude, error rate, and error acceleration); he then transferred to a partially aided display (error amplitude plus error rate), and finally he tracked with a nonaided display (error amplitude only). Operators trained with complete and partial aiding performed at a higher level on the nonaided display than did operators who lacked this prior training (positive transfer of augmented F). However a group receiving an intermediate amount of training was superior on the nonaided task to groups receiving relatively larger or smaller amounts of training.

In a second study the operator received auditory clicks when tracking error was within specified limits; he then transferred to a situation in which augmented F occurred only if both error and error rate were within bounds. Next he received augmented F only if error, error rate, and error acceleration were within bounds. Finally, all augmented F was removed. Augmented F groups were inferior in performance to control groups without F, perhaps because the increased number of criteria reduced the frequency with which F was presented. In a third experiment, however, augmented F based on the most complex criterion (all three error indices) did lead to superior performance.

Case 188. The hypothesis underlying this study was that given clearly discernible input and intrinsic F cues, augmented F is supplementary and is not an essential feature of the task. Therefore deletion of the augmented F after training should not result in performance deterioration.

The hypothesis was tested with a one-dimensional CRT tracking situation. Two types of auditory F were provided; one noisy, the other not. An experimental group and a control group experienced visual noise during training trials; two other groups had a noise-free display. Following training, the augmented F was removed during transfer. Results tended to confirm the hypothesis. The noise-free groups, for which intrinsic F was not deliberately obscured, experienced less deterioration when augmented F was removed.

Conclusions. Some of the discrepant results of the previous studies and the one described in Case 189 suggest that it is possible to overload the operator with excessively complex F information.

Illustrative Studies

Case 189. In a tracking task augmenting F was provided in the form of a visual display (criterion circle surrounding the primary error display; the diameter of the former varied as a function of the error criterion applied to the task). Two criteria were employed: simple "average error," and a much more complex time-weighted error in which the relative importance of the error increased as a linear function of time. The criterion circle improved performance with the simple criterion but reduced performance with the complex one. The supplemental information demanded the tracker's attention and interfered with his primary task performance.

Somewhat comparable results were found by Williams (1974) in a simple linear-positioning task. Augmented F was disadvantageous under conditions of minimal practice because the complexity of the augmented F cues apparently impeded the learner's ability to select the criterion position. Augmented F proved to be beneficial after more practice.

Conclusions. Under certain circumstances and strictly for training purposes, cueing the operator in advance of an input can be as effective as or even more effective than F.

Illustrative Studies

Feedback functions after the initiating action has been completed. Thus for F to have an effect, it must act on the immediately following action. In that obvious sense the F acts as guidance or a *cue* to the performance of subsequent actions.

Taking this as a vantage point, a number of researchers have attempted to contrast the usefulness of F with that of information presented to the operator *before* the response occurs (cueing) in an effort to influence that response. The question is: does cueing have as much effect as F in learning?

The major proponent of cueing in contrast to F is Annett (Annett and Clarkson, 1964; Annett and Paterson, 1966, 1967; Annett, 1969), who points out that F is required only when a response is necessary to extract information from the environment (e.g., manual tasks depending mainly on kinesthetic information). Other types of tasks (e.g., those of perception, identification, and discrimination) could as readily be taught by cueing and prompting methods. Cueing is essentially a form of direct instruction; it avoids the pre-F delay and should, therefore, be even more effective than F.

The case for cueing as opposed to KOR is equivocal. In vigilance Annett and Clarkson (1964) found cueing superior to KOR in transfer, but Annett and Paterson (1966) obtained no differences. Annett (1966) found cueing superior to KOR in another study. More recently, Wiener and Attwood (1968) compared four training conditions in a vigilance task: KOR, cueing, KOR plus cueing, and a control group. The KOR-trained group detected significantly more signals during transfer than did the control, but the cueing group did not perform better. Nor was the combination of KOR and cueing superior to KOR alone.

The author's point of view is that there is nothing mutually exclusive between F and cueing *as far as training is concerned.* Both should be employed in training when appropriate. For the regulation of operational systems, however, and particularly for indeterminate situations in which Rcn is necessary to guide selection of response alternatives, cueing cannot be utilized and F becomes essential.

IDIOSYNCRATIC FACTORS

It is a reasonable presumption (although unfortunately not demonstrated by very substantial evidence) that social personality and attitudinal (idiosyncratic) factors influence F effectiveness. Since F is always (ultimately) received by an operator, his ability to receive and interpret that F in part determines F consequences.

One might, therefore, hypothesize that the weaker the F stimuli, the more idiosyncratic factors would come into play. In the same way, in indeterminate systems and tasks, where the probability of the correctness of any given response is not very high, these factors would have a greater effect on response selection.

At the other end of the continuum, the nature of the F received in any group situation should influence the social (non-task-oriented) characteristics of that situation (how one member of a team feels about the team, how the team feels about itself, etc.). If all this seems somewhat tenuous, it is because idiosyncratic factors are themselves inherently tenuous.

It is assumed that any complex MMS becomes in part a social situation and as such is influenced by these factors. This may be more characteristic of indeterminate than of determinate systems, although the paucity of data compels us to express the notion tentatively. In all probability, the effect of such factors is minor compared with the effect of task-oriented factors such as system requirements or operating procedures.

The HF specialist's interest in this topic is purely pragmatic. These non-mission-oriented factors are likely to detract from the performance of the task, and it is to the system developer's advantage to reduce their influence. It is first necessary, however, to understand how these factors influence F.

Conclusions. There is some evidence to indicate that F can affect the operator's attitudes and his perception of the situation in which he works (see also the discussion of goal setting, pp. 333–336). How these effects are translated into improved performance is much less clear, however.

Illustrative Studies

Several studies of non-task-oriented group behavior suggest that F may affect "social" learning (i.e., learning to interact with the group). The ambiguity of the terms employed makes it difficult to evaluate the meaningfulness of this work.

Kelman (1950) manipulated F in an autokinetic situation to study the effects of motivational factors on suggestibility. Using three experimental F conditions and a control (no-F) condition, he measured the extent to which subjects changed their judgments in the presence of an experimental confederate who judged differently. Subjects given no F showed more variability of response in the presence of the confederate. Subjects given F indicative of success had lower suggestibility scores than the control group, and those who were given failure F manifested higher suggestibility. Suggestibility scores of the group receiving ambiguous F were not significantly different from those of the control group.

Leavitt and Mueller (1951) concluded that "free feedback" is an aid to accuracy in interpersonal communication and that the presence or absence of F affects the sender-receiver relationship. Zero F is presumably accompanied by low confidence and hostility; free F is accompanied by high confidence and amity.

Gibb et al. (1955) suggested that positive F produced significantly less defensive feeling within a group than did negative F. Feedback concerning feelings produced significantly less defensive feeling than did task F. Significantly higher task efficiency occurred under positive F and also under feeling F compared to task F.

French (1956) reported an interaction between F and motivation on task performance. Her results showed that achievement-motivated subjects obtain higher performance scores under task-relevant F, whereas affiliation-motivated subjects performed better under "feeling" F.

Case 190. One study investigated the hypothesis that group effectiveness increases with increased member awareness of group satisfaction with task results. The tasks presented were paper and pencil problems, differing in task difficulty; three F conditions were presented (no F, overt F, and covert F). Subjects were permitted to express their degree of satisfaction to the group by operating a bank of indicator lights. Group performance scores were higher with F than with no F, and higher with covert (as opposed to overt) F on two tasks. The F effects appeared most prominently in connection with more difficult tasks.

Conclusions. One of the problems in the development of new equipment is influencing users to accept that equipment. Feedback can be used to change operator attitudes toward utilization of new equipment.

Illustrative Studies

Case 191. This study was designed to induce pilots to accept a new altimeter display that differed radically from the conventional altimeter. All pilots in one condition received KOR favoring performance on the new display. Pilots in this group were required to take readings from the old and new altimeters, and the F they received arbitrarily favored the new display.

Another group of pilots was given printed information handouts summarizing much of the evaluation data on the new and old displays and demonstrating the superiority of the new display. The third group, a control, received no information about the new display except what members had gathered from looking at a photograph of the new equipment. The performance measure was subjects' rating of their preference for the two types of display.

Figure 6-13 indicates that F resulting from trials with successful performance significantly altered preferences for the new altimeter. The informational F provided was no more effective than no F at all. Thus the

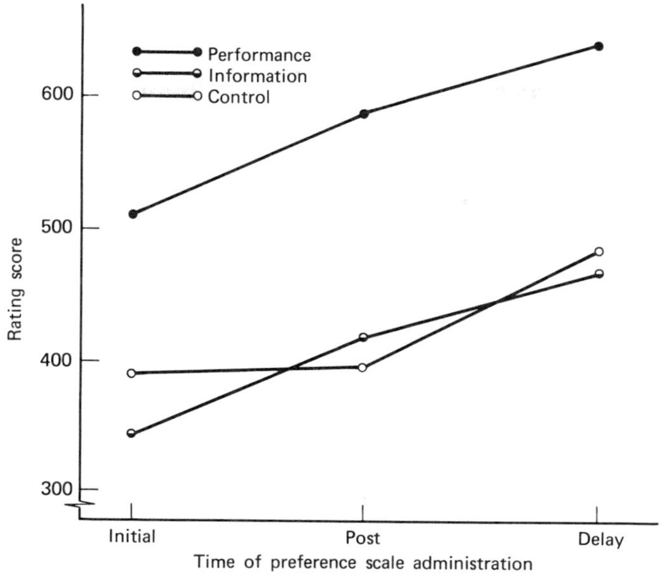

Figure 6-13 Preference ratings for new equipment as a function of type of KOR. With permission, from Berger et al., 1966, Case 191.

usefulness of F appears to be at least partly a function of the involvement of the operator with the activities leading to the F.

A similar technique was used to learn whether preferences would shift from an initially preferred display to one initially not preferred as a result of informing operators that they had attained higher proficiency with the nonpreferred equipment. Two new radar altimeter displays were used.

One group of pilots, after expressing initial preference for one of the two displays, was given an opportunity to perform static readouts on both displays. Each subject received F indicating that his performance was better with the display he initially did not prefer. Pilots in a second group were told that their performance was the same with both displays.

The results of this experiment supported the effectiveness of the testing technique, although statistical significance could not be achieved because of the small number of subjects. Although 8 of the 17 subjects in the trial or successful performance group changed their stated preferences, only 4 of the 17 in the control group changed. It is probable that pilots in this experiment based their evaluations of overall performance on their preconceptions as well as on the experimental F. This may have produced a conflict that attenuated F effects.

A third experiment was performed to determine the effect on preference change of varying the delay in F presentation. In the previous studies F had been given the subject after he had completed an entire series of stimulus presentations. Since F effectiveness might depend on how rapidly F is given to the subject after a response (pre-F interval), in this experiment F was given under three conditions: *(a)* immediately after a subject response, *(b)* after 18 responses, and *(c)* after 36 responses. The results of the study indicated that providing F after each response or after smaller blocks of responses did not increase F effectiveness. This is entirely in accordance with results of Bourne and Bunderson (1963).

In line with Case 191, empirical case studies of attitudinal factors in the acceptance of innovations in the Navy (Mecherikoff and Mackie, 1970) indicated that "successful performance with a new system leads to positive attitudes more readily than other forms of introduction." (See also Mackie et al., 1972.)

SUMMARY OF CONCLUSIONS

1. The general (but not invariable) effect of F is to improve performance both in learning and in system operation. However it is easier to demonstrate this improvement in the laboratory than in real world tasks.
2. The available studies supply no evidence to demonstrate that F effectiveness decreases with delay in receiving F. There is some indication that the longer the interval between F and the next stimulus, the greater the F effect, because of the opportunity for the operator to "study" the F.
3. Any F is more effective than none. At least in simple situations, F frequency need not be very great (no more 20–30% of all responses) for the operator to make effective use of that F.
4. Although little reliable evidence is available, it appears that performance improves with increasing F as long as *(a)* the F is task relevant, *(b)* the operator needs the additional information to perform the task, and *(c)* he can discriminate the number of information categories included in the F.
5. Operator performance tends to improve as F becomes more specific, provided the added specificity is task relevant and the operator needs it to perform the task.
6. Quality of operator performance is generally in inverse relation to the amount of misinformation provided, but incorrect F is more useful than no F at all. The negative effects of F misinformation can be reduced by making the operator aware that he is receiving incorrect F. Misinformation based on transformations of F data does not reduce performance as much as grossly inaccurate F.
7. Conclusions with regard to medium of presentation are confusing,

and studies derive only from vigilance situations. Verbal F is more effective than automatically produced F, perhaps because of the motivation inherent in direct contact between the subject and the experimenter. Auditory F is more effective than visual displayed F, perhaps because the former is more attention demanding.

8. Operator performance differs significantly as a function of F source level. Feedback is most effective when it describes the operator's output directly than when it describes the output as processed by operators at more remote system levels.

9. Feedback that directly represents the consequences of the individual operator's performance (direct F) is almost always more effective than F representing the team output (confounded F). However team F is more effective than no F at all. Team F affects individual performance to the degree that it accurately describes the individual operator's performance. Team F is particularly helpful when team coordination processes must be learned. Postperformance (debriefing) F can be highly effective in improving team performance in complex tasks. Consensus F can also be useful in highly indeterminate situations.

10. Feedback has both motivational (stimulating, goal setting, and criteria selection) and directive properties. One of the effects of F is to cause the operator to adjust his performance goals, which may in turn produce performance improvement. Monetary rewards interact with informational F to improve performance. The effect of F on a stressful task appears to be positive, although the mechanisms are obscure.

11. Augmented F in training is almost always effective. However the usefulness of augmented F in transferring to the operational situation (i.e., after augmented F has been removed) has not been clearly established. It is possible to overload the operator with excessively complex F. Under certain circumstances cueing the operator in advance of an input can be as effective as or even more effective than F in training.

12. Social and attitudinal factors can influence F effectiveness and can in turn be affected by F characteristics. Feedback appears to influence the operator's job satisfaction and perception of system performance, but acceptable evidence on this point is sparse. Feedback can be used to change operator attitudes toward utilization of new equipment.

DEVELOPMENTAL IMPLICATIONS

Because the F literature does not supply precise quantitative values to be incorporated in system design, its primary value lies in pointing out F variables that should be kept in mind in analyzing and developing the system configuration (particularly its training subsystem).

The data suggest a balanced view of F. A little F goes a long way, and one need not be overly concerned about precise temporal and quantitative factors. Even the correctness of F, though important, is not critical.

At the same time it is apparent that there are limits to F usefulness: higher-order F is not invariably effective in system operations, and augmented F may be valuable only insofar as it helps shorten training time.

The major design questions that need to be answered are as follows.

1. How does the HF specialist include F augmentation in the design of the task?

2. What are the most effective ways for F to be included in system status displays?

The designer has always included some F in hardware form (e.g., meters and indicators to show the status of the equipment being operated or the extrasystem environment being manipulated). The F included in this way is usually relatively obvious, and an HF specialist may not be needed to point it out.

The literature provides certain general rules that can aid the designer. For example, any F to be incorporated in hardware design should be as specific as possible; it is inadvisable to give the operator F about which he can do nothing; giving the operator more F than he can handle (whose categories he cannot discriminate) will only overload.

These are, however, only general principles. Since F is always highly task specific, the key to effective F implementation in system design lies in more adequate analysis of system and task requirements. The following procedures may be helpful.

1. Determine what information the operator needs to do his job. This will probably unearth all the relatively obvious F he needs.

2. Determine what F is intrinsic to the job. If a task has considerable intrinsic F, it may be that no supplementary F is necessary. If, however, successful performance of the task depends on cues that are not highly visible, it may be advisable to amplify the cues, either by incorporating an F display in the equipment or by intensive training of intrinsic cues to make them more visible. In deciding whether to provide supplementary system F or augmented F in training, remember that if the former is properly incorporated in the system, it is probably more effective in regulating operator performance than the latter.

3. In developing system F the HF specialist should consider the following points:

(a) Concentrate on qualitative and quantitive needs for F. Although F frequency, delays, specificity, and precision factors should not be ignored, they should be considered to be secondary variables.

(b) Make sure that any supplementary F will not overload the operator by requiring him to perform additional discrimination tasks.

(c) Ensure that the operator receives only that F which is compatible with task requirements (what he needs for performing his job) and that he does not receive higher-order system F unless required by task demands.

(d) In dull, repetitive jobs, provide supplementary F for motivational purposes as long as it does not distract the operator from his task.

GENERAL REFERENCES

Adams, J. A. and Humes, J. M. Monitoring of complex visual displays: IV. Training for vigiliance. *Hum. Factors,* 1963, **5,** 147–153.

Ammons, R. B. Effects of knowledge of performance: A survey and tentative theoretical formulation. *J. Gen. Psychol.,* 1956, **54,** 279–299.

Annett, J. *The role of knowledge of results in learning: a survey.* Report NAVTRADEVCEN 342–3, U.S. Naval Training Device Center, Port Washington, N.Y., May 1961. (AD 262–937.)

Annett, J. Training for perceptual skills. *Ergonomics,* 1966, **9,** 459–468.

Annett, J. *Feedback and Human Behavior.* Baltimore: Penguin Books, 1969.

Annett, J. and Clarkson, J. K. *The use of cueing in training tasks.* Report NAVTRADEVCEN 3143–1, U.S. Navy Training Device Center, Port Washington, N.Y. 1964. (AD 601–352.)

Annett, J. and Kay, H. Knowledge of results and skilled performance *Occup. Psychol.,* 1957, **31,** 69–79.

Annett, J. and Paterson, L. *The use of cueing in training tasks: Phase II.* Report NAVTRADEVCEN 4119–1, U.S. Navy Training Device Center, Port Washington, N.Y., 1966. (AD 630–260.)

Annett, J. and Paterson, L. *The use of cueing in training tasks: Phase III.* Report NAVTRADEVCEN 4717–1, U.S. Navy Training Device Center, Port Washington, N.Y., 1967 (AD 657–367.)

Antonelli, D. C. and Karas G. G. Performance on a vigilance task under conditions of true and false knowledge of results. *Percept. Mot. Skills,* 1967, **25,** 129–138.

Arps, G. F. Work with knowledge of results versus work without knowledge of results. *Psychol. Monogr.,* 1920, **28,** No. 3 (whole no. 125).

Baker, C. H. *Three minor studies of vigilance.* Report 234–2, Defense Medical Research Laboratory, Toronto, Canada, 1958.

Baker, C. H. Maintaining the level of vigilance by means of artificial signals. *J Appl. Psychol.,* 1960, **44,** 336–338.

Bergum, B. O. and Lehr, D. J. Monetary incentives and vigilance. *J. Exp. Psychol.,* 1964, **67,** 197–198.

Bilodeau, E. A. *Some effects of various degrees of supplemental information given at two levels of practice upon the acquisition of a complex motor skill.* USAF Human Resources Research Center Research Bulletin 52–15, 1952.

Bilodeau, E. A. Speed of acquiring a simple motor response as a function of the systematic transformation of knowledge of results. *Am. J. Psychol.,* 1953, **66,** 409–420.

Bilodeau, I. Mc. D. Information feedback. In E. A. Bilodeau (Ed)., *Acquisition of Skill.* New York: Academic Press, 1966, pp. 255–296.

Bolin, S. F. et al. *Team procedures in image interpretation.* Technical research note 164, U.S. Army Behavioral Science Research Laboratory, Washington, D.C., December 1965.

Bourne, L. E. Effects of delay of information feedback and task complexity on the identification of concepts. *J. Exp. Psychol.,* 1957, **54,** 201–207.

Bourne, L. E. Long term effects of misinformative feedback upon concept identification. *J. Exp. Psychol.,* 1963, **65,** 139–147.

Bourne, L. E. and Bunderson, C. V. Effects of delay of informative feedback and length of postfeedback interval on concept identification. *J. Exp. Psychol.,* 1963, **65,** 1–5.

Bourne, L. E. and Pendleton, R. B. Concept identification as a function of completeness and probability of information feedback. *J. Exp. Psychol.,* 1958, **56,** 413–420.

Bourne, L. E. et al. Concept identification: The effects of varying lengths and informational components of the intertrial interval. *J. Exp. Psychol.,* 1965, **69,** 624–629.

Brown, I. D. An asymmetrical transfer effect in research on knowledge of performance. *J. Appl. Psychol.,* 1966, **50,** 118–120.

Chapanis, A. Knowledge of performance as an incentive in repetitive, monotonous tasks. *J. Appl. Psychol.,* 1964, **48,** 263–267.

Chinn, R. M. and Alluisi, E. A. Effect of three kinds of knowledge-of-results information on three measures of vigilance performance. *Percept. Mot. Skills,* 1964, **18,** 901–912.

Clarke, D. E. The effects of simulated feedback and motivation on persistence at a task. *Organ. Behav. Hum. Performance,* 1972, **8,** 340–346.

Colquhoun, W. P. Training for vigilance: A comparison of different techniques. *Hum. Factors,* 1966, **8,** 7–12.

Cummings, L. L. et al. Performance and knowledge of results as determinants of goal setting. *J. Appl. Psychol.,* 1971, **55,** 526–530.

Doten, G. W. et al. The use of teams in image interpretation: Information exchange, confidence and resolving disagreements. *Hum. Factors,* 1968, **10,** 107–116.

Egerman, K. et al. *Increasing team proficiency through training. 3. Detrimental effects of reinforcement in teams with redundant members.* Technical report AIR-B64-6/62-TR, American Institute for Research, Pittsburgh, June 1962. (AD 276–428.)

Egerman, K. et al. *Increasing team proficiency through training. 4. A learning-theoretic analysis of the effects of team arrangement on team performance.* Technical report AIR-B64-9/63-TR, American Institute for Research, Pittsburgh, September 1963. (AD 422–328.)

Elwell, J. L. and Grindley, G. C. The effect of knowledge of results on learning and performance. I. A coordinated movement of the two hands. *Brit. J. Psychol.,* 1938, **29,** 39–53.

Eriksen, C. W. Effects of practice with or without correction on discrimination learning. *Am. J. Psychol.,* 1958, **71,** 350–358.

French, E. G. Effects of the interaction of feedback and motivation on task performance. *Am. Psychol.,* 1956, **11,** 395 (abstract).

Gagne, R. M. and Rohwer, W. D. Instructional psychology. *Ann. Rev. Psychol.,* 1969, **20,** 381–418.

Garvey, W. D. et al. *The use of "artificial signals" to enhance monitoring performance.* NRL Report 5259, Naval Research Laboratory, Washington, D.C., 1959.

Geldard, F. A. *Conference on cutaneous communication systems and devices.* Austin, Texas: Psychonomic Society, 1974.

Gibb, J. R. et al. Effects of positive and negative feedback upon defensive behavior in small problem-solving groups. *Am Psychol.,* 1955, **10,** 335 (abstract).

Gibbs, C. B. and Brown, I. D. *Increased production from the information incentive in a repetitive task.* No. 230, Medical Research Council, Applied Psychology Research Unit, Great Britain, March 1955.

Goldman, M. A comparison of indivdual and group performance for varying combinations of ability. *J. Pers. Soc. Psychol.,* 1965, **1,** 210–216.

Goldstein, I. L. et al. Effect of percentage and specificity of feedback on choice behavior in a probabilistic information-processing task. *J. Appl. Psychol.,* 1968, **52,** 163–168.

Goldstein, I. L. et al. Feedback in a complex multiman-machine system. *J. Appl. Psychol.,* 1967, **51,** 346–351.

Green, R. F. et al. *The effects of varying degrees of knowledge of results on knob-setting performance.* SPECDEVCEN 241-6-20, U.S. Navy Training Device Center, Port Washington, N.Y., 1955. (AD 109–352.)

Hall, R. L. Group performance under feedback that confounds responses of group members. *Sociometry,* 1957, **20,** 297–305.

Higgins, N.C. *Feedback in group instruction.* Report AFOSR-TR-73-0016, Instructional Resources Laboratory, Arizona State University. February, 1972. (AD 754–979.)

Horrocks, J. E. et al. *Team training II: Individual learning and team performance.* NAVTRADEVCEN 198–2, U.S. Naval Training Device Center, Port Washington, N.Y., August 1960. (AD 247–147.)

Howell, W. C. and Kreidler, D. L. Information processing under contradictory instructional sets. *J. Exp. Psychol.,* 1963, **65,** 39–46.

Howell, W. C. and Kreidler, D. L. Instructional sets and subjective criterion levels in a complex information-processing task. *J. Exp. Psychol.,* 1964, **68,** 612–614.

Johanson, A. M. The influence of incentive and punishment upon reaction-time. *Arch. Psychol.,* 1922, **8** (54).

Johannsen, W. J. Concept identification under misinformative feedback and subsequent informative feedback conditions. *J. Exp. Psychol.,* 1962, **64,** 631–635.

Johnston, W. A. Individual performance and self-evaluation in a simulated team. *Organ. Behav. Hum. Performance,* 1967, **2,** 309–328.

Johnston, W. A. and Nawrocki, L. H. Effect of simulated social feedback on individual tracking performance. *J. Appl. Psychol.,* 1967, **51,** 145–151.

Judd, C. H. Practice without knowledge of results. *Psychol. Res. Monogr. Suppl.,* 1905, **7,** 185–199.

Karlin, L. *Psychological study of motor skills: Phase I.* Report NAVTRADEVCEN 558-1, U.S. Naval Training Device Center, Port Washington, N.Y. 1960. (AD 243–372.)

Karlin, L. and Mortimer, R. G. Effect of verbal, visual and auditory augmenting cues on learning a complex motor skill. *J. Exp. Psychol.,* 1963, **65,** 75–79.

Kelman, H. C. Effects of success and failure on suggestibility in the autokinetic situation. *J. Abnorm. Soc. Psychol.,* 1950, **45,** 267–285.

Kim, J. S. and Hamner, W. C. Effect of performance feedback and goal setting on productivity and satisfaction in an organizational setting. *J. Appl. Psychol.,* 1976, **61,** 48–57.

Kimble, G. A. *Hilgard and Marquis' Conditioning and Learning.* New York: Appleton-Century Crofts, 1961.

Kinkade, R. G. *Augmented feedback and tracking skill.* Report NAVTRADEVCEN 508–3, U.S. Naval Training Device Center, Port Washington, N.Y., 1959. (AD 230–999.)

Kinkade, R. G. *A differential influence of augmented feedback on learning and on performance.* Report AMRL-TDR-63–12, Aerospace Medical Research Laboratories, Wright-Patterson AFB, Ohio, February 1963. (AD 402–973.)

Klaus, D. J. and Glaser, R. *Increasing team proficiency through training. 8. Final summary report.* Report AIR-E1-6/68-FR, American Institute for Research, Pittsburgh, May 1968.

Leamon, T. B. An investigation into the effects of knowledge of results on operator performance. *Ergonomics,* 1974, **17,** 639–650.

Leavitt, H. J. and Mueller, R. A. H. Some effects of feedback on communication. *Hum. Relat.,* 1951, **4,** 401–410.

Lindahl, L. G. Movement analysis as an industrial training method. *J. Appl. Psychol.,* 1945, **29,** 420–436.

Locke, E. A. Toward a theory of task motivation and incentives. *Organ. Behav. Hum. Performance,* 1968, **3,** 157–189.

Locke, E. A. and Bryan, J. F. Cognitive aspects of psychomotor performance: The effects of performance goals on level of performance. *J. Appl. Psychol.,* 1966a, **50,** 286–291.

Locke, E. A. and Bryan, J. F. The effects of goal-setting, rule learning and knowledge of score on performance. *Am. J. Psychol.,* 1966b, **79,** 451–457.

Locke, E. A. and Bryan, J. F. Performance goals as determinants of level of performance and boredom. *J. Appl. Psychol.,* 1967, **51,** 120–130.

Locke, E. A. et al. Goals and intentions as mediators of the effects of monetary incentives on behavior. *J. Appl. Psychol.,* 1968, **52,** 104–121.

Locke, E. A. and Bryan, J. F. Knowledge of score and goal level as determinants of work rate. *J. Appl. Psychol.,* 1969, **53,** 59–65.

Loeb, M. and Schmidt, E. A. *Influence of time on task and false information on efficiency of responding to pure tones.* Report 426, U.S. Army Medical Research Laboratory, Fort Knox, Ky., 1960.

Mackie, R. R. et al. *Factors leading to the acceptance or rejection of training devices.* Report NAVTRAEQUIPCEN 70-C-0276-1, Naval Training Equipment Center, Orlando, Fla., 1972.

Manzer, C. W. The effect of knowledge of output on muscular work. *J. Exp. Psychol.,* 1935, **18,** 80–90.

Matthews, L. Subject-determined postfeedback interval in auditory concept identification. *Psychonomic Sci.,* 1972, **26,** 193–194.

McCormack, P. D. A two factor theory of vigilance. *Brit J. Psychol.,* 1962, **53,** 357–363.

McCormack, P. D. and McElheran, W. G. Follow-up of effects on reaction time with partial knowledge of results. *Percept. Mot. Skills,* 1963, **17,** 565–566.

McCormack, P. D. et al. Effects on reaction-time of knowledge of results of performance. *Percept. Mot. Skills,* 1962, **14,** 367–372.

McCormack, P. D. et al. Effects on reaction time of partial knowledge of results of performance. *Percept. Mot. Skills,* 1963, **17,** 279–281.

McKeachie, W. J. Instructional Psychology. *Ann. Rev. Psychol.,* 1974, **25,** 161–193.

Mackworth, J. F. The effect of true and false knowledge of results on the detectability of signals in a vigilance task. *Can. J. Psychol.,* 1964, **18,** 106–117.

MacPherson, S. J. et al. The effect of knowledge of results on learning and performance: II. Some characteristics of very simple skills. *Q. J. Exp. Psychol.,* 1948, **1,** 68–78.

Metcherikoff, M. and Mackie, R. R. *Attitudinal factors in the acceptance of innovations in the Navy.* Technical report 784-1, Human Factors Research, Inc., Goleta, Calif., June 1970.

Montague, W. E. and Webber, C. E. Effects of knowledge of results and differential monetary reward on six uninterrupted hours of monitoring. *Hum. Factors,* 1965, **7,** 173–180.

Namikas, G. and Carey, R. N. Concept identification as a function of the conceptual rule and length of postinformative feedback interval. *Psychonomic Sci.,* 1971, **23,** 417–419.

Payne, R. B. and Hauty, G. T. Effect of psychological feedback upon work decrement. *J. Exp. Psychol.,* 1955, **50,** 343–351.

Peretti, P. O. Effects on reaction time of random-false knowledge of results of performance. *Percept. Mot. Skills,* 1970, **30,** 838.

Pishkin, V. Effects of probability of misinformation and number of irrelevant dimensions upon concept identification. *J. Exp. Psychol.,* 1960, **59,** 371–378.

Pressey, S. L. Development and appraisal of devices providing immediate automatic scoring of objective tests and concomitant self-instruction. *J. Psychol.,* 1950, **29,** 417–447.

Rosenberg, S. The maintenance of a learnt response in controlled interpersonal conditions. *Sociometry,* 1959, **22,** 124–138. See also, Rosenberg, S. A laboratory approach to interpersonal aspects of team perforance. *Ergonomics,* 1959, **2,** 335–348.

Rosenberg, S. and Hall, R. L. The effects of different social feedback conditions upon performance in dyadic teams. *J. Abnorm. Soc. Psychol.,* 1958, **57,** 271–277.

Roweton, W. E. and Davis, G. A. Effects of preresponse interval, postinformative feedback interval , and problem difficulty on the identification of concepts. *J. Exp. Psychol.,* 1968, **78,** 642–645.

Sadacca, R. et al. Image interpretation task—status report. Technical research report 1129, U.S. Army Behavioral Science Research Laboratory, Washington, D.C., June 1962.

Saunders, P. S. and Neimark, E. D. Effects of misinformative feedback and awareness of misinformation on concept attainment strategy. *Proceedings,* American Psychological Association Convention, 1967, pp. 49–50.

Sipowicz, R. R. et al. The effects of reward and knowledge of results on the performance of a simple vigilance task. *J. Exp. Psychol.,* 1962, **64,** 58–61.

Smith, R. L. et al. Effects of monetary rewards and punishments on vigilance performance. *J. Appl. Psychol.,* 1967, **51,** 411–416.

Smode, A. F. Learning and performance in a tracking task under two levels of achievement information feedback. *J. Exp. Psychol.,* 1958, **56,** 297–304.

Sulzer, J. E. and Levy, C. M. A test for the effect of numerical magnitude of knowledge of results. *Percept. Mot. Skills,* 1966, **22,** 311–316.

Taubman, R. E. The effect of practice with and without reinforcement on the judgment of auditory number. *J. Exp. Psychol.,* 1944, **34,** 143–151.

Thorndike, E. L. The law of effect. *Am. J. Psychol.,* 1927, **39,** 212–222.

Wallis, D. and Samuel, J. A. Some experimental studies of radar operating. *Ergonomics,* 1961, **4,** 155–168.

White, R. M. and Schmidt, S. W. Preresponse intervals versus postinformative feedback intervals in concept identification. *J. Exp. Psychol.,* 1972, **94,** 350–352.

Wiedenfeller, E. W. et al. Effects of knowledge of results (true and false) on vigilance performance. *Percept. Mot. Skills,* 1962, **14,** 211–215.

Wiener, E. L. Knowledge of results and signal rate in monitoring: A transfer of training approach. *J. Appl. Psychol.,* 1963, **47,** 214–222.

Wiener, E. L. Training for vigilance: Repeated sessions with knowledge of results. *Ergonomics,* 1968, **11,** 547–556.

Wiener, E. L. An adaptive vigilance task with knowledge of results. *Hum. Factors,* 1974, **16,** 333–338.

Wiener, E. L. and Attwood, D. A. Training for vigilance: combined cueing and knowledge of results. *J. Appl. Psychol.,* 1968, **52,** 474–479.

Williams, A. C. and Briggs, G. E. On-target versus off-target information and the acquisition of tracking skill. *J. Exp. Psychol.,* 1962, **64,** 519–525.

Williams, I. D. Practice and augmentation in learning. *Hum. Factors,* 1974, **16,** 503–507.

Zajonc, R. B. The effects of feedback and probability of group success on individual and group performance. *Hum. Relat.,* 1962, **15,** 149–161.

Ziller, R. C. Toward a theory of open and closed groups. *Psychol. Bull.,* 1965, **64,** 164–182.

CASE REFERENCES

Case 149. Johnson, E. M. and Payne, M. C. Vigilance: Effects of frequency of knowledge of results. *J. Appl. Psychol.,* 1966, **50,** 33–34.

Case 150. Bergum, B. O. and Lehr, D. J. *Vigilance performance as a function of task and environmental variables.* Research report 11, Human Resources Research Office, U.S. Army Air Defense Human Research Unit, Fort Bliss, Tex., May 1963.

Case 151. McCormack, P. D. et al. Effects on reaction time of partial knowledge of results of performance. *Percept. Mot. Skills,* 1963, **17,** 279–281.

Case 152. McCormack, P. D. and McElheran, W. G. Follow-up of effects on reaction time with partial knowledge of results. *Percept. Mot. Skills,* 1963, **17,** 565–566.

Case 153. Goldstein, I. L. et al. Effect of percentage and specificity of feedback on choice behavior in a probabilistic information-processing task. *J. Appl. Psychol.,* 1968, **52,** 163–168.

Case 154. Bourne, L. E. and Pendleton, R. B. Concept identification as a function of completeness and probability of information feedback. *J. Exp. Psychol.,* 1958, **56,** 413–420.

Case 155. Goldstein, I. L. et al. Feedback in a complex multiman-machine system. *J. Appl. Psychol.,* 1967, **51,** 346–351.

Case 156. Smode, A. F. Learning and performance in a tracking task under two levels of achievement information feedback. *J. Exp. Psychol.,* 1958, **56,** 297–304.

Case 157. Hunt, D. P. The effect of the precision of informational feedback on human tracking performance. *Hum. Factors,* 1961, **3,** 77–85.

Case 158. Naylor, J. C. et al. *Long-term skill transfer and feedback conditions during training and rehearsal.* Report AMRL-TDR-63-136, Behavioral Science Laboratory, 6570th Aerospace Medical Research Laboratories, Wright-Patterson AFB, Ohio, December 1963. (AD 431–222.)

Case 159. Cotterman T. E. *Effects of variations in specificity of knowledge of results on the improvement of a perceptual skill.* Behavioral Science Laboratory, Aerospace Medical Research Laboratories, Wright-Patterson AFB, Ohio, August 1960. (AD 246-349.)

Case 160. Morin, R. E. Factors influencing rate and extent of learning in the presence of misinformative feedback. *J. Exp. Psychol.*, 1955, **49,** 343–351.

Case 161. Kinkade, R. G. et al. *A study of tactical decision making behavior* (experiment II). Report ESD-TR-66-61, Decision Sciences Laboratory, Hanscom Field, Mass., November 1965. (AD 478–769.)

Case 162. Monty, R. A. Auditory perception of numerosity as affected by number and by correct and incorrect knowledge of results. *Hum. Factors,* 1962, **4,** 193–199.

Case 163. Federico, P. A. Some effects of inflating information feedback on the acquisition of a discrete lever-positioning response. *Percept. Mot. Skills,* 1970, **31,** 447–459.

Case 164. Noble, C. E. and Broussard, I. G. Effects of complex transformations of feedback upon simple instrumental behavior. *J. Exp. Psychol.*, 1955, **50,** 381–386.

Case 165. Ware, J. R. and Baker, R. A. Effects of method of presentation, modes and response category knowledge of results on detection performance in a vigilance task. *J. Eng. Psychol.*, 1964, **3,** 111–116.

Case 166. Hardesty, D. et al. Influence of knowledge of results on performance in a monitoring task. *Percept. Mot. Skills,* 1963, **16,** 629–634.

Case 167. Feallock, J. B. and Briggs, G. E. *A multiman-machine system simulation facility and related research on information-processing and decision-making tasks.* Report AMRL-TDR-63-48, Behavioral Science Laboratory, 6570th Aerospace Medical Research Laboratories, Wright-Patterson AFB, Ohio, June 1963. (Study 2, 84–109.) (AD 411–552.)

Case 168. Hall, R. L. Group performance under feedback that confounds responses of group members. *Sociometry,* 1957, **20,** 297–305.

Case 169. Rosenberg, S. and Hall, R. L. The effects of different social feedback conditions upon performance in dyadic teams. *J. Abnorm. Soc. Psychol.*, 1958, **57,** 271–277.

Case 170. Zajonc, R. B. The effects of feedback and probability of group success on individual and group performance. *Hum. Relat.*, 1962, **15,** 149–161.

Case 171. Alexander, L. T. et al. The effectiveness of knowledge of results in a military system-training program. *J. Appl. Psychol.*, 1962, **46,** 202–211.

Case 172. Cockrell, J. T. *Maintaining image interpreter proficiency through team consensus feedback.* Technical research note 195, U.S. Army Behavioral Science Research Laboratory, Washington, D. C., April 1968. (AD 833–583.)

Case 173. Cockrell, J. T. *Maintaining target detection proficiency through team consensus feedback.* Technical research note 219, U.S. Army Behavioral Science Research Laboratory, Arlington, Va., December 1969. (AD 707–376.)

Case 174. Rosenberg, S. Cooperative behavior in dyads as a function of reinforcement parameters. *J. Abnorm. Soc. Psychol.*, 1960, **60,** 318–333.

Case 175. Horrocks, J. E. et al. *Team training II: Individual learning and team performance.* Report NAVTRADEVCEN 198-2, U.S. Naval Training Device Center, Port Washington, N.Y., August 1960. (AD 247–147.)

Case 176. Gibbs, C. B. and Brown, I. D. *Increased production from the information incentive in a repetitive task.* No. 230, Medical Research Council, Applied Psychology Research Unit. Great Britain, March 1955. Described in Case 177.

Case 177. Chapanis, A. Knowledge of performance as an incentive in repetitive monotonous tasks. *J. Appl. Psychol.*, 1964, **48,** 263–267.

Case 178. Hundal, P. S. Knowledge of performance as an incentive in repetitive industrial work. *J. Appl. Psych.*, 1969, **53,** 224–226.

Case 179. Locke, E. A. and Bryan, J. F. Knowledge of score and goal level as determinants of work rate. *J. Appl. Psychol.,* 1969, **53,** 59–65.

Case 180. Leamon, T. B. An investigation into the effects of knowledge of results on operator performance. *Ergonomics,* 1974, **17,** 639–650.

Case 181. Denny, N. R. and Gruemmer, H. C. *Event frequency memory and categorization performance as a function of conditions of feedback.* Report, Contract DAAD05-69-C-0102, Texas Technical University, Lubbock, March 23, 1970. (AD 703-848.)

Case 182. Karsh, R. et al. Effects of knowledge of results and method of payoff on keeping-track performance. *J. Psychol.,* 1970, **75,** 73–79.

Case 183. Coulter, X. *The effect of performance relevance and feedback upon resistance to anticipatory stress.* Report NAMRL-1114, USAARL No. 69-10, Joint Army-Navy Report, U.S. Army Aeromedical Research Laboratory, Naval Aerospace Medical Research Laboratory, August 14, 1970.

Case 184. Wiener, E. L. Knowledge of results and signal rate in monitoring: a transfer of training approach. *J. Appl. Psychol.,* 1963, **47,** 214–222.

Case 185. Micheli, G. *Augmenting feedback and transfer of training.* Report NAVTRADEV-CEN 1H-41, Naval Training Device Center, Port Washington, N.Y., March 1966. (AD 631–405.)

Case 186. Briggs, G. E. *Research on augmented feedback and the acquisition and transfer of skill.* Report NAVTRADEVCEN 836-3, U.S. Naval Training Device Center, Port Washington, N.Y., September 5, 1962. (AD 293–884.)

Case 187. Briggs, G. E. *On the scheduling of training conditions for the acquisition and transfer of perceptual-motor skills.* Report NAVTRADEVCEN 836-1, U.S. Naval Training Device Center, Port Washington, N.Y., December 1961. (AD 419–362.)

Case 188. Kinkade, R. G. *A differential influence of augmented feedback on learning and on performance.* Report AMRL-TDR-63-12, Aerospace Medical Research Laboratories, Wright-Patterson AFB, Ohio, February 1963. (AD 402–973.)

Case 189. Ward, J. L. and Senders, J. W. *Methodological studies of tracking behavior: The effect of various supplemental information feedbacks.* Report 1458, Final report Nonr 4605(00), Bolt, Beranek, and Newman, Inc., Cambridge, Mass., May 31, 1966. (AD 649–574.)

Case 190. Shaw, M. E. and Blum, J. M. Group performance as a function of task difficulty and the group's awareness of member satisfaction. *J. Appl. Psychol.,* 1965, **49,** 151–154.

Case 191. Berger, P. K. et al. *The role of trial in the acceptance and adoption of new equipment, a review and summary.* Report LS/TR 66-1, Life Sciences, Inc., Fort Worth, Tex., August 1966.

VII CONCLUDING COMMENTS

SUMMARY. Most behavioral research does not constitute investigation of the system because the studies are directed primarily at the individual. The questions they answer are not very relevant to system design, and they do not supply required quantitative data. A number of quasi-system studies have been performed, however, and from these it appears that man in the system does not merely react but is a highly active modifier of system functioning.

INTRODUCTION

This is a short chapter. More might be said about topics discussed previously and about certain subjects that have not been touched on, such as extended operations, fatigue, and work-rest cycles; but an end must be made at some point.

In this chapter we attempt to summarize the characteristics of the research described in previous chapters. What have we learned about the way behavioral research has been and should be conducted on the system?

Viewed as a totality, the behavioral literature summarized in this book is impressive; researchers have no need to feel ashamed of what they have learned. Having said this, however, we must not ignore the gaps in the literature. Many of the conclusions reached can be only tentative until they are verified with additional tasks and more varied subject populations. If the author had unlimited financial resources and time and free choice of experimenters, his first major research task would be to "beef up" what is known, to nail down present data against any doubt.

RESEARCH CHARACTERISTICS

It should come as no great surprise to the reader that most of the studies reviewed in this book are not "true" system studies because they do not satisfy the system criteria specified in the preface. Many of these investigations might be termed "man-machine" experiments because their context is that of equipment operation or is directed toward answering questions of operator behavior. Even though these studies attempt to represent a system context by including the outward hardware features of a system (e.g., aircraft instruments or their analogues), their primary interest is in the operator's responses to stimuli, in what happens to and within the individual, rather than how operator responses are transformed into system outputs.

Naturally there are examples of studies that come close to the system criteria listed in the preface. The distinguishing characteristics of these studies are described later.

Collectively, the mass of potentially relevant behavioral research may seem overwhelming to the reviewer. Yet when this research is analyzed in terms of specific questions to be answered, it is surprising how few data we have for many topics. This is perhaps understandable in view of the numerous variables influencing system operation; it is also possible that with so few studies performed in a system framework, the amount of relevant data must be minimal.

Another characteristic of the studies reviewed (again with a few significant exceptions) is that the questions they address are not very relevant to system design or to understanding how the system functions. We cannot therefore answer the research questions posed in Chapter 1. The studies performed are preoccupied with determining whether variable X or Y is or is not significant. It should be obvious that every variable is both significant and insignificant (from a practical as well as a statistical standpoint), depending on the way it interacts with other variables simultaneously affecting the system. From that standpoint, the questions these studies typically address are essentially irrelevant. The real question is: *under what conditions* is a variable significant? Or, rather, what quantitative value does the variable have when it does significantly influence operator and system output?

The experimenter's task is greatly complicated when he is obliged to provide *numbers* that represent the point along the continuum at which the system developer can expect a significant change in operator and system functioning. This is not a new requirement. Christensen addressed the same point when he wrote ". . . the design engineer does not need to know that 'a' is significantly better than 'b'; but he needs to know how much greater and the consequences in terms of system effectiveness of selecting points along a dimensionalized 'a' " (Meister, 1971, p. 388).

In this context Chapanis (1967, 1971) has pointed out that little basic research is useful for HF even when the research was undertaken to solve a real world problem. Even worse, very little is known, as Christensen and

Mills (1967) found, about operator activities in real world systems performed under operational (field) conditions.

Another reason why it is difficult to supply the system developer with usable data is that researchers create a great variety of task situations for their studies (Chapter 3). This makes it extremely difficult to combine data readily from different studies. It is possible that task-specific characteristics account for a large proportion of performance variance. Perhaps then we should not expect conclusions to generalize easily from one type of task to another. On the other hand, it is difficult to have confidence in conclusions derived solely from research on a single task type. The question of how task differences affect operator/system performance, which not only urgently needs answering but is an opportunity for future research, has not yet been answered.

Available behavioral research is valuable insofar as it provides general principles that can guide the system designer. But values specific to a particular design application must be acquired either logically (by analysis, if that is possible) or experimentally (by the so-called quick and dirty study). Since opportunity for experimentation (of any type) is somewhat limited in system development, the HF specialist must increase the emphasis he places on use of analytic methods. He must develop methods to permit the translation of system/equipment data into behavioral consequences (e.g., the deduction of training requirements from system characteristics) and vice versa (e.g., the deduction of desired equipment configurations based on operator aptitudes). Unfortunately, there appears to have been little original thinking about task analysis methods since Miller's initial work (Miller, 1953). Some innovations in this area would be welcome.

Another characteristic of the research reviewed is that subject factors are accorded little significance. One must ask whether the traditional subject in behavioral research—the college student—is actually representative of the operator population. The author suspects (there are few data on this point) that the average system operator differs substantially from the average experimental subject, not necessarily in terms of intellectual capability but in terms of attitudes and performance strategies.

Ultimately we must ask why HF research so often fails to address the system. There are, of course, obvious answers. HF research is more costly; it requires more subjects and more time; it is more difficult to develop experimental situations that reproduce real world systems. Beyond these explanations, however, one suspects that many HF researchers have been "overtrained" in the psychological tradition that concentrates on the individual. This tends to produce a "homocentric" concept of the system; that the individual within the system is considered to be *the* critical system element. The system, therefore, tends to be conceptualized in terms of individual rather than interactive system variables.

The individual operator is indeed critical to the MMS, but by overemphasizing his individuality, the researcher may tend to ignore *other* system characteristics—in particular, relationships among subsystems and between system personnel and the total system.

What are the consequences of failing to include the system in HF studies? It is evident that the HF specialist working in system development receives only part of the data support he requires from research. Beyond this, the question most critical to HF (i.e., the relationship of the human to the system) cannot be effectively answered. It is therefore difficult to demonstrate that HF activities in system development and functioning make any real difference to overall system performance. Unfortunately, although this discipline does contribute usefully to system development, potential users of HF research remain unconvinced of its utility in system terms.

All these difficulties and inadequacies are not, of course, the fault of the researcher alone. He is often limited by funding and by ill-informed demands made on him by employers who are not themselves behavioral scientists.

EFFECTIVE SYSTEM RESEARCH CHARACTERISTICS

One reason for the case approach adopted in this book was to permit us to discern study patterns that could suggest effective ways of attacking system questions. Although it would be foolish to attempt to rank the vast amount of research in terms of qualities like "goodness" or "effectiveness," certain studies—rather, series of studies—are particularly impressive: the work of Kidd and his colleagues (Cases 2, 3, 4, 5, and 137); Briggs and his co-workers (Cases 71, 75, 133, 143, and 148); Horrocks et al. (Case 127); Glaser and his co-workers (Cases 130, 138, and 140); Lanzetta and Roby (Cases 6, 68, 123, and 144); the *DM* research described by Howell in Case 113.

All the studies cited met in some degree the criteria specified in the preface. In addition, the following characteristics ought particularly to be marked.

Continuity

Each citation represents a *series* of studies dealing with a single topic or several interrelated topics that are studied on the basis of the same task (or very similar ones). Unlike a single study that illuminates one or two points but is not necessarily followed up by the same investigator, these studies permit one to examine the same variable from different viewpoints. This enhances their value. For example, during the four years of team training studies performed at Ohio State University . . ."the topics progressed from the initial question of team versus individual training to a detailed treatment of task organization, task fidelity, system criteria, communications channels, and feedback effects on team training. The results are significant because they represent an orderly attack . . ." (Williges, 1975).

Task Representativeness

In almost all cases the exemplary research was performed with real world systems or realistic analogues of these, featuring the multiplicity of functions found in the real world. This is not to say that effective research cannot be performed with simpler functions like reaction time, or laboratory tasks like vigilance or tracking, but it is hard to generalize the results of the latter to the real world with its greater complexity and functional richness.

In order to utilize tasks in their system context, more research may have to be performed in the field environment. In investigating highly molar system variables, e.g., comparing total ship against individual crew performance, it may be impossible to use traditional laboratory or even controlled field experiments. Historical-descriptive techniques making use of non-experimental data such as the system records described in Case 121 may be required.

Even though human performance is studied in a system context, the results have implications for individual aspects of human performance, such as information processing. The most effective system-oriented research, such as that of Briggs and his co-workers, can, through preplanning and advance problem definition, provide meaningful data for both individual performance and system applications.

Subject Experience

Other investigations ought to be cited because they used subjects who were experienced in the tasks they had to perform. Many such studies have been performed in a military context [see Parsons (1972) for a history of these] but there is nothing to prevent other investigators from doing the same. Behavioral research has been and is largely a discipline based on the reactions of naïve college students, when the behavior it is desired to predict is that of trained professionals.

SYSTEM THEORY

One of the characteristics of HF research is that it is more pragmatic than theoretical. Researchers like Kidd and his colleagues, or Briggs and Howell and their co-workers, set themselves questions of fact—what happens, for example, if one enlarges the size of the team? (Kidd, Case 2)—and then proceeded to answer these questions. There are, of course, theoretical concepts implied in the research of Glaser et al. and the *DM* research summarized by Howell (Case 113); but the theory making is not very elaborate compared with theories of the individual, such as those of vigilance (e.g., Tanner and Swets, 1954). Whatever theory one finds in HF research is relatively low level. This is not meant pejoratively. The pragmatic emphasis is perhaps dictated by necessity. It may even be desirable; many problems can be attacked more effectively by direct methods.

In any event, most HF theories do not stem from or describe any unitary, highly elaborated set of system concepts; they usually describe the performance of the individual who is in the system but not part of it. For example, Kidd (1959) considers that the major factor influencing the performance of his experimental air traffic controllers is short-term memory, a parameter that describes the behavior of the individual but not that of the system.

There has, in fact, been a good deal of theory construction in HF, but it has not been recognized as such because we are conditioned by our psychological training to think of theory primarily in terms of individual performance variables and by our HF experience to think of answers to problems rather than of constructs.

Yet the terms in which HF is phrased are theoretical constructs in themselves. Such terms as task inputs/outputs, the crew, system requirements, work load and man-machine interface, imply a theoretical structure that is meaningful only in systems and cannot be found in theories of individual behavior. The theoretical implications of these terms have not, however, been systematically elaborated by HF researchers.

MAN IN THE SYSTEM

One can examine HF research not only from the narrower perspective of the scientist but also from the viewpoint of what that research has contributed to society in general. The effect of a behavioral discipline is manifested not only in data but also in terms of its influence on culture, in terms of viewing ourselves and our environment. The outstanding example is psychoanalysis.

From that viewpoint HF research has emphasized the symbiotic relationship between man and his technology. It teaches us that man is embedded in one system or the other—civilian, military, social, political, or the one of greatest interest to us, the *MMS,* and is not merely a passive, reactive element at the mercy of the system as a whole: he examines his environment critically; he develops a cognitive model of that system; he not only adapts his behavior to system requirements but often attempts to modify the system to his own needs. The system becomes part of the phenomenal reality he is constantly exploring and seeking to change.

The active striving of the man in the system may appear to be obvious from real life and from artistic representations of that reality (novels, plays, etc.). Perhaps, then, HF researchers have simply demonstrated experimentally what many of us had suspected from our own perceptions of reality.

It may appear that the conclusions derived from HF research do little more than confirm the obvious, what we as individuals apprehend as "reality" from everyday experience or the communicated experiences of others. For example, it is "obvious" that instructional feedback improves learning, that inputs that appear too rapidly are difficult to monitor, that people have difficulty making probabilistic decisions, and so on, and so forth. Conclusions such as these may seem trivial to the reality-oriented reader.

Experiential phenomenology—direct apprehension of behavior, that makes many behavioral conclusions appear obvious—is, however, more characteristic of individual than of system contexts. When we study the relationship of operator behavior to the overall system, we have much less experience to suggest which variables are important or how they function. People have difficulty directly apprehending macro-systems. From that standpoint it is more difficult to accuse system-oriented studies of being "obvious."

The ultimate challenge to the HF researcher is of course to assist in constructing reality by helping to develop the system.

GENERAL REFERENCES

Chapanis, A. The relevance of laboratory studies to practical situations. *Ergonomics,* 1967, **10,** 557–577.

Chapanis, A. The search for relevance in applied research. In Singleton, W. T. et al. (Eds.), *Measurement of Man at Work.* London: Taylor and Francis, Ltd., 1971, pp. 1–14.

Christensen, J. M. and Mills, R. G. What does the operator do in complex systems? *Hum. Factors,* 1967, **9,** 329–340.

Kidd, J. S. *A summary of research methods, operator characteristics, and system design specifications based on the study of a simulated radar air traffic control system.* WADC Technical report 59–236, Aerospace Medical Research Laboratories, Wright-Patterson AFB, Ohio, July 1959.

Meister, D. *Human Factors: Theory and Practice.* New York: Wiley, 1971.

Miller, R. B. *Some working concepts of systems analysis.* American Institute for Research, Pittsburgh, February 1953.

Parsons, H. M. *Man-Machine System Experiments.* Baltimore: Johns Hopkins Press, 1972.

Tanner, W. P. and Swets, J. A. A decision-making theory of visual detection. *Psychol. Rev.*, 1954, **61,** 401–409.

Williges, R. C. *A tribute to George: reflections from colleagues. Part 2—Lessons in system research.* Paper presented at the American Psychological Association annual meeting, Chicago, Ill., September 1975.

INDEX